The Biology of Peatlan

f ser
Fine

THE BIOLOGY OF HABITATS SERIES

This attractive series of concise, affordable texts provides an integrated overview of the design, physiology, and ecology of the biota in a given habitat, set in the context of the physical environment. Each book describes practical aspects of working within the habitat, detailing the sorts of studies which are possible. Management and conservation issues are also included. The series is intended for naturalists, students studying biological or environmental science, those beginning independent research, and professional biologists embarking on research in a new habitat.

The Biology of Rocky Shores
Colin Little and F. A. Kitching

The Biology of Polar Habitats
G. E. Fogg

The Biology of Lakes and Ponds
Christer Brönmark and Lars-Anders Hansson

The Biology of Streams and Rivers
Paul S. Giller and Bjorn Malmqvist

The Biology of Mangroves
Peter F. Hogarth

The Biology of Soft Shores and Estuaries
Colin Little

The Biology of the Deep Ocean
Peter Herring

The Biology of Lakes and Ponds, Second ed.
Christer Brönmark and Lars-Anders Hansson

The Biology of Soil
Richard D. Bardgett

The Biology of Freshwater Wetlands
Arnold G. van der Valk

The Biology of Peatlands
Håkan Rydin and John K. Jeglum

The Biology of Peatlands

Håkan Rydin
Department of Plant Ecology,
Uppsala University

and

John K. Jeglum
Department of Forest Ecology,
Swedish University of Agricultural Sciences

with contributions from

Aljosja Hooijer, Beverley R. Clarkson, Bruce D. Clarkson,
Dmitri Mauquoy, and Keith D. Bennett

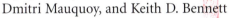

OXFORD
UNIVERSITY PRESS

OXFORD

UNIVERSITY PRESS

Great Clarendon Street, Oxford OX2 6DP

Oxford University Press is a department of the University of Oxford.
It furthers the University's objective of excellence in research, scholarship,
and education by publishing worldwide in

Oxford New York

Auckland Cape Town Dar es Salaam Hong Kong Karachi
Kuala Lumpur Madrid Melbourne Mexico City Nairobi
New Delhi Shanghai Taipei Toronto

With offices in

Argentina Austria Brazil Chile Czech Republic France Greece
Guatemala Hungary Italy Japan Poland Portugal Singapore
South Korea Switzerland Thailand Turkey Ukraine Vietnam

Oxford is a registered trade mark of Oxford University Press
in the UK and in certain other countries

Published in the United States
by Oxford University Press Inc., New York

British Library Cataloguing in Publication Data

Data available

Library of Congress Cataloging-in-Publication Data

Rydin, Håkan.
  The biology of peatlands / Håkan Rydin and John K. Jeglum; with
contributions from Aljosja Hooijer . . . [et al.]
    p. cm. — (The biology of habitats)
    ISBN-13: 978–0–19–852872–2 (978–0–19–852871–5: alk. paper)
    ISBN-10: 0–19–852872–8 (0–19–852871–X: alk. paper)   1. Peatland
plants.   2. Peatland ecology.   3. Peatlands.   I. Jeglum, J. K.   II.
Hooijer, Aljosja.   III. Title.   IV. Series.
    QK938.P42R93 2006
    577.68'7—dc22   2006006151

Typeset by Newgen Imaging Systems (P) Ltd., Chennai, India
Printed in Great Britain
on acid-free paper by
Biddles Ltd., King's Lynn

ISBN 0–19–852871–X   978–0–19–852871–5
ISBN 0–19–852872–8 (Pbk.)   978–0–19–852872–2 (Pbk.)

10 9 8 7 6 5 4 3 2

Preface

The study of peatland biology and habitats encompasses a broad range of subjects and disciplines. The task of providing a comprehensive overview of such a broad subject is challenging and somewhat daunting, and we undertake this with a good deal of humility. We are fully aware of the enormous worldwide variation among the peatlands, and we therefore decided to focus our presentation on the temperate and boreal peatlands of the northern hemisphere. This admitted bias is partly a reflection of our own competence, but is also a reflection of the fact that this is where the main peatland areas and the main peat stores are to be found. Many of the concepts and processes we describe are, however, generally applicable, and we were fortunate to be able to include contributions from colleagues to cover tropical and southern hemisphere peatland regions (Chapter 11).

The book is organized into 13 chapters in a logical sequence from biological themes, through geophysical and developmental themes, to world distribution and human impacts. In the first chapter we introduce the subject of peatland habitats with basic terminology and classification of wetlands, with the ambition to bridge the differences in terminology in different countries. Chapters 2–4 present an overview of the biodiversity of peatlands, and describe the innumerable adaptations of organisms to these peculiar habitats. Particular attention is paid to the biology of one of the most important species groups, the peat mosses of the genus *Sphagnum*, because of its dominating influence on peatland development in many parts of the world. Peat, the material in which the organisms live, and which they also form themselves, is characterized in Chapter 5. This is followed by two chapters treating the subject of development of peatlands and their use as historical archives for vegetation and climate. After this we deal with environmental factors affecting peatland organisms. First, the quantity and quality of water, so crucial for any wetland, is dealt with in the hydrology chapter, and this is followed by discussions on nutrients, light, temperature, and climate. In Chapters 10 and 11 we provide an overview of the extent, distribution and incredible variation of forms and patterns found in peatlands over the world, including three detailed treatments of quite dissimilar peatlands in southeast Asia, New Zealand, and Tierra del Fuego. Chapter 12 deals with the highly topical subject of peat and carbon accumulation and global warming. Finally, Chapter 13 gives an overview of the uses, values, management, and restoration of peatlands.

The Biology of Peatlands is intended as an introduction both for students and for professionals in the fields of biology, forestry, conservation, etc. Our aims have been to present a modern view of peatlands, with references to the recent literature and research trends. In addition we have provided links to the rich historical literature with its immense accumulated knowledge of natural history, much of which we strongly recommend for further reading.

Many friends and colleagues have helped us by providing data, tables, and photographs. Others have kindly reviewed texts and helped us to improve the chapters. For their various efforts we thank Per Alström, Dick Andrus, Taro Asada, Ingvar Backéus, Lisa Belyea, Viktor Boehm, Suzanne Campeau, Martha Carlson, Dicky Clymo, Nils Cronberg, René Doucet, David Gibbons, Paul Glaser, Urban Gunnarsson, Tomas Hallingbäck, Richard Hebda, Stephen Heery, Barrie Johnson, Hans Joosten, Erik Kellner, Zhao Kuiyi, Raija Laiho, Jill Lancaster, Stefani Leupold, Lars Lundin, Kalle Mälson, Sean McMurray, Edward Mitchell, Mats Nilsson, Juhani Päivänen, Jonathan Price, Line Rochefort, Jonathan Shaw, Andrey Sirin, Göran Thor, Sebastian Sundberg, Eeva-Stiina Tuittila, Gert van Wirdum, Peter Uhlig, Harri Vasander, Henrik von Stedingk, and Douglas Wilcox. In addition numerous people have patiently answered our detailed questions, assisted with hints on literature, or been helpful in many other ways.

Special thanks to our colleagues who contributed sections to Chapter 11: Beverley Clarkson, Bruce Clarkson, Dmitri Mauquoy, Keith Bennett, and Aljosja Hooijer. We also thank Ian Sherman, Stefanie Gehrig, and Anita Petrie, our patient editors at Oxford University Press.

It will be noted that we give a few references to 'Hugo Sjörs, personal communication', but in fact there are scores of items in this book that he augmented or modified from his immense field experience. In particular, his contributions to Chapters 11 and 12 were substantial, and for this we are extremely grateful.

October 2005   Håkan Rydin, Uppsala
                                                John K. Jeglum, Umeå

Contents

List of Contributors

Keith D. Bennett — Palaeobiology Program, Department of Earth Sciences, Uppsala University, Uppsala, Sweden

Beverley R. Clarkson — Landcare Research, Hamilton, New Zealand

Bruce D. Clarkson — Centre for Biodiversity and Ecology Research, Department of Biological Sciences, University of Waikato, Hamilton, New Zealand

Aljosja Hooijer — River Basin Management, WL | Delft Hydraulics, Delft, The Netherlands

John K. Jeglum — Department of Forest Ecology, Swedish University of Agricultural Sciences, Umeå, Sweden

Dmitri Mauquoy — Department of Geography and Environment and the Northern Studies Centre, University of Aberdeen, Aberdeen, UK

Håkan Rydin — Department of Plant Ecology, Uppsala University, Uppsala, Sweden

1 Peatland habitats

This book deals with the diverse, beautiful, and fascinating world of peatlands. They represent very special kinds of transitional, amphibious ecosystems with habitats between uplands and water, where organic matter tends to accumulate because of the waterlogged, often poorly aerated conditions. Here we encounter *Sphagnum* peat mosses with an infinite variety of colours – greens, reds, browns; insect-eating plants and beautiful orchids; reeds, sedges, and cotton grasses; low, often evergreen shrubs; floating plants and emergents at the water's edge; quaking mats; vast wetlands with spectacular surface patterns; springs and soaks; thickets, sparsely treed woodlands, and tall forests. As a consequence of their rich vegetation, peatlands also house a multitude of microorganisms, insects, birds, and other animals.

Peat accumulations can be tens of metres deep and provide material that can be harvested and used as fuel and for horticulture. After drainage, large areas have been converted to arable land, meadows, or forests. The peats are also valuable archives of past vegetation and climate, where we may find the buried remains of ancient settlements, trackways, fields, and even preserved humans – the so-called 'bog people' of northern Europe (Coles and Coles 1989; Turner and Scaife 1995).

The aim of this chapter is to provide the reader with an understanding of the main terms and concepts used in peatland science, and a general appreciation of the main peatland habitats. It is essential at the outset to provide a basic language of peatlands which, even if not universally agreed upon, will define the usage for this book. The variation in terminology reflects the great diversity and complexity of habitats and ecosystems. Unfortunately several terms are not consistently used, even in the same country or language. This reflects traditional differences in understanding and comprehension among specialists, and differences between geographical areas. Table 1.1 lists a select set of peatland types in several languages. Many glossaries and definitions are available; particularly useful are IPS (1984) and Joosten and Clarke (2002).

Table 1.1 Peatland terminology. It is difficult to find exact translations, and the terms are sometimes used inconsistently, even within one language

English	German	Russian	French	Finnish	Swedish
Wetland	Nassboden, vernässter Boden, Feuchtgebiete	Заболоченная местность; З. земля	Milieux humides	Kosteikko	Våtmark
Peat[a]	Torf	Торф	Tourbe	Turve	Torv
Peatland	Torfmoor	Торфяник, торфяное болото	Tourbière	Turvemaa	Torvmark
Mire	Moor	Болото	Tourbière, tourbière vivante	Suo	Myr
Bog	Regenmoor, Hochmoor	Болото атмосферного питания,Верховое болото	Tourbière ombrotrophe, tourbière haute,	Ombrotrofinen suo, rahkasuo	Mosse
Fen	Niedermoor	Низинное болото	Tourbière minérotrophe, tourbière basse, bas-marais	Sarasuo, minerotrofinen suo	Kärr
Marsh	Marschmoor	Марш[b]	Marais	Marskimaa	–[c]
Swamp forest	Bruchwald, Moorwald	Болото лесное	Marécage, forêt marécageuse	Korpi	Sumpskog

[a] An English word related to the word used for peat in other languages is 'turf' (often used in old literature, e.g. King 1685).
[b] Rarely used term, usually used for translation and in plural form (марши).
[c] No commonly used term, since marshes are classified as fens (when peaty) or shore vegetation.

Wetlands, peatlands, and mires

The three main terms used in the current literature to encompass the subject are wetlands, peatlands, and mires. These terms are defined somewhat differently, although there is considerable overlap (Fig. 1.1). The broadest concept is that of wetlands.

Wetlands

Wetlands include shore, marsh, swamp, fen, and bog. Scientifically, we can characterize wetland by the following points:

- The water table is near the ground surface.
- As a consequence, the substrate is poorly aerated.
- Inundation lasts for such a large part of the year that the dominant plants and other organisms are those that can exist in wet and reducing conditions.

The Ramsar Convention provides a very wide definition:

For the purpose of this Convention wetlands are areas of marsh, fen, peatland or water, whether natural or artificial, permanent or temporary, with water that is static or

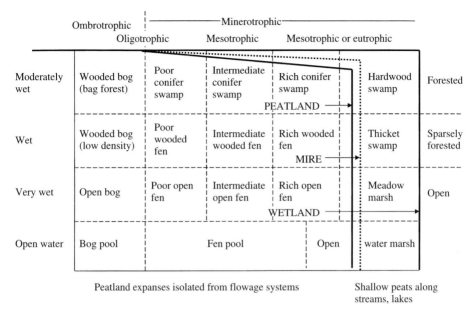

Fig. 1.1 A general scheme to define the position of broad wetland types in an ordination based on the two most important environmental gradients. Wetness, or distance between vegetation surface and water table, varies along the vertical axis, and the complex gradient with variation in pH, base saturation, and nutrient status is depicted along the horizontal axis. Wetland is an even broader category than shown here, since it includes various habitats of shore and shallow waters.

flowing, fresh, brackish or salt, including areas of marine water the depth of which at low tide does not exceed six metres (Ramsar 1987).

The Canadian Wetland Classification has a narrower definition, which emphasizes the extent and duration of high water levels to separate a wetland from upland ecosystems:

A wetland is defined as: land that is saturated with water long enough to promote wetland or aquatic processes as indicated by poorly drained soils, hydrophytic vegetation and various kinds of biological activity which are adapted to a wet environment (National Wetlands Working Group 1997).

Comprehensive reviews of the multitude of wetland definitions used for scientific and regulatory purposes in the USA and some other countries are given by Tiner (1998; 1999).

Peat and peatland

In order to define peatland we must first define peat. Peat is the remains of plant and animal constituents accumulating under more or less water-saturated conditions owing to incomplete decomposition. It is the result of anoxic conditions, low decomposability of the plant material, and other complex causes. Peat is organic material that has formed in place, i.e. as sedentary material, in contrast to aquatic sedimentary deposits. Quite different plant materials may be involved in the process of peat formation, for instance, woody parts, leaves, rhizomes, roots, and bryophytes (notably *Sphagnum* peat mosses). Much of the material originates below ground.

Peatland is a term used to encompass peat-covered terrain, and usually a minimum depth of peat is required for a site to be classified as peatland. In Canada the limit is 40 cm (National Wetlands Working Group 1997), but in many countries and in the peatland area statistics of the International Mire Conservation Group it is 30 cm (Joosten and Clarke 2002).

Mire

Mire is a term for wet terrain dominated by living peat-forming plants (e.g. Sjörs 1948). Both peatland and mire are narrower concepts than wetland, because not all wetlands have conditions that allow peat to accumulate.

In one sense mire is a slightly broader concept than peatland, because peat accumulation can occur on sites that do not have the required depth of peat to qualify as peatland. In another sense peatland is broader – a drained site, for instance a site being used for peat harvesting, is still a peatland, but having lost its original peat-forming vegetation it is no longer a mire. A reason for the seemingly conflicting definitions is that the terms are used for different purposes. Mire is a term often used in botanical and ecological investigations of the vegetation types or the process of peat formation. Peatland is often used in forestry and land management, which makes the peat depth limit crucial. There are also traditional differences among countries: in North America peatland has been more widely used, and in Scandinavia mire (*myr*) is more common.

The concept of 'peat-forming plants' is problematic. Even if some species more commonly give rise to peat than others, peat formation is a *process* that can befall most plant materials.

Peatland habitats along wetness and chemical gradients

Two complex environmental gradients are responsible for the distinction of main peatland types. One is linked to wetness and aeration and the other is a combination of pH, calcium (Ca) content, and base saturation. The latter is to some degree linked to nutrient availability, but is not a universal predictor of productivity.

Variation in wetness and aeration

The overriding physical condition controlling peatlands is the high water table. The initial formation of peat is related to wet conditions at the surface. Oxygen moves very slowly in stagnant water, and is used up rapidly by microorganisms in saturated peat, creating anoxic conditions.

Among and within peatlands the position of the water table varies in time and space, so some parts of the peatland are below, some at, and some slightly raised above the water level permanently or temporarily. This variation creates a moisture–aeration regime, which depends not only on the position of the water table, but also on pore structure of the peat, the fraction of the total pore spaces filled with water versus air, and the oxygen content of the water. This regime can be segregated for analytic purposes into the moisture factor and the aeration factor, although they are related.

The lack of oxygen influences the rate of decomposition of organic matter that is laid down by the peatland plants. In virgin mires with actively growing surface vegetation there is a net gain of organic matter and hence active growth of the peat layer. Different organisms respond differently to the depth of, or depth to, the water table. Limitations for vascular plants relate primarily to lack of oxygen in the rooting environment (discussed further in Chapter 3).

Variation in pH, base richness, and nutrient availability

The accumulation of peat usually causes increasingly more acid and nutrient-poor conditions, as the influence of the cations derived from mineral soil decreases with time. The organic matter has a high cation exchange capacity (CEC), and tends to take up (adsorb) cations in exchange for hydrogen ions. Therefore, most chemicals, notably cations, are adsorbed on the peat particles, and only a minor – but important – fraction is actually free in solution.

The chemical regime can be segregated into two factor groups. One is the variation in pH, linked also to electrical conductivity, Ca content, and base richness. The other is availability of plant nutrients. As in most terrestrial ecosystems nitrogen (N) is a key nutrient, but the scarcer phosphorus (P) and potassium (K) are more often limiting in peatlands than in mineral soil.

Origin of groundwater and trophic classes

From early scientific work the importance of the origin of the mire water as a major controlling factor has been recognized (e.g. Du Rietz 1954). Peat accumulation begins on wet mineral soils or as quaking mats encroaching on open water. At this stage the water in the peat surface is connected with, or has passed over or through, mineral parent materials. Such sites are termed *minerogenous* (or *geogenous*) to indicate that water is added to the peatland from the surrounding mineral soil. However, as the peat layer grows higher, the vegetation may become progressively more isolated from the mineral soil water. Peatlands with a surface isolated from mineral-soil-influenced groundwater will receive water only by precipitation. These peatlands are called *ombrogenous*. To emphasize the chemical effects on the site we refer to minerogenous peatlands as *minerotrophic*, nourished by mineral soil groundwater. Correspondingly we refer to ombrogenous peatlands as *ombrotrophic*, nourished by precipitation (and airborne dust). The terms minerogenous and ombrogenous stress the hydrological regime, whereas the terms minerotrophic and ombrotrophic focus on the way nutrients are provided.

Many mire ecologists follow the simple convention of using the term *fen* for minerotrophic mires and *bog* for the ombrotrophic ones. The idea of giving vernacular terms a strict meaning in ecological literature was probably first introduced in Sweden (*kärr* = minerotrophic, *mosse* = ombrotrophic, *myr* for both), Germany, and Finland. Later these terms were adopted into English as *fen, bog,* and *mire*. This is a useful convention, not least when it comes to communicating peatland science in popular form, and we follow it in this book.

Minerogenous peatland is further divided into three major hydrologic systems (von Post and Granlund 1926; Sjörs 1948):

- *Topogenous peatlands* have flat (or virtually flat) water tables, and are located in terrain basins with no outlet, a single outlet, or both inlets and outlets.
- *Soligenous peatlands* are sloping, with directional water flow through the peat or on the surface.
- *Limnogenous peatlands* are located along lakes, streams, or intermittent stream channels that are flooded periodically by waters carried in these channels.

Another set of terms is *oligotrophic, mesotrophic,* and *eutrophic*. These terms are commonly used in limnology in relation to plankton productivity, which is often explained in terms of levels of N or P. The terms have also been adopted for use in peatland work, and the sequence oligo-, meso-, eutrophic is explained as a gradient of increasing productivity and nutrient availability. The oligotrophic class is portrayed as somewhat broader than the ombrotrophic class (Fig. 1.1), and it includes weakly minerotrophic sites with low pH. However, there are also oligotrophic sites (with low productivity) in minerotrophic conditions that have very high pH and Ca content, because the P has become unavailable by binding with Ca.

The main ecosystems: marsh, swamp, fen, bog

In a long-term programme of wetland studies in Canada, four high-level ecosystem classes were identified – *marsh, swamp, fen,* and *bog* (National

Wetlands Working Group 1997). These four terms, and the ecosystem classes that they represent, are among the most common used in the wetland literature. A basic principle of classification is that it is purposive; that is, it is done for a specified purpose or use. It is difficult to find 'natural' classifications, because there are often no sharp boundaries between ecosystems. A physiognomic and dominance type approach is useful, because it is based on structure and form of vegetation, and on the main dominants that control the appearance of the vegetation. Such a classification can be used by people who are not specialists in flora, and also from the point of view of vegetation mapping and remote sensing. In order to clarify the divisions between the four main peatland ecosystems and some of their physiognomic groups, we present a key in Table 1.2, and the key features of marsh, swamp, fen, and bog in Table 1.3. Generally the physiognomic groups are not sharply distinguished in the field, and they are separated by rather arbitrary cut-off levels for the main vegetational features.

Marsh

Marshes (Fig. 1.2) are characterized by standing or slowly moving water with submergent, floating leaved, or emergent plant cover. They are permanently flooded, or seasonally flooded and intermittently exposed. Nutrient-rich water generally remains within the rooting zone for most of the growing season. Bottom surfaces may be mineral glacial drift, aquatic sedimentary deposits, or precipitates of inorganic compounds or organics. Initial root mats of peat may be developing over the mineral or sedimentary deposits.

Many marsh habitats are not peatlands since they have only little peat, which means that most vascular plants are rooted in the underlying mineral soil from which they can take up nutrients. However, marshes often have some mineral-rich organic deposits, or shallow accumulations of true peat, developing over mineral or aquatic sedimentary deposits. The deep beds of *Phragmites* peat beneath Irish bogs are a case of peat development in marshes, but it is questionable whether one should call the *Phragmites* community a marsh or a fen. Under semi-arid or tropical conditions, marshes are often the predominant kind of wetlands (for example, *Papyrus* marshes in Africa).

The main physiognomic groups of marsh are *open water marsh, emergent marsh* (including reedswamp which is actually a marsh, or sometimes rather a fen), and *meadow marsh*. These types are often arranged as zones beside open waters including lakes, ponds, pools, rivers, streams, and drainage ways. The main complex factors within marshes are water level (floodings, drawdowns) and in some places disturbance by wave or current energy.

Table 1.2 Key to the main peatland classes – bog, fen, marsh and swamp – and physiognomic groups within the classes. Separation of the physiognomic groups is augmented by physiography, hydromorphology, and floristics (modified from Harris *et al.* 1996). The key was developed for northern Ontario and is valid for most boreal regions, but can be adapted even to tropical peatlands

1 Beside open channels or bodies of water, standing or flowing water, < 2 m deep; submergent, floating, emergent plants; or flooded seasonally; mostly mineral substrate, sometimes with peat. Open vegetation..................Marsh

  2 Standing or flowing water with emergent plant cover <25%. Submergent and/or floating-leaved plants cover normally > 25%..................Open water marsh

  2 Relatively open cover of graminoids and herbs, >25% cover of emergents

    3 Flooded for most of the growing season. Relatively open cover of graminoids and herbs, >25% cover of emergent species, interspersed with pools or channels, sometimes with submerged and floating plants..................Emergent marsh

    3 Little or no standing water for most of the growing season, but flooded seasonally. Closed cover of graminoids, sometimes with herbs. Often tussocks or hummocks. Transitional types to open fen can occur..................Meadow marsh

1 Not beside open channels or lakes, not flooded seasonally or if so, by shallow overland sheet surface flow; on peat with open, wooded, or forested vegetation

  4 Woody vegetation (height >2 m) with canopy cover >25%

    5 Conifer trees dominant.

      6 Indicators of minerotrophy present. Trees dense and large enough to be merchantable (height often >10 m)..................Conifer swamp forest

      6 Indicators of minerotrophy absent. Trees generally smaller and more sparse..................Wooded bog (Bog forest)

    5 Broad-leaved species dominant

      7 Hardwood trees dominant, height usually > 10 m and large enough to be merchantable..................Hardwood swamp forest

      7 Tall shrubs (height > 2 m) dominant. Includes species of *Alnus* and *Salix*..................Thicket swamp forest

  4 Woody vegetation (height > 2 m) absent or scanty with canopy cover < 25%.

    8 Indicators of minerotrophy present

      9 Woody vegetation (height > 2 m) with cover < 10%. Lawn level or carpet level tend to be dominant, hummock level in minority..................Open fen

      9 Woody vegetation (height > 2 m) with cover 10–25%. Often with hummock level mounds which support low-growing trees or tall shrubs. The hummock level and lawn-carpet level are of similar magnitudes..................Wooded fen

    8 Indicators of minerotrophy absent

      10 Woody vegetation (height > 2 m) absent or with cover < 10%. Dominated by lawn or carpet level, *Sphagnum* and low sedges, with hummock level often in minority, but the hummock level may also dominate and dwarf shrubs or lichens become abundant..................Open bog

      10 Woody vegetation (height > 2 m) with 10–25% cover of small conifers. Lacks trees taller than 10 m. Vegetation mainly *Sphagnum* and dwarf shrubs with some low sedges. Dominated by hummock level with low-growing trees..................Sparsely wooded bog

Table 1.3 Key features of marsh, swamp, fen, and bog

Peatland attribute	Marsh	Swamp	Fen	Bog
Vegetation	Submergents, floating-leaved, reeds, tall sedges	Forests, tall shrub thickets, herbs, graminoids, bryophytes	Open or sparse cover of low trees, low shrubs, graminoids, herbs, bryophytes	Open or with low trees, dwarf shrubs, low cyperaceous plants, bryophytes
Soils/peats	Mineral, organic-rich mineral, or shallow peat	Mineral, organic-rich mineral, shallow to deep peat	Usually >40 cm peat	Usually >40 cm peat
Origin of water	Limno-, topo-, soligenous	Limno-, topo-, soligenous	Topo-, soligenous, sometimes limnogenous	Ombrogenous
Shape	Flat, concave, or sloping	Flat, concave, or sloping	Flat, concave, or sloping	Flat, convex, or sloping
Moisture regime	High water level fluctuation, with flooding and drawdowns common	Groundwater below surface, hummock level dominates providing aerated peat to support trees or tall shrub dominance	Groundwater below to above surface; in open fen lawn, carpet and mud-bottom most abundant; in wooded fen hummocks occur as islands in a matrix of lawn, carpet and mud-bottom level	Groundwater below to above surface; in open bog lawn, carpet and mud bottom levels are abundant; in wooded bog the hummock level is dominant
Ground surface microtropography	Level or tussocky	Irregular, with high hummocks; logs and tree bases provide support for hummocks	Level, or with low hummocks, or patterned with low or high ridges alternating with depressions (flarks)	Level, or with low hummocks and hollows, or patterned with hummocks or ridges alternating with hollows
Nutrient regime	Minerotrophic, eu- to mesotrophic	Minerotrophic, eu-to oligotrophic	Minerotrophic, eu- to oligotrophic	Ombrotrophic, oligotrophic

Swamp or swamp forest

Swamp forests (Fig. 1.3) are forested or sometimes thicketed wetlands (in vernacular English a swamp could refer to almost any kind of wetland). They have minerogenous water that may come from watercourses or the underlying soil or lateral groundwater throughflow. They have standing or gently flowing water in pools or channels, or subsurface flow. The water table is usually well below the surface, so that the surface layer is aerated and supports the roots of trees or other tall woody plants. Substrates are organic–mineral mixtures, or shallow to deep peat (in which wood is a large component).

The main physiognomic groups of swamp forest are *conifer swamp forest*, *hardwood swamp forest* (deciduous or evergreen), and *thicket swamp*. These

Fig. 1.2   Tussock meadow marsh with *Calamagrostis canadensis* and *Carex stricta* along a slow-flowing stream that overflows the marsh periodically. Ontario, Canada.

Fig. 1.3   Swamp forests: (A) Herb-rich alder (*Alnus glutinosa*) swamp forest, eastern Sweden. (B) Eastern white cedar (*Thuja occidentalis*), herb-rich swamp. Ontario, Canada.

Fig. 1.4   Fens: (A) Topogenous poor fen dominated by *S. papillosum*. In the foreground *Carex rostrata*, towards the background with increasing cover of *Eriophorum vaginatum*. The shelf to the right is the transition to the bog. Sweden, midboreal zone. (B) Extremely rich calcareous fen, with shallow water with greyish marl deposition. Hummocks made up of *Schoenus ferrugineus, Carex lasiocarpa*, and brown mosses in low mounds, with rich-fen indicator *Sphagnum warnstorfii* in higher hummocks. Sweden, midboreal zone.

types are located either along open waters, in drainage ways, near springs, or as a part of larger peatlands where they sometimes form a zone between the peatland and the upland forest. The thickets are often somewhat wetter than the swamp forests. The main complex factors within the swamp forests are nutrient regime, pH-base richness, moisture–aeration, and light.

Fen

These are minerotrophic peatlands with water table slightly below, at, or just above the surface (Fig. 1.4). Usually there is slow internal drainage by seepage, but sometimes with oversurface flow. Peat depth is usually greater than 40 cm, but sometimes less (for instance adjacent to the mineral edges). Two broad types are *topogenous* (basin) fen and *soligenous* (sloping) fen.

The main physiognomic groups of fen are open fen and wooded fen (with tree cover, or a sparse tall shrub cover, sometimes called *shrub carr*). The latter indicates that the distinction between fen and peaty swamp forest is diffuse. In Table 1.2 we set the limit at 25% tree cover, but there is no universal agreement on this. The dominant fen vegetation could be bryophytes, graminoids, or low shrubs. Ground surfaces may be relatively firm, or they may be floating mats (quaking mats) occurring in basins or invading out over open water. The main complex factors are nutrient regime, pH-base richness, and moisture–aeration (and to some extent light).

Bog

Bogs (Fig. 1.5) are ombrotrophic peatlands with the surface above the surrounding terrain or otherwise isolated from laterally moving mineral-rich

Fig. 1.5   Bogs: (A) Open bog with dwarf-shrub-dominated hummocks (dark areas) and lawns with *Eriophorum vaginatum* and *Scirpus cespitosus* (light-coloured). Western Sweden, boreal zone. (B) Wooded bog with *Pinus sylvestris* and a dense shrub understory dominated by *Vaccinium uliginosum* and *Rhododendron tomentosum* (*Ledum palustre*). Eastern Sweden, boreonemoral zone.

soil waters. Some bogs are convex in shape (raised bogs), but bogs can also be quite flat or sloping, with slight rises at the margin that isolate them from incoming minerogenous water. The peat is usually more than 40 cm deep.

The main physiognomic groups are *open bog* and *wooded bog* (bog forest). In bogs with a pattern of hummocks and hollows, there is large variation in wetness. Ground surfaces may be relatively firm, but quaking mats occur along internal water bodies (bog pools) or in hollows at wet centres of raised bogs. The main complex factors influencing biotic variation are moisture–aeration and light. Since they are nourished only through precipitation there is less local chemical variation than among the fens. Bogs are extremely nutrient poor and strongly acidic; surface water pH is usually around 4 or even lower, but in some coastal areas the pH and the content of some minerals may be higher as a result of sea spray influence.

A word of caution

Bog and fen are old, vernacular words which were only given a strict scientific meaning during the last century. Even if the term bog is used in a scientific paper to describe a site, it may not be a bog according to the definition above. For instance, in North American literature until the 1960s the term 'bog' was used rather loosely for a *Sphagnum*-dominated peatland. After the influential publications of Sjörs (1959; 1963) and Heinselman (1963; 1970) the term bog became reserved for ombrotrophic peatlands, following the Scandinavian usage. However, the non-specific words continue to live on in names on maps, for instance, and botanical usage is still variable. In botanical literature, mire (*myr* in Scandinavia) is sometimes used as a collective term for fen and bog. In one sense this is logical since not all marshes and swamp forests are peat producing, but with the definition we use it is clear that those marshes and swamp forests that produce peat should also be considered as mires.

Environmental ordination

As indicated above, the two most important regimes influencing vegetational variation are moisture–aeration and pH-base richness. Since these are the principle determinants of biological variation, these two regimes are used to provide the reader with a simple, two-dimensional environmental ordination (see Fig. 1.1). The main ecosystem classes and physiognomic groups are positioned in this scheme. The wetlands occupy the total area of the ordination (and even go beyond the scheme), whereas the mires and peatlands are more restricted. We portray the mires as slightly broader than the peatlands, because the mires can exist on less than 30 cm of peat, expanding out over mineral substrates or aquatic sedimentary deposits. Ombrotrophic bogs occupy the left-hand side of the model, the nutrient poorest and most acid, whereas the minerotrophic fens, swamp forests, and marshes are on the right-hand side of the model, which has higher pH and usually more nutrients. The minerotrophic peatlands encompass a much broader range of nutrient and pH variation, and hence also a good deal more abiotic and biotic variation than the bogs. The general relationship between oligo-, meso-, and eutrophic classes can also be seen, but this portrayal disguises the fact that some high-pH sites are rather oligotrophic. To the right we find the peatlands on shallower peats or adjacent to streams and lakes. The vertical axis shows increasing dryness, progressing from open water through open types, wooded or thicketed types, and forested types.

From this scheme it is clear that the peatland category includes the bog and fen ecosystems, and covers a substantial part of the swamp forests. However, most of the marshes are not peatlands, and we do not deal with them in any detail in this book.

Frameworks for finer classification of peatlands

The Swedish system

A system developed in Sweden by Du Rietz and elaborated by Sjörs (1948) has been adopted in Norway (e.g. Økland 1990a), and has also had a large impact in other countries. In this system there are three main lines of vegetational variation, related to primary environmental regimes, but distinguished in the field on the basis of vegetation composition.

The environmental factors causing vegetational variation within site are the same as those governing the separation into the main ecosystems:

- the *bog–poor fen–rich fen* series, related to pH and base richness
- the *hummock–mud-bottom* series, related to moisture–aeration regime
- the *mire margin–mire expanse* series, related to the distance from the upland mineral soil.

The bog–poor fen–rich fen series

Ombrotrophic equates with bog, and minerotrophic with fen. The minerotrophic sites are further divided into poor fen and rich fen (and sometimes with a finer grouping into extremely poor, intermediate, rich, and extremely rich). In this system, 'rich' implies that the type is rich in a floristic sense and has higher pH and base richness (as reflected for instance by electric conductivity of the mire water). It is a common but unfortunate mistake to equate 'rich' as used here with 'nutrient rich' or eutrophic. Bogs are always oligotrophic, but rich fens can be either quite productive or oligotrophic (since P becomes unavailable at high Ca concentration).

There is considerable overlap, but the approximate pH ranges of the mire types are:

- bog, 3.5–4.2 (higher in oceanic areas)
- poor fen, 4–5.5
- intermediate and moderately rich fen, 5–7
- extremely rich fen, 6.8–8.

Northern temperate and boreal bogs are extremely acid because of the acidifying effect of the dominating *Sphagnum* mosses (Chapter 4) and the low buffering capacity of the incoming rainwater. In fens pH depends on the properties of the soil and bedrock that the water has passed, with rich fens occurring in areas with calcareous soil. In practice, plant indicators are used to recognize the levels of richness. In a first subdivision, bogs are recognized by the *absence of sensitive indicators of minerotrophy*, which are plants that cannot exist under the paucity of some mineral cations in the bogs. They have to be defined for a region, since they differ for different regions. Then the principle of sensitive indicators of successively richer sites is used to separate poor and rich fens. The various indicators extend only so far down into poorer conditions; for example, poor fens are defined by the absence of rich fen indicators.

Most of Sjörs's work has been in the open and sparsely wooded mires, bogs, and fens. A similar sequence can be recognized for the treed sequence: bog forest–poor swamp–intermediate swamp–rich swamp (see Fig. 1.1). This requires a different set of indicator species from the open mires (Jeglum 1991; Økland *et al.* 2001b).

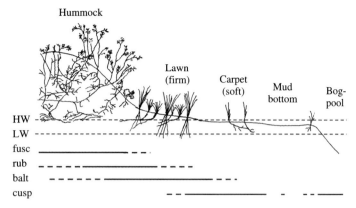

Fig. 1.6   Schematic presentation of the microtopographic gradient in a bog. The hummocks have aerated peat, which allows for the growth of dwarf shrubs. Lawns often have a dense cover of graminoids (e.g. *Scirpus cespitosus, Eriophorum vaginatum*) with dense rhizomes and roots making them firm. Carpets have a sparse cover of cyperaceous plants, whereas mud-bottoms are often inundated and lack plants almost totally. Pools may have some floating *Sphagnum* at the edge. Approximate levels for high water (HW) and low water (LW) are indicated. The distribution is indicated for the peat mosses *Sphagnum fuscum, S. rubellum, S. balticum*, and *S. cuspidatum*.

The hummock–mud-bottom series

Many workers have noted that the simple measure of depth to the water table from the ground surface has one of the strongest relationships with vegetational gradients in peatlands, and mire ecologists have followed the practice of Sjörs (1948) in dividing mires into microtropographic or microstructural levels along the water table gradient: hummock–lawn–carpet–mud-bottom–pool (Fig. 1.6).

- *Hummocks* are raised 20–50 cm above the lowest surface level and are characterized by dwarf shrubs. The lower limit of the hummock can be rather well defined in regions where there is a good correlation between the lowest level occupied by a certain species and the duration of flooding. An example is the lower limit of *Calluna vulgaris* used by Malmer (1962a,b).
- *Lawns* are most of the time 5–20 cm above water table; graminoids (cyperaceous plants, grasses, etc.) are dominant. Because of their strong rooting systems, lawns are so firm that footprints rapidly disappear. The moss cover is very diversified, and lawns seem to have the greatest richness in this respect.
- *Carpets* are often from 5 cm below to 5 cm above the water table. They often have a sparse cover of cyperaceous plants, and their bryophyte dominance make them so soft that a footprint remains visible for a long time.

- *Mud-bottoms* are often inundated and may almost totally lack vascular plants. They are often covered incompletely by horizontally growing mosses or by liverworts, but otherwise have exposed bare peat, often with a thin covering of algae.
- *Pools* are permanently water-filled basins, often with some vegetation at their edges.

This variation was described mostly from the perspective of fens and bogs, but one can apply these levels to swamps and marshes to obtain comparable characterizations of ground surfaces relative to water tables.

These terms were originally coined in his English summary by Sjörs (1948). They are not literal translations of his original Swedish terms, and this has caused some confusion, especially in British literature. In the original, Sjörs (1948) stressed that lawns are firm structures (*fastmatta*), whereas carpets are softer (*mjukmatta*). Mud-bottoms have a loose consistency (*lösbotten*) and they are usually covered by bare peat, in contrast with the common use of 'mud', which usually indicates a wet substrate with high mineral content.

These levels give many peatlands a characteristic patterning. In bogs we speak about a hummock–hollow microtopography and then use 'hollow' as a rather general term for the depressions between the hummocks (encompassing lawns to mud-bottoms). Sometimes the hummock level dominates, with hollows as narrow, wet pits without any segregation of distinct lawn–carpet–mud-bottom–pool phases. This is common in drier, continental, often wooded bogs. In British literature, 'hollow' is sometimes restricted to the carpet and mud-bottom levels (Lindsay *et al.* 1985). In patterned fens the wet mud-bottom dominated structures are termed *flarks* (in Finnish *rimpi*), as described in Chapter 10.

The mire margin–mire expanse series

This is the gradient one sees when walking from upland forests down into forested peatlands (e.g. swamp forests), then into sparsely wooded peatlands (e.g. wooded fens or bog forests), then into open mires. This gradient is related to the water table, deeper at the margin and shallower towards the open mires. It is a common gradient in boreal regions where wooded and forested peatlands are abundant, covering as much as half of the total peatland area. This gradient was recognized in Sweden in the work of Sjörs (1948), and more recently discussed for instance by Økland (1990a). It is also implicit in the Finnish system (see below). Most of the species of the mire margins also grow on dry ground. The differentiation of the mire margin communities is probably caused by simultaneous variations in several abiotic factors. The peat is normally thin, and deeply rooted vascular plants can reach the mineral soil beneath. The relatively dense tree and shrub layers create shade and shed litter that contains nutrients for the ground flora. The mire margins are in most cases swamps or wooded fens, although some ecologists include pine bogs among them.

Peatland classification in Finland

The system used in Finland is one of the most detailed and complex of peatland classifications. Its origin is based on early work of Cajander (1913) who recognized sites mainly on the basis of understorey vegetation and tree cover (since the system was developed largely for the purpose of drainage for forestry). The basis is water level and nutrient status, with the addition of information of supplementary nutrient influence, such as seepage and spring water (Eurola and Holappa 1985). The schematic presentation in Fig. 1.7 is from Ruuhijärvi (1983). As in Fig. 1.1, the trophic types are indicated on the horizontal axis together with the corresponding ombrotrophic–minerotrophic gradient. Wetness is depicted on the vertical axis with the main division based on degree of forest cover: treeless, sparsely covered by trees, and forested types. The types are grouped into four main types of peatland – *letto* (rich open peatlands), *neva* (other open wet peatlands), *korpi* (spruce and birch dominated peatlands), and *räme* (pine-dominated or hummocky peatlands). These types are represented by the letters L, N, K, and R, which can be combined, and also combined with other letters to denote various subgroups. Ruuhijärvi and Reinikainen (1981) list the Finnish abbreviations and give English translations for all types.

Wetland classification in Ontario

In pioneering work by Sjörs (1959; 1963) in the Hudson Bay Lowland in Ontario, Canada, the Swedish approach was used to characterize the peatlands. This has had a large impact on the peatland work in Ontario and elsewhere in Canada and the USA. Subsequently, since the early 1970s wetland classification in Canada has been greatly advanced by the efforts of an informal committee. This classification is hierarchical with four levels: class, form, type, and specialized needs (National Wetlands Working Group 1997). There are five classes: shallow open water, marsh, swamp, fen, and bog. These are divided into peatland forms, based upon the morphology of the peat body and its physiographic location (see Chapter 10). The finest classification, specialized needs, represents special purposes such as floristic associations, soil types, nutrient levels, and so on. A detailed classification of vegetation types was proposed in the wetland classification system for Northwestern Ontario (Harris *et al.* 1996), and this is one of the more comprehensive treatments to date in Canada of the total range of variation of wetland vegetation for a particular region. In this classification, 36 wetland vegetation types were recognized for the whole range of wetlands, including marsh, fen, bog, and swamp. For simplicity, these 36 types were combined by Racey *et al.* (1996) into 17 ecosites, on the basis of similarities in moisture regime, nutrient regime, soil, or substrate. These ecosites can be ordered in the same way as the Finnish and Swedish systems (Fig. 1.8).

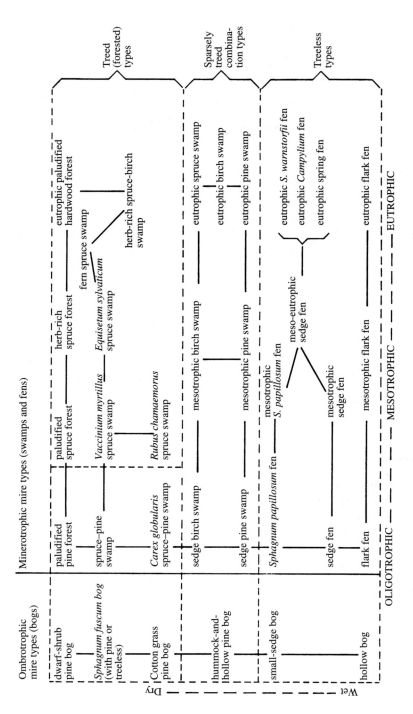

Fig. 1.7 A scheme of the Finnish peatland types (Ruuhijärvi 1983). Because the conifer species differ, this is not valid in North America.

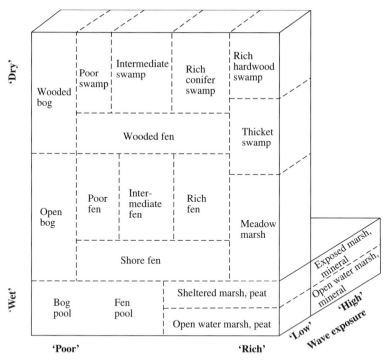

Fig. 1.8   A scheme showing the position of 17 wetland ecosites, recognized on the basis of similarities in moisture regime, nutrient regime, soil or substrate (Racey *et al.* 1996). The
scheme was developed for northwestern Ontario, and the ecosites are further subdi-
vided into 36 vegetation types. Largely valid also in boreal Europe.

Plant community classification

Phytosociological classifications are based solely on plant species composition. In large parts of Europe (except the Nordic countries and the British
Isles) the Braun–Blanquet system for classification of vegetation, including
peatlands, is widely used. We refer to Dierssen's work for an introduction
(1996) and in depth treatment (1982). Another phytosociological approach
is the National Vegetation Classification (NVC), which was developed in the
1980s as a standard tool for all vegetation types in Britain (Rodwell 1991;
Elkington *et al.* 2001).

2 Diversity of life in peatlands

Peatlands host a wide diversity of species and life forms. The plants are the most obvious, since they provide the structural foundation and are the source of organic compounds derived from photosynthesis for the sustenance of animals and microorganisms. In this chapter we give an overview of important life forms and the associated diversity of main groups within different types of peatlands, mainly in temperate and boreal areas.

Fungi and microorganisms

Heterotrophic microorganisms, that is, those microbes that obtain energy from degradation of organic material, consist mainly of bacteria and fungi. These organisms have high diversity in the peats of marsh, swamp, fen, bog, and their subtypes. We have already introduced in Chapter 1 the importance of the pH – base richness – nutrients complex factor and the moisture–aeration complex factor in determining the main lines of variation and diversity for higher vegetation, and these complex factors also underlie the diversity of microorganisms. However, for microorganisms a key factor is the nature of the peat, specifically its botanical composition and its quality (i.e. the amount of easily decomposed organic matter). In fact, the peat type and amount of quality organic matter are undoubtedly the most important factors influencing rates of decomposition, biomass growth, and diversity. Another key factor influencing both taxonomic and functional diversity of microorganisms is the degree of oxygen saturation, and this is the basis for classifying microorganisms into *aerobic* and *anaerobic*. An overview of the taxonomic and functional diversity of microorganisms in peatland was provided by Dickinson (1983).

Fen, swamp, and marsh peats often contain sedge material which influences the organic matter composition of the peat, and these peats have higher pH and nutrient contents than bog peats. Table 2.1, taken from Waksman and

Table 2.1 Occurrence of microorganisms at different depths of a probably somewhat drained fen ('low moor') peat profile from New Jersey, USA (Waksman and Stevens 1929). Numbers refer to counts per gram fresh mass

Depth (cm)	pH	Water content (%)	Bacteria (aerobic and facultative aerobic) and actinomycetes	Actinomycetes (%)	Fungi	Aerobic cellulose decomposing bacteria	Nitrifying bacteria	Anaerobic bacteria
Surface	5.9	61.1	6 000 000	90	1 050 000	++	+++	+
30	6.0	72.5	350 000	40	250	+	++	++
45	6.2	82.3	450 000	25	175	0	++	++
60	6.3	87.5	40 000	20	150	0	+	++
75	6.3	87.1	35 000	25	33	0	+	++
90	6.4	80.8	20 000	15	0	0	0	++
120	6.7	83.6	100 000	2	0	0	0	+++
150	6.8	84.5	500 000	0	0	0	0	++++
165	8.0	64.8	200 000	0	0	0	0	++++

+ designates a few; ++ a fair number; +++ abundance of organisms; ++++ numerous (about 250 000 or more colonies formed by 10 g of material).

Stevens (1929), presents some basic properties of the peat and microorganisms of a fen peat profile. There was an abundance of bacteria throughout the whole profile, which was 165 cm deep. Below 90 cm the number of bacteria actually increased, owing to the increase of anaerobic forms. *Actinomyces* and fungi were very numerous only at the surface of the peat, but diminished rapidly with depth. The same was true for aerobic cellulose-decomposing and nitrifying bacteria. Waksman and Stevens (1929) noted that this peatland had undergone a certain amount of drainage, as shown by the low moisture within the upper 30 cm of peat. The occurrence and abundance of microorganisms in *Sphagnum* bog peat with low pH presented a different picture. Nitrogen-fixing *Azotobacter* was not present, and nitrifying and aerobic cellulose-decomposing bacteria were completely lacking. The fungi were present only at the very surface and actinomycetes were almost entirely lacking.

Peatland bacteria include the genera *Bacillus*, *Pseudomonas*, *Achromobacter*, *Cytophaga*, *Micrococcus*, *Chromobacterium*, *Clostridium*, *Streptomyces*, *Actinomyces*, *Mycobacterium*, *Micromonospora*, and *Nocardia* (Williams and Crawford 1983). In a Scottish raised bog the aerobic bacteria were domin-ated by spore-forming *Baccillus* (50–60%). Other main groups were Gram-negative non-sporing rods (30%), and *Arthrobacter* (5%) (Wheatley *et al.* 1996). Aerobic bacteria were present down to 3 m, but numbers decreased drastically at the level where conditions started to be reducing. Facultative aerobic bacteria are normally found throughout the entire peat profile. A rather limited number of bacteria are actually obligate anaerobes.

Fungi are abundant in the aerated surface layers of the peat (see Table 2.1). Fungi characteristic of ombrotrophic peat bogs are of the genera *Penicillium*, *Cladosporum*, *Trichoderma*, *Mucor*, *Mortierella*, *Cephalosporium*, and *Geotrichum* (review in Bélanger *et al.* 1988), and several types of yeasts have been isolated (Nilsson *et al.* 1992). Outbreaks of a skin disorder (sporotrichosis) have been reported among workers handling peat, for instance in tree nurseries. However, it appears that the fungus causing the disease, *Sporothrix schenckii*, does not grow in living *Sphagnum*, but probably thrives during or after harvest (Zhang and Andrews 1993).

There is probably greater microorganism diversity in minerotrophic than in ombrotrophic peat. Nitrogen-fixing species, nitrifying bacteria, cellulose-decomposing bacteria, *Actinomyces*, moulds, and yeasts (e.g. *Streptomyces*) are all significant in minerotrophic peatlands. Fungi include those listed for bogs above plus *Fusarium*, *Cylindrocarpon*, *Arthrinium*, *Volutella*, and *Pseudeurotium* (review in Bélanger *et al.* 1988).

It is difficult to assess the species diversity of macrofungi (species that form fruiting bodies), since the production of fruiting bodies is erratic, and dependent on moist weather. A large Finnish study which spanned four seasons gives a good picture (Salo 1993). The peatlands included were mostly pine bogs, wooded fens, and herb-rich swamp forests. A total of 130 macrofungus species were listed from peatlands, about 50% of which were mycorrhizal (see Chapter 3). The rest were saprophytes (species that live on dead plants or animals), 35% living on plant litter and 15% on wood. Compared with the whole flora, the peatlands have a lower proportion of wood saprophytes and a higher proportion of mycorrhizal species. On the specialized side, nine species were saprophytes on *Sphagnum*, and at least one of these (*Lyophyllum palustre*) can parasitize live *Sphagnum* (see Chapter 4). In a database of Swedish macrofungi including about 4000 species, only 70 (2.2%) had bog or fen as their main habitat (Rydin *et al.* 1997), even though these habitats comprise about 11% of the country's total area.

Functional diversity of microorganisms

It is often more relevant to present the microorganisms in terms of their function than by species names. The functional classification can be based on what they feed on and which chemical transformations they perform in the peat. All fungi and most bacteria are heterotrophs, that is, they live on organic compounds produced through the photosynthesis of green plants.

The most important autotrophic microorganisms in peatlands are the photosynthesizing cyanobacteria ('blue-green algae' in older literature). There are free-living species of several genera (Table 2.2), and in addition there are species growing inside *Sphagnum* cells (Chapter 4).

Table 2.2 Number of taxa of microalgae and cyanobacteria per site in ombrotrophic bogs along a coastal-continental gradient from Manitoba to Newfoundland (n is the number of sites). Data from Yung et al. (1986)

	Coastal $n = 19$	Maritime $n = 3$	Continental $n = 9$	Common taxa
Desmids	29	23	11	Actinotaenium cucurbita, Cylindrocystis brebissonii, Netrium digitus, Penium silvae-nigrae
Other green algae	24	19	15	Spirogyra, Mougeotia, Binuclearia
Diatoms	8	4	5	Eunotia exigua, Pinnularia viridis, Tabellaria fenestrata, Frustulia rhomboides, Navicula subtilissima
Cyanobacteria	7	8	6	Anabaena, Nostoc, Calothrix, Microchaete, Scytonema, Stigonema Hapalosiphon
Other	7	9	5	Chrysophyta, Cryptophyta, Pyrrhophyta, Rhodophyta
Total	75	63	42	
Cl^- in mire water (μeq L^{-1})	>250	100–250	<60	

Diversity of chemical transformations by microorganisms

Oxygen is normally depleted under water-saturated conditions, since the oxygen consumption exceeds the rate of diffusion. Key factors influencing the functional diversity of the microorganisms are the degree of oxygen saturation and the related reduction–oxidation (redox) status of the peat. The degree of oxic condition is mainly determined by the degree of water saturation and temperature. At water saturation a number of alternative inorganic compounds may be used as terminal electron acceptors by the microorganisms. As reduction intensity increases, the electron acceptors will be used in the order oxygen (O_2), nitrate (NO_3^-), ferric iron (Fe^{3+}), sulfate (SO_4^{2-}), and carbon dioxide (CO_2) (see Chapter 5).

Bacteria have different roles in the nitrogen (N) cycle, and depending on the availability of oxygen they transform N between forms that are more or less available to plant uptake (Chapter 9). The functional diversity includes N-fixing species such as cyanobacteria, and the bacteria (actinomycetes) associated with *Alnus* and *Myrica* trees and shrubs. Many bacteria are aerobic decomposers, and their consumption of organic matter transforms organic N compounds to ammonium (NH_4^+). Under less acid conditions, aerobic nitrifying bacteria transform NH_4^+ to nitrite (NO_2^-) (e.g. *Nitrosomonas*) and further to NO_3^- (e.g. *Nitrobacter*). If nitrate is available this will be the first electron acceptor to be used after oxygen depletion; this anaerobic process is known as *denitrification* and results in production of either nitrous oxide (N_2O) or atmospheric nitrogen (N_2).

There are anaerobic Fe-reducing bacteria transforming Fe^{3+} to soluble Fe^{2+} in the anoxic zone. Where the water reaches the surface the Fe^{2+} is oxidized back to Fe^{3+} and then becomes insoluble. This process is common in fens and swamps, but not in bogs. The oxidation is promoted by bacteria such as *Leptothrix* species which gain energy from the process and build cases and a reddish, gelatinous deposit called *ochre*. Bacteria may also be responsible for the crust of Fe hydroxides sometimes seen around roots that leak oxygen in the otherwise anoxic zone.

Another typical process for waterlogged environments is that SO_4^{2-} is reduced to sulfide (S^{2-}) by anaerobic bacteria (e.g. *Desulfovibrio*), giving off a 'rotten egg' odour. Conversely, there are aerobic chemoautotrophic bacteria (e.g. *Thiobacillus*) that oxidize S^{2-} to S and further to SO_4^{2-}. As with several bacterial processes this occurs just at the anoxic–oxic border: higher up, S^{2-} would not be available, and further down oxygen is absent.

Two ecologically important groups of microorganisms are the methanogenic (CH_4-producing) Archaea (a different kingdom from Bacteria) living in anoxic peat, and a diverse group of CH_4-consuming bacteria, the methanotrophs, which oxidize CH_4 to CO_2. Transformations of CH_4 and CO_2 are important in the C cycle of peatlands, as we see in Chapter 12. For more details on anaerobic processes, see Sikora and Keeney (1983).

Protozoans

Under this heading we find several unicellular organisms. Important peatland groups are ciliates (exemplified by *Paramecium*) and rhizopods. Of the rhizopods, the testate amoebae are common in peatlands. They are amoebae with a shell; some species form the shell themselves, whereas others build it with material taken up from the habitat (Fig. 2.1). They can

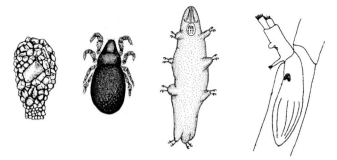

Fig. 2.1   Examples of the variety of microscopic life found in peatlands. From left to right: testate amoeba (*Difflugia oblonga*, 0.1–0.4 mm), mite (*Hermannia gibba*, c. 0.6 mm), tardigrade (*Macrobiotus hufelandii*, c. 0.5 mm), rotifer (*Habrotrocha roeperi*, c. 200 μm, with upper part of the body extending out of a pore on a *Sphagnum* stem). Illustrations from Hingley (1993), which is an excellent introduction to the diversity of small organisms found in peatland focusing on organisms living in association with *Sphagnum* mosses. Reproduced with kind permission from The Naturalist Handbooks and The Company of Biologists.

creep by extending the cell into pseudopodia, and they also feed as the pseudopodium encapsulates food particles. Testate amoebae occur in high diversity in *Sphagnum* vegetation, and there may be over 2000 individuals per cm³ (Tolonen *et al.* 1992), many of them with very precise habitat requirements. In a study of bogs and near-ombrotrophic peatlands across Europe, a total of 54 species was recorded. Each site had 16–40 species in samples taken from similar and very uniform *Sphagnum* lawns (Mitchell *et al.* 2000).

Testate amoebae as indicators of habitat

Many species of testate amoebae have well-defined ecological preferences for ecological conditions, such as water table depth and mineral concentrations, and they can be used as fine-scale ecological indicators. Since they have shells that are preserved in peat they can also be used to reconstruct previous conditions in the peatland. Most peatland species are cosmopolitan, and they are therefore useful when comparing conditions of peatlands from different biogeographic regions. An example to show how depth to water table can be predicted from the composition of amoebae species is given in Fig. 2.2.

Algae

Nowadays, algae are placed in the kingdom Protista. In ecological studies, cyanobacteria are often sampled and discussed together with microalgae. Peatlands are rich in microalgae, and their diversity seems to be related to

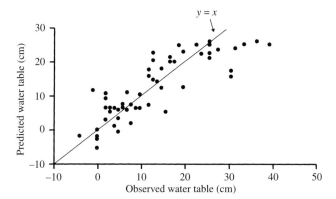

Fig. 2.2   Amoebae as indicators of water table in mires on Newfoundland. The optimum water level was determined for 40 species of testate amoebae, and a mathematical model developed to established the relationship between the abundance of these species and the water table at a sampling point. Each dot represent a sampling point, and water level predicted by the model corresponds well with the measured water table. From Charman, D. J. and Warner, B. G. (1997). The ecology of testate amoebae (Protozoa: Rhizopoda) in oceanic peatlands in Newfoundland, Canada: Modelling hydrological relationships for palaeoenvironmental reconstruction. *Écoscience*, **4**, 555–62, with permission.

the rich–poor gradient in much the same way as vascular plants. Species of *Euglena* and *Chlamydomonas* are very numerous, occurring in tens of thousands per cm^2 in *Sphagnum* mats (Hingley 1993). Of the Desmidiaceae a boreal rich fen could contain about 100 species, a poor fen about 35, and a bog area about 15 species (Flensburg 1965). In the same habitat diatoms may be more numerous, but there are fewer species (Hingley 1993), especially in the silica-deficient bogs.

Most microalgae have a wide geographical distribution, and largely the same species are found in Europe and North America. The diversity is not only related to the rich–poor gradient. Yung *et al.* (1986) found 252 taxa (including Cyanobacteria) in samples from 31 bogs in a stretch from Manitoba to Newfoundland. Their study included only 3 minerotrophic sites and these contained 118 taxa (89 taxa were common to bogs and fens). Within bogs there are more species of green algae (but not of diatoms or cyanobacteria) in coastal than in continental bogs (see Table 2.2). The reasons could be differences in water chemistry, but also that oceanic bogs have more patches with permanent open water. Deposits rich in Fe are often very rich in desmids.

Mud-bottoms are sometimes covered by large algal mats, notably (in bogs) by the violet-coloured filamentous green alga *Zygogonium ericetorum*, but species diversity is higher among submerged plants, such as *Sphagnum* in pools. *Sphagnum* hummocks are species poor; the diatom *Eunotia exigua* is one of the few found here.

Lichens

A lichen is a mutualistic association between a fungus (the mycobiont) and a species of green microalgae or cyanobacteria (the photobiont). The mycobiont makes up most of the thallus, and creates a suitable environment for the photobiont which in turn performs photosynthesis and thus feeds both itself and its fungal partner. In species with a cyanobacterial photobiont the lichen can also provide itself with N through fixation from the air. Lichens are normally included with the plants in ecological studies. There are three main life forms among lichens: *crustose* species grow firmly attached to the substrate, *foliose* species form a leaf-like thallus, and *fruticose* species look like miniature shrubs.

Some lichens have an outstanding ability to endure desiccation, and are prominent on rocks, tree trunks and other exposed habitats. Once they are dry, photosynthesis and growth cease completely. They can cycle between the active and passive states repeatedly, even within a day, but the overall growth rate of lichens is very low. Being particularly adapted to dry habitats, some lichens can become dominant on exposed high hummocks of treeless bogs, especially further north towards the northern boreal and tundra zones and also in some blanket bogs. There may be as many as 20 species of *Cladonia*, but most of them do not have peatlands as their main habitat. There are several large fruticose species known as *Reindeer lichen*. The Canadian lichen bogs have such a high cover of *Cladonia* that they appear white on remote sensing. *Cetraria islandica* is another large species that can achieve rather high covers in northern peatlands.

Crustose lichens typically grow on rocks and tree trunks, but there are some peatland species that grow on *Sphagnum*, such as *Absconbditella sphagnorum* and *Icmadophila ericetorum*. An almost aquatic lichen, *Siphula ceratites*, is found in shallow pools in highly oceanic bogs, both in North America and in Europe.

Plants

The focus here will be on the description of plant life forms that characterize various types of peatlands. Life forms are based on general morphology and size. Some life form classes coincide with taxonomic groups, such as bryophytes and lichens, whereas the life forms within vascular plants often include species that are not taxonomically related. Many bryophytes occurring in peatlands are wetland specialists, whereas a majority of the vascular plants also grow on mineral soil.

Since the species differ among regions, only a few typical examples will be mentioned to indicate similarities and differences between different parts of the world (especially north-west Europe and North America). Non-flowering plants are to a larger degree the same species in different parts of the world than are flowering plants.

Bryophytes

In systematics, the two large groups of bryophytes are the mosses and the liverworts. In peatland ecology, a useful life form classification is peat mosses, brown mosses, liverworts, and feathermosses.

Peat mosses are a taxonomically well-defined group, the genus *Sphagnum*, which is also easily recognized in the field. Because of their importance, the whole of Chapter 4 is devoted to their biology. They are dominant on the poor, low-pH side of the peatland range, and there are good grounds for dividing the mires into *Sphagnum* dominated (bogs and poor fens) and brown moss dominated (rich fens). The diversity of *Sphagnum* is often highest in poor fens, when counting either number of species or the proportion of *Sphagnum* to all bryophytes (Table 2.3). Even though *Sphagnum* takes a subordinate role in rich fens, there are several species that are

Table 2.3 Number of *Sphagnum* and other bryophyte species in peatlands in central Sweden (Sjörs 1948) and continental western Canada (Vitt and Belland 1995)

	Central Sweden			Continental Western Canada			
	Intermediate and rich fen	Poor fen	Bog	Extremely rich fen	Moderately rich fen	Poor fen	Bog
Sphagnum	23	27	10	9	7	16	12
Other bryophytes	51	35	26	58	28	28	41
Total	74	62	36	67	35	44	53

Fig. 2.3   Examples of brown mosses. Left *Scorpidium cossonii*; right, *Calliergon giganteum*.

restricted to such mires and are actually indicators of relatively high pH (e.g. *S. warnstorfii*, *S. contortum*, *S. teres*). High hummocks in rich fens are often dominated by *Sphagnum* species more typical of poor fens or bogs (e.g. *S. fuscum* or *S. rubellum*); the reason is that the height of these hummocks isolates them from the influence of the high pH of the groundwater, at the same time as the *Sphagnum* mosses further acidify the surface peat.

Brown mosses (Fig. 2.3), by contrast, are not a taxonomic entity, but rather an ecological group mostly characteristic of rich fens. However, the species of *Warnstorfia* and *Straminergon stramineum* occur under less rich conditions. Although species of the Amblystegiaceae family dominate (e.g. *Scorpidium* spp., *Calliergon* spp., *Loeskypnum badium*, *Calliergonella cuspidata*, *Campylium stellatum*), species from other families are also included (e.g. *Tomentypnum nitens* and *Paludella squarrosa*). Most but not all species have some degree of brown, reddish or yellowish brown, or even golden colour. The brown mosses indicate somewhat different degrees of richness; for instance, Kooijman and Hedenäs (1991) showed that in Sweden *Scorpidium cossonii* (previously *Drepanocladus intermedius*) indicates higher pH and electrical conductivity than *S. revolvens*. It is also an indication of higher pH when more species of brown mosses are encountered in a site and when the cover of brown mosses relative to *Sphagnum* increases.

Feathermosses are another ecological category. They are most abundant in fresh to moist coniferous forests, but they also occur in wet swamp forests. In peatlands they occur under forest shade, where they may share abundance with forest-dwelling *Sphagnum* species. The highest hummocks in treed bogs often hold feathermosses, and they expand especially in drained sites. Common species in Eurasia and North America are *Hylocomium splendens*, *Pleurozium schreberi*, and *Ptilium crista-castrensis*.

Albinsson (1996) listed 43 liverwort species from peatlands in southern Sweden. There are leafy species and thalloid ones, and in peatlands the former are the most diverse and abundant. However, to the eye they are rarely conspicuous. Typically, they occur as individual shoots in dense *Sphagnum* mats, and *Mylia anomala* is present in most mats. Occasionally, liverworts cover larger patches; examples are *Cladopodiella fluitans*, *Kurtzia setacea*, *Calypogeia sphagnicola*, and *Gymnocolea inflata* which sometimes can expand over mud-bottoms in bogs, appearing as dark green to black mats.

There are some peatland bryophytes that do not fall into the categories above. High hummocks in oceanic blanket bogs are often covered by *Racomitrium lanuginosum*. In swamps and in fens bordering mineral soil *Polytrichum commune* is common, and *P. strictum*, *Aulacomnium palustre*, and *Dicranum* spp. also form small hummocks in open mires.

Graminoids

Graminoids include grasses (Poaceae) and other groups with a grass-like morphology, such as sedges (*Carex*), cotton grasses (*Eriophorum*) and other Cyperaceae, rushes (Juncaceae), *Scheuchzeria palustris* (the only species in its family), and in the southern hemisphere some Restionaceae. Horsetails (*Equisetum*; a non-flowering plant) are also sometimes included. In northern boreal and alpine areas a large number of *Carex* species are encountered.

Marshes are characterized by a dense sward of graminoids, often with rather tall species and lush swards of sedges at the water's edge in well-nourished situations. Fens are also characterized by graminoids, but the swards are often shorter, and can have a wide range of densities. A few species of sedges (e.g. *Carex lasiocarpa* and *C. rostrata*) and grasses (*Molinia caerulea*) can cover large fen areas, and these have been utilized for grazing and haymaking (Øien and Moen 2001). In fens bordering open water, advancing and quaking sedge mats are formed by, for instance, *C. rostrata*, *C. lasiocarpa*, *C. limosa*, *C. chordorrhiza*, and *Phragmites australis*.

In *Sphagnum*-dominated peatlands the diversity of graminoids is quite low, especially in bogs, where, however, *Eriophorum vaginatum*, its American equivalent *E. spissum*, and often *Scirpus cespitosus* are frequently dominant. Wooded fens and thicket swamps have varying densities of graminoids, depending on the proportion of the lawn phase and tree cover. Also in swamp forests a number of grasses and *Carex* can be represented, but often as scattered plants with low abundance.

Herbs

Although the widespread *Menyanthes trifoliata* and *Rubus chamaemorus* are circumboreal, peatlands generally have a rather low diversity of herbs, but intermediate and rich fens hold a considerable number of species. Amphiatlantic examples are the carnivorous *Pinguicula vulgaris*, the orchid *Liparis loeselii*, and the 'spikemoss' *Selaginella selaginoides* (a non-flowering vascular plant). In Europe rich fens are often characterized by orchids, but in North America several orchid species occur in *Sphagnum* mires (*Arethusa bulbosa*, *Calopogon tuberosus*, *Cypripedium acaule*, *Habernia clavellata*, and *Pogonia ophioglossoides*).

Especially on thin peat, at the transition between rich fen and calcareous wet meadow, the herb diversity may be high. More species can reach the mineral substrate with their roots, and escape some of the nutrient limitation. Diversity may further be augmented by mowing or cattle grazing. This keeps large graminoids in check (their shade and litter will otherwise be detrimental to low herbs), and the trampling promotes seed germination. Sloping fens with their generally thin peat and moving water are sometimes rather herb-rich. Quaking sedge mats often hold a few herb species, for instance *Menyanthes trifoliata* and *Potentilla palustris*.

Bogs have few herbs (examples being *Rubus chamaemorus* and the carnivorous *Drosera*). In poor fens several species can occur, and the number of species increases at the mire margin (e.g. *Potentilla erecta*, *Trientalis*). Swamp forests can be herb-rich, especially if the peat layer is thin and the underlying soil is nutrient-rich or if soil aeration and nutrient uptake is improved by moving groundwater close to the surface.

Aquatic vascular plants

Some vascular plants grow in standing water. These species are graminoids or herbs, and many of them also grow on peat. Aquatic plants are grouped according to the location in relation to the water surface:

- *Submergents* have their leaves beneath the water surface (but may have flowers that emerge). Examples are *Utricularia*, *Potamogeton*, *Ceratophyllum*, *Myriophyllum*, and the peculiar stoneworts (green algae in the Characeae family) sometimes found in water bodies of calcareous fens.

- *Floating-leaved species* have their leaves at the water surface (e.g. *Nuphar*, *Nymphaea*, and some *Potamogeton* species).
- *Emergents* have their leaves above the surface; they are usually tall or very tall graminoids including *Phragmites australis*, *Schoenoplectus* spp., *Typha* spp., and *Equisetum fluviatile*, and a number of herbs such as *Calla palustris*, *Lysimachia thyrsiflora*, *Menyanthes trifoliata*, and *Potentilla palustris*.

Aquatic plants are obvious in marshes. A few species may occur where there is permanently open water in fens – both emergents and floating-leaved species occur along the edges of quaking mats in fens that border lakes. In flarks there are a few scattered emergent graminoids and herbs, often the same species (e.g. *Menyanthes trifoliata*) that are found on the peat surrounding the flark. Bog pools rarely have these aquatics, but when they do it could mean some weak connection of the pool to minerotrophic peats below.

Shrubs

Evergreen dwarf shrubs are typical for wooded bogs and hummocky parts of open bogs, and also for fen hummocks which can become high enough to form 'miniature bogs'. Examples are *Andromeda polifolia*, *Calluna vulgaris*, *Chamaedaphne calyculata*, *Erica tetralix*, *Empetrum nigrum*, *Kalmia polifolia*, and *Vaccinium* spp. (both deciduous and wintergreen species). In the shade of wooded bogs some species may be common: European pine bogs have *Rhododendron tomentosum* and *Vaccinium uliginosum* along with several of the aforementioned species, and North American *Picea mariana* bogs ('black spruce muskeg') host their relatives *Rhododendron groenlandicum*, *Vaccinium angustifolium*, and *V. myrtilloides*.

A few species such as *Andromeda polifolia*, *A. glaucophylla* (eastern North America) and *Vaccinium oxycoccos* grow in *Sphagnum* at levels somewhat below the hummock. Another bog shrub is the northern dwarf birch species *Betula nana* (Europe, Asia, and subarctic North America). In thicket swamps and in the transition between fen and swamp, several *Salix* species occur abundantly. Numerous low to intermediate-height shrubs characterize fens, especially in North America: *Aronia prunifolia*, *Betula pumila* *Myrica gale* (also in bogs in oceanic locations), *Potentilla fruticosa*, and *Salix* spp. In thicket swamps and understories of rich swamp forests there are numerous tall shrub species such as *Salix* spp. (Europe and North America), *Alnus incana* var. *rugosa*, *Ilex verticellata*, and *Nemopanthus mucronatus* (North America).

Trees

Tree species diversity is lowest in the boreal region, and increases through temperate, subtropical, and tropical. Many species overlap from mainly upland locations onto peatlands, adding to the diversity. If we consider only the main dominant peatland trees, we have one for northwestern China and western Siberia (*Larix gmelinii*); two for Hokkaido, Japan (*Alnus* spp., *Picea glehnii*); four species in the main boreal forest in North America (*Picea mariana, Larix laricina, Thuja occidentalis, Fraxinus nigra*); three species in western British Columbia, Canada (*Thuja plicata, Pinus contorta, Chamaecyparis nootkatensi*); four species in north-west Europe (*Betula pubescens, Picea abies, Pinus sylvestris, Alnus glutinosa*); about four in New Zealand (e.g. *Dracophyllum arboretum, Dacrycarpus dacrydioides*); about five in north temperate North America (e.g. *Acer rubrum, Fraxinus nigra, Picea rubrum, Ulmus americana*); about ten in the southeastern USA (e.g. *Liquidambar styraciflua, Nyssa sylvatica, Pinus elliottii, Taxodium distichum*); and scores of species in subtropical and tropical peatlands (see Chapter 11).

Species and life form diversity among peatland types

The relative role of the plant groups in boreal fens and bogs is shown in Fig. 2.4. The total species richness increases dramatically in the order bog < poor fen < rich fen. In relative terms (proportion of total species number), lichens and liverworts are important in bogs, and herbs and graminoids in fens. However, in absolute numbers there are roughly equally many liverwort species in all three mire types. Even though bogs and poor fens are dominated by *Sphagnum*, the number of species is very low in bogs. Rich fens have almost as many *Sphagnum* species as the poor fens, largely because small bog-like mounds are present also in many fen types. For the same reason the dwarf shrubs occur over the whole range of mires.

In Chapter 1 we introduced the wetland classification system for northwestern Ontario (Harris *et al.* 1996), and Table 2.4 summarizes species richness

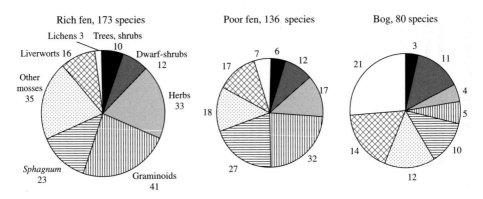

Fig. 2.4   Diversity of plants of different life forms in rich fens, poor fens and bogs, based on Sjörs's (1948) detailed investigations of a mire area in central Sweden. The sectors indicate the relative contribution to species richness, with labels showing the number of species.

Table 2.4 Vegetation profile for the 17 wetland ecosites of northwestern Ontario (Racey et al. 1996). The ecosites consist of combinations of finer vegetation types from the wetland survey of Harris et al. (1996) and the Ecosystem Classification for Northwestern Ontario (Sims et al. 1997). For each life form, the first number is the highest number of species found in a sample plot (plot size for forested type 10 × 10 m, for open types 5 × 5 m), and the second number is their average cover (% of area). Data provided by Peter Uhlig

	Submerged aquatics	Floating aquatics	Bryophytes, lichens	Herbs, graminoids	Trees, shrubs <0.5 m	Trees, shrubs 0.5–2 m	Trees/ shrubs 2–10 m	Subdominant trees >10 m	Dominant trees >10 m
Marshes									
ES49 Open water marsh on peat	10/57	4/30	1/1	5/4					
ES47 Sheltered marsh	8/18	5/9	2/4	10/80	2/<1	12/3	2/<1	1/<1	
ES46 Meadow marsh	4/3	3/<1	5/6	19/93	4/1				
Fens									
ES45 Shore fen	5/2	2/1	10/21	24/64	10/12	11/32	4/5	1/<1	
ES43 Open rich fen	2/<1		10/86	32/19	8/6	12/20	3/3	1/<1	
ES42 Open moderately rich fen	2/<1	1/<1	9/70	31/38	7/4	11/17	3/1	1/<1	
ES41 Open poor fen	1/1		9/86	16/23	7/2	9/18	3/3		
ES40 Treed fen	1/<1		10/88	31/19	8/3	10/27	3/20		
Swamps									
ES44 Thicket swamp		1/<1	7/24	26/63	8/7	9/28	5/67	3/6	1/1
ES38 Rich hardwood swamp			20/19	39/56	2/1	18/19	9/29	4/23	4/43
ES37 Rich conifer swamp			26/35	34/39	6/2	20/23	7/19	4/14	6/40
ES36 Intermediate swamp			28/77	33/27	5/2	25/33	8/20	3/9	4/35
ES35 Poor swamp			18/92	27/13	9/5	11/40	4/8	2/3	2/22
Bogs									
ES39 Open bog			13/97	10/17	7/3	8/26	4/4		
ES34 Treed bog			12/94	14/13	5/3	10/40	3/22	1/<1	2/2

among plant life forms in the 17 ecosites covering the whole range of wetlands, including marshes, swamps, fens, and bogs. In terms of the structural categories used, the most diverse ecosites are those in the swamp forests. These ecosites are more complex as they contain several members of all life forms. The least diverse are the marshes, and also the pools in fen and bogs. The most species rich was ES43 – open, extremely rich fen – and the swamps were also species rich.

In Chapter 1 we noted that two gradients, depth to the water table (the hummock to mud-bottom gradient) and pH (the rich fen to bog gradient) largely determine the plant species composition in northern peatlands. There are numerous studies and reviews describing this, both in Eurasia and in North America (Sjörs 1948; Malmer 1962b; Jeglum 1971; Vitt and Slack 1975; Vitt and Slack 1984; Økland 1990c, to mention just a few). Figure 2.5 illustrates this variation with species from northern Europe (Rydin *et al.* 1999). The variation in life forms is similar in North America, and many species (especially bryophytes) are the same.

Plants as indicators of habitat

As noted in Chapter 1, the absence of sensitive indicators of minerotrophy can be used to distinguish bogs and fens in the field, and other indicators are used to further divide the minerotrophic sites into poor fens and rich fens, and even finer divisions, keeping in mind that there are no sharp limits between these units. Many studies have shown that plant species can be effective predictors of ecological conditions (e.g. Sjörs 1948; Jeglum 1971; Gignac *et al.* 1991; Nicholson and Gignac 1995; Rydin *et al.* 1999; Tiner 1999). It is, for example, quite possible to make an assessment of pH from the presence of some plant species. It is also possible to draw conclusions about the normal wetness or water level even if a site is visited during an unusually wet or dry period.

There are a number of pitfalls in the use of indicators. Even if the environmental conditions are within the range of a certain species, this species may not necessarily be present. Species that are poorly dispersed are therefore not very useful, and for this reason bryophytes with their often almost circumboreal distribution are often better indicators than vascular plants. Another reason why bryophytes are good indicators is that they do not have roots – the whole plant grows in close contact with the immediate abiotic environment. Vascular plants may have an extensive root system, and collect nutrients over many square metres and from deeper, richer layers of peat or mineral soil.

Perhaps the greatest risk in the use of indicator species is when the relationship between species and environment has been established in one region, and then uncritically applied in another region. For example, *Eriophorum angustifolium* and *Sphagnum papillosum* are 'sensitive indicators of minerotrophy' in northern and eastern Sweden, but in the southwestern

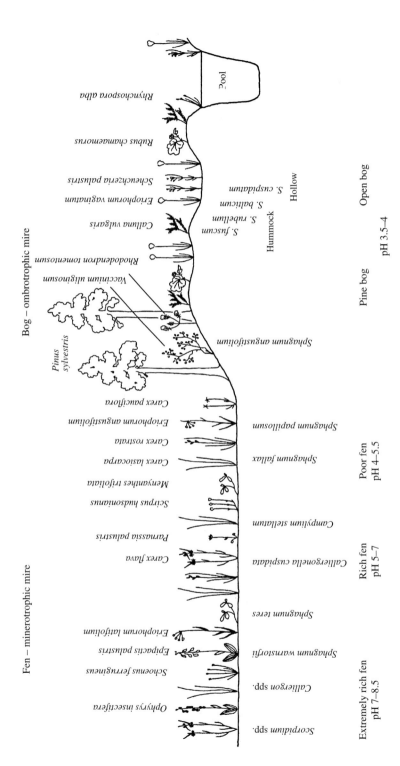

Fig. 2.5 Change in characteristic plant species along the gradient from extremely rich fen to bog. The species are exemplified from Northern Europe, but the bryophyte flora is very similar in North America. For the bog vegetation the difference between the wooded and open part and the variation from hummock to hollow is indicated. Illustration by C. Lind from an original by H. Sjörs.

parts of the country they occur in bogs (Rydin *et al.* 1999). Even more species enter the bogs in oceanic areas such as the British Isles (for example *Myrica gale* and *Menyanthes trifoliata*). The reasons for these shifts in distribution are not completely clear, but different temperature regimes, rain water chemistry, and higher and more evenly distributed precipitation in oceanic areas have been invoked as explanations. Owing to the coastal salt spray, bogs near the ocean are distinctly less acid.

Animals

The peatland animals broadly fall into three compartments: the aquatic component of the open water pools, the terrestrial fauna confined to the peatland (including the invertebrates that live in peat), and the terrestrial fauna that range beyond the peatland itself. Many insects do not directly prefer the peatland habitat, but as herbivores or pollinators they may instead be strongly linked to a certain host plant species. In addition, many upland mammals, birds, and insects regularly visit peatlands for food or water.

To exemplify the diversity of peatland animals, Table 2.5 shows the abundance of invertebrates in a Finnish pine bog, with well over a million individuals per square metre (Vilkamaa 1981).

Table 2.5 Density of invertebrates in peat in a dwarf shrub dominated pine bog in Finland, based on data in Vilkamaa (1981)

Phylum		Individuals m^{-2}
Tardigrada		5000
Arthropoda, Insects		
	Thysanoptera	60
	Collembola	4200
	Blattodea	1
	Psocoptera	1
	Hemiptera	158
	Lepidoptera (larvae)	11
	Diptera (larvae and adults)	137
	Coleoptera (larvae and adults)	92
	Hymenoptera, ants	164
	Other Hymenoptera	8
Arthropoda, Spiders		
	Oribateid mites	106 100
	Other mites	22 300
	True spiders	169
	Pseudoscorpionida	4
Arthropoda, Centipedes		9
Arthropoda, Millipedes		66
Nematodes		886 000
Enchytraeidae		4400
Rotatoria (Rotifers)		33 000

Arthropods

The arthropods include insects, spiders, centipedes, millipedes, and crustaceans. There are several ecological groups of peatland arthropods. Aquatic arthropods living in permanent bog or fen pools, or temporary waters, include larvae of species that are flying or living on the surface as adults. The species lists seen in the literature to some extent reflect methods of catching as well as ecological grouping. Arthropods may be collected from water in pools and ponds, extracted from the *Sphagnum* carpet, or caught with various trap types.

An example showing the diversity of arthropods sampled with pan traps is given in Table 2.6 (Blades and Marshall 1994). Pan traps are small beakers, 10–20 cm wide and a few centimetres deep, that are inserted into the peat surface. They are filled with water and a few drops of liquid detergent (to reduce surface tension, so that the insects that fall into the trap cannot escape) and sometimes painted yellow (to be more attractive to insects). The catch was grouped by trophic level, and the numbers of predatory and parasitic species are striking. Spiders and carabid beetles were most common and most rich in species among the predators. Predators are indeed common on mires, but the result is also an effect of their behaviour: they move around a lot on the peat surface and are more likely to be trapped than other species. The table shows that peatland specialists are in a minority; many species also occur in wet grasslands, heaths, and similar habitats.

Insects

The pantrap example (see Table 2.6) shows the enormous species diversity of peatland insects (see also Spitzer and Danks 2006). Diptera, Hymenoptera, and Coleoptera were by far the most species-rich orders. Some insects occur in very high densities, and peatlands are infamous for their mosquitoes and midges. A peatland specialist which is unpleasantly abundant is the Scottish or Highland midge, *Culicoides impunctatus* – densities exceeding 20 million per hectare have been reported.

A number of carabid beetles occur in mires. Only few of them are found in the completely open *Sphagnum*-dominated microhabitats, but with some shading by dwarf shrubs on hummocks or in wooded bogs there could be more than a dozen species, many of which are generalists of shaded habitats and forests. Among the mire specialists there are Old World species (bog specialist *Agonum ericeti*; fen specialist *Pterostichus rhaeticus*), New World species (bog specialists *A. mutatum*, *Bembidion quadratulum*; fen specialist *A. darlingtoni*) and also Holarctic species (fen specialist *Platynus mannerheimi*) (Främbs 1996).

Aquatic Coleoptera (water beetles) live in bog pools. In pools in blanket bogs in the British Isles 15–20 species can be trapped, and they share the habitat with Ephemeroptera (mayflies), several Corixidae (lesser water boatmen), and other Hemiptera (water bugs and water skaters, *Gerris* spp.) High

Table 2.6 Number of species of terrestrial arthropods collected in pan traps in peatlands in Ontario. Based on Blades and Marshall (1994), with permission from the Entomological Society of Canada and the authors

	Rich fen	Poor fen	Bog
Systematic groups			
Cl Arachnoidea; spiders, mites	100	78	78
Cl Crustacea, Isopoda; hog louse	2	0	0
Cl Diplopoda; millipedes	8	0	1
Cl Chilopoda; centipedes	0	1	1
Cl Insecta with following orders:	760	791	654
Ephemeroptera; mayflies	3	1	1
Odonata; dragonflies, damselflies	4	8	5
Plecoptera; stoneflies	5	0	0
Dictuoptera; cockroaches, etc.	0	1	0
Dermaptera; earwigs	1	0	0
Grylloptera; crickets	2	2	2
Orthoptera; grasshoppers	6	1	2
Psocoptera; bark lice	7	4	2
Hemiptera; true bugs	26	21	23
Homoptera; aphids, etc.	58	52	40
Neuroptera; lace wings	1	2	0
Coleoptera; beetles	193	174	189
Diptera; flies, mosquitoes	198	259	195
Siphonaptera; fleas	2	1	0
Trichoptera; caddisflies	8	5	4
Hymenoptera; ants, bees, wasps, etc.	246	260	191
Trophic groups			
Herbivores	155	171	131
Predators	194	184	166
Omnivores	120	100	97
Parasitoids	199	211	162
Detritivores	49	77	59
Fungivores	32	28	28
Epiphytic grazers	27	12	11
Others and unknown	94	87	80
Habitat preference groups			
'Bog species'	37	43	55
'Non-bog species'	65	45	29
'Widespread species'	342	338	279
'Unknown preference'	426	444	371

abundances can be reached by mayflies (e.g. *Leptophlebia vespertina*) and Diptera (e.g. *Chaoborus* and Chironomidae) (Downie *et al.* 1998; Standen *et al.* 1998). Table 2.7 illustrates the aquatic invertebrate diversity in bog pools.

An indication that the peatland fauna has been poorly studied is a report on chironomid midges in *Sphagnum*-dominated fens in Ontario (Rosenberg *et al.* 1988). In total 84 species were collected; 23 of them were new records

Table 2.7 Diversity of aquatic invertebrates. The total number of Canadian species and the numbers reported for bogs and fens is based on Danks and Rosenberg (1987). The diversity and density in Newfoundland bog pools is based on data in Larson and House (1990). Here, not all groups could be identified to the species level, so the real number of species is somewhat higher than the tabulated number of taxa. Densities are averages for all pools, ranging in size from <1 m² to >100 m². With permission from the Entomological Society of Canada and the authors

	No. of aquatic species in Canadian fauna			Newfoundland bog pools	
	Total	Bogs	Fens	No of taxa	Density
Arthropoda, Insects					
Ephemeroptera	301	1	0	0	
Plecoptera	250	0	0	0	
Odonata	195	63	22	13	66.6 m^{-2}
Hemiptera	138	33	61	10	3.2 m^{-2}
Coleoptera	579	107		27	13.4 m^{-2}
Trichoptera	546	9	2	9	20.2 m^{-2}
Diptera[1]	866	84	60	31	135.6 m^{-2}
Arthropoda, crustacean	NA	NA	NA	16	16.5 L^{-1}
Annelida, mostly Enchytraeidae	NA	NA	NA	4	13.5 m^{-2}

[1] The 866 dipteran species include only families Culicidae, Tabanidae, Ceratopoginadae, and Chironomidae.
NA = data not available.

for North America or Canada, and 10 of these were previously undescribed species. As many as 37 of these chironomids appear to be peatland specialists. The chironomids are understandably a good food source for birds – between 500 and 2000 individuals emerge per square metre each year (Rosenberg *et al.* 1988).

Among butterflies and moths (Lepidoptera), species richness is often higher on mineral land than on adjacent peatland. In mires these species as adults have their highest diversity and abundance in lagg fens and the nearby sparsely treed part of bogs (Väisänen 1992), and become more rare on open mire. This is a matter of habitat selection, and also an effect of individuals moving in from the surroundings. Väisänen (1992) caught 85 species in the pine bog and 32 in the open bog. Larvae of a number of butterflies, for example, *Papilio machaon* and several species of *Boloria*, feed on mire plants.

The dragonflies (Odonata) are represented by big and colourful species which are striking elements especially in productive fens and marshes.

Several ant species (*Formica* spp., *Myrmica* spp.) live in bogs and fens where they make nests by tunnelling in *Sphagnum*, *Polytrichum*, and *Dicranum* hummocks and other natural mounds, or building mounds with plant material, such as *Calluna* short shoots.

Spiders and mites

There are two main groups, the true spiders (Araneida) and the mites and ticks (Acarina). The true spiders are a conspicuous group on open mires, and it is likely that a large part of their food is insects that blow in from nearby richer habitats such as forests and grasslands. A Finnish study of open and treed bogs reported 21 species of wolf spiders (Lycosidae) which hunt on the ground without webs. Many of them showed a clear preference for different bog microhabitats, even if they were not restricted to peatlands (Itämies and Jarva-Kärenlampi 1989).

There could be more than 100 000 mites per square metre of *Sphagnum* in a mire, and the dominant group is the oribatid mites (Fig. 2.2; Table 2.5). Both diversity and abundance increase from the wet hollows to higher hummocks: Donaldson (1996) found 7–11 species with *S. cuspidatum* in the deepest hollows, 17–24 with *S. fallax* in intermediate levels, and 31–35 species with *S. magellanicum* in the higher hummocks.

Crustaceans

These are mainly Cladocera (water fleas), but also some copepods and ostracods in areas rich in nutrients or Ca. These taxa are important components of the aquatic fauna and crucial to the bog-pool food web (see Table 2.7). Most of the Cladocera are herbivores or filter feeders, and are found on the bottom sediments rather than in the water column. One exception is *Polyphemus* – a predatory cladoceran with huge eyes and good swimming ability.

Tardigrades, leeches, nematodes, and enchytraeid worms

The peculiar tardigrades, or 'water bears', look like miniature bears (<1 mm in size) and are quite common in moss carpets, as seen in Table 2.5. Leeches, *Hirudo medicinalis*, occur in small water bodies in fens. Enchytraeid worms are much more important in *Sphagnum* peatlands. In blanket peat it has been estimated that enchytraeid worms constitute up to 70% of the faunal biomass, and 90% of their individuals belong to *Cognettia sphagnetorum* (Springett and Latter 1977). Small nematodes are very common; they can be seen under the microscope crawling through the pores of *Sphagnum* leaves.

Flatworms

Dendrocoelum lacteum and *Polycelis tenuis* are macroscopic flatworms (from the group Tricladida) that occur in several types of limnic environments and also in fen waters. A more typical peatland group are the Catenulida, but it is somewhat uncertain whether they actually should be classified as flatworms. They are 0.5–3 mm in size and very abundant in *Sphagnum* – several hundreds can be extracted from a handful of moss, and there are some 20 species in peatlands. They have rather wide geographical distribution, and several species occur both in Europe and in North America (K. Larsson, pers. comm.).

Rotifers

In a survey across nutrient-poor mire types, Pejler and Berzins (1993) recorded as many as 328 rotifer taxa (see Fig. 2.2). Most of them have a wide amplitude, but 20–30 are probably bog specialists. As with microalgae, the diversity is highest in pools with submerged plants (about 60 species with *Utricularia* or *Sphagnum fallax*), but as many as 14 species were found on the highest bog hummocks with *Cladonia* lichens. In Newfoundland Bateman and Davis (1980) likewise found more species in hollows with *Sphagnum pulchrum*, *S. subnitens*, or *S. cuspidatum*. The abundance, however, was higher in the hummock, where the upper 5 cm of *S. fuscum* could contain over 400 individuals per squre centimetre. There are a number of species that attach to plants in ponds and pools. Some of them are rather host-specific, and important hosts are *Utricularia*, *Nymphaea*, filamentous green algae, and cyanobacteria (Wallace 1977).

Amphibians and reptiles

Salamanders occur in bog pools, for instance the four-toed salamander (*Hemidactylum scutatum*) in the USA (Tiner 1998), and the palmate newt (*Triturus helveticus*) and common newt (*T. vulgaris*) in British blanket bogs (D. Gibbons, pers. comm.).

Many frogs, toads, and salamanders depend on wet habitats, but wetlands other than peatlands are more important for most species. The pool frog (*Rana lessonae*) at the northern limit of its distribution in Sweden is dependent on small water bodies in rich fens for breeding (its winter quarters are cavities in the forest ground). This is an area of land uplift, and new fens are created close to the coast at the same time as the ponds farther inland are filled in and become inhospitable (Sjögren-Gulve 1994). Each population is therefore doomed to extinction, and local extinctions also occur randomly in sites close to the coast. There is some dispersal between these local populations (such a system of dispersal-connected populations is referred to as a *meta-population*), and the ability of the frogs to migrate to new sites is a prerequisite for the long-term regional survival of the species. It is very likely that other animals occurring in the same localities (for example the great crested newt, *Triturus cristatus*) depend on dispersal among the fens in the same way.

Among reptiles there are a few wetland species that frequent fens, bogs, or swamps. Examples in the eastern USA are spotted turtle (*Clemmys guttata*), bog turtle (*C. muhlenbergii*), painted turtle (*Chrysemus picta*), eastern ribbon snake (*Thamnophis sauritus*), and common garter snake (*T. sirtalis*) (Tiner 1998). In northwestern Europe, both the small lizard (*Lacerta vivipara*) and the adder (*Vipera berus*) occur in bogs, although they are essentially forest dwellers.

Birds

Northern peatlands have a wide diversity of birds, with several spectacular species. The common crane (*Grus grus*) in Europe and the sandhill crane (*G. canadensis*) in North America are the most conspicuous ones. It is an exciting experience to watch the lek (courting display) of the Eurasian black grouse (*Tetrao tetrix*) on the frozen bog on early spring mornings. Even if they mostly breed in forests, they need an open space, such as a bog, for the lek.

Among the treed peatlands, nutrient-rich deciduous swamps have a particularly rich bird fauna, mostly with passerine birds (a characteristic northern European species is the rustic bunting, *Emberiza rustica*), but also including for instance woodpeckers. These swamp forests are often more species rich than adjacent upland forests, because of the abundance of insects and availability of tree holes for breeding. The American spruce grouse (*Falcipennis canadensis*) and the European hazel grouse (*Bonasa bonasia*) are often found in swamp forests, as are the American woodcock (*Scolopax minor*) and the Eurasian woodcock (*S. rusticola*).

In open boreal mires, richness of bird species increases northwards. One reason for the high diversity of migratory birds in northern mires is the high productivity of insects (Järvinen and Väisänen 1978), and the long days available for foraging in the summer. A large number of waders have a decidedly northern distribution. Many of them are bound to the peatlands and other open habitats, and they also increase in population density towards the north (Boström and Nilsson 1983; Desrochers 2001). Another reason for the high diversity is that farther north the peatlands are bigger, more open, have more and larger water bodies, and overall contain more habitat types including large sedge fens, flark systems, willow thickets, and bog areas with pools. Waders and passerine birds with various habitat requirements are thereby attracted to large peatlands, as are raptors and other birds that require a certain territory size (Fig. 2.6). Examples of predators are the merlin (*Falco columbarius*), the hen harrier (northern harrier; *Circus cyaneus*), and the short-eared owl (*Asio flammeus*).

Examples of peatland waders occurring in both North America and Eurasia are the red-necked phalarope (*Phalaropus lobatus*) and the whimbrel

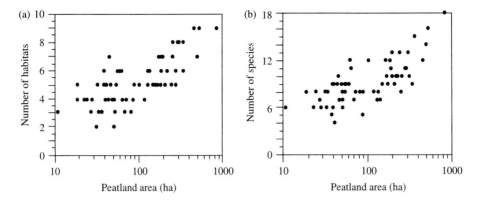

Fig. 2.6 Species – area relationships for birds on peatlands in eastern Canada. Bird diversity increases with the size of the peatland, which reflects habitat diversity. Note the logarithmic area scale. Modified from Desrochers (2001).

(*Numenius phaeopus*). Other Eurasian waders are Temminck's stint (*Calidris temminckii*), European golden plover (*Pluvialis apricaria*), broadbilled sandpiper (*Limicola falcinellus*), jack snipe (*Lymnocryptes minimus*), greenshank (*Tringa nebularia*), spotted redshank (*T. erythropus*), wood sandpiper (*T. glareola*), and ruff (*Philomachus pugnax*). For North America we can note the short-billed dowitcher (*Limnodromus griseus*), solitary sandpiper (*Tringa solitaria*), and lesser yellowlegs (*Tringa flavipes*).

Among passerine birds sparrows and warblers are typical for North American open peatlands, for example, Lincoln sparrow (*Melospiza lincolnii*), swamp sparrow (*Melospiza georgiana*), savannah sparrow (*Passerculus sandwichensis*), Harris's sparrow (*Zonotrichia querula*), white-throated sparrow (*Z. albicollis*), palm warbler (*Dendroica palmarum*), and common yellowthroat (*Geothlypis trichas*). Euroasian passerines are yellow wagtail (*Motacilla flava*), citrine wagtail (*M. citreola*), meadow pipit (*Anthus pratensis*), petchora pipit (*A. gustavi*), and dusky warbler (*Phylloscopus fuscatus*).

The dependence on mire area for bird diversity contrasts with vascular plants, which have high diversity in specific types of peatlands, such as small calcareous fens. An important effect of peatlands is that they add mosaic diversity to the landscape in areas with otherwise quite uniform areas of agriculture or forestry. In such cases even small swamp forests or fens can be important for bird life. Marshes are also rich in birds, but peatland marshes much less so than marshes with mineral soil.

Mammals

In *Sphagnum* peatlands, grazing animals have little to feed on, as the vegetation is dominated by bryophytes, dwarf shrubs, and sedges of low palatability. In contrast, thicket swamps, nutrient-rich swamps, and the fen – forest ecotone offer plenty of food for grazers and browsers, and also water during dry seasons. Important food items are grasses with broad and thin leaves, herbs, and deciduous shrubs (e.g. *Salix*). Swamps and fens are important feeding grounds for small rodents and for the largest animal, the moose (*Alces alces*). An example of a large animal even more dependent on peatlands is the white-tailed deer (*Odocoileus virginianus*) in parts of the northeastern USA. Its critical winter habitats are evergreen white cedar (*Thuja occidentalis*) swamps (Tiner 1998). Muskrat (*Ondatra zibethica*) is a North American species introduced into Europe in early 1900s. Muskrat and beavers (*Castor canadensis* in North America and *C. fiber* in Eurasia) are uniquely adapted to life in wetlands. They reshape their habitat by building dams and they can stay under water for many minutes, but largely these large rodents live in non-peaty wetlands, although they sometimes inhabit marshes and swamps with some peat. The damming activity of beavers may affect hydrology in surrounding peatlands, and as they fell trees they have large impacts on swamp forests surrounding the stream they occupy. The dams may also turn upland forests into swamp forests (see Chapter 7).

3 Adaptations to the peatland habitat

Peatland organisms live in an environment which is typically wet and nutrient-poor, partly anoxic, often acid, and in large parts exposed to wind and sun. In this chapter, we consider the many adaptations in plants and animals that enable them to inhabit peatlands successfully. For the fens and bogs in the northern hemisphere, many of the conditions described in this chapter are caused by the growth and accumulation of peat by the peat mosses, *Sphagnum*. Their peculiar adaptations are described in the next chapter.

Plant adaptations to flooding and anoxic conditions

Root growth and uptake of minerals and nutrients from the peat are processes that require oxygen. A high water table leads to anoxic conditions, especially in stagnant water. How far this will cause problems for the plants depends on how high the water table reaches, and the duration of the flooding. Some species are confined to sites with moving water where the aeration is better, such as sloping fens and springs.

Morphological adaptations to flooding

Aerenchyma – providing the roots with oxygen

Vascular plants which have their roots for longer periods of time in anoxic conditions must be able to transport oxygen to the below-ground parts. Such plants have *aerenchyma* – widened intercellular spaces leading from the leaves, through the stem and down to rhizomes and roots (Fig. 3.1). In addition to providing channels for the transport of air, the aerenchyma also increases the buoyancy of the plants, helping to keep them afloat. One can demonstrate the presence of aerenchyma by cutting off a rhizome or root section and blowing air bubbles into water with it. Aided by aerenchyma,

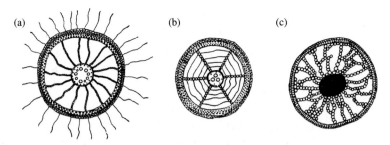

Fig. 3.1   Cross section of roots (c. 1 mm in diameter) of peatland plants with air-conducting tissue: (a) *Phragmites australis*, (b) *Eriophorum vaginatum*, (c) *Rubus chamaemorus*. From Metsävainio (1931).

peatland plants can have live roots deep down in the anoxic zone. The classic source of information for root morphology in peatland plants (Metsävainio 1931) mentions several deep-rooted species: *Calla palustris* (down to 70 cm), *Carex lasiocarpa* (75), *Equisetum fluviatile* (80), *Eriophorum angustifolium* (80), and *Phragmites australis* (100). *Carex rostrata* can transport gas from above ground to depths of at least 230 cm (Saarinen 1996). Even so, 50–80% of all roots in peatlands are found in the top 10 cm (Metsävainio 1931), where the living biomass is often much bigger than above ground.

Well developed aerenchyma also occurs in many other wetland plants, e.g. *Peucedanum palustre*, *Pedicularis sceptrum-carolinum*, *P. palustris*, *Pinguicula vulgaris, Caltha palustris, Molinia caerulea, Equisetum* spp., *Sparganium* spp., *Scheuchzeria palustris, Rhynchospora alba, Carex* spp., and *Scirpus* spp. In *Menyanthes*, the rhizome is formed above ground but penetrates down into the surface peat with long vertical adventitious roots. Peatland plants that lack aerenchyma, such as trees, dwarf shrubs, and many orchids, must therefore have superficial roots, or grow on hummocks were there is aerated peat above the water table.

Growing in water

Some plant species live entirely submerged. The leaves are often very thin, with a large surface area, and lack stomata. Some of these plants are rooted in the bottom, but others have no roots at all (for instance *Utricularia* spp.). Waters around these plants can be still, slowly moving, or rapidly mixing as in case of rivers and lakes with peatland margins. Such plants take up carbon dioxide (CO_2) and nutrients directly into the leaves from the water, just in the way that bryophytes do. Carbon dioxide is rarely limiting, but in waters with very high pH (as in calcareous fens) the availability of CO_2 is much reduced, and some plants have the ability to take up bicarbonate (HCO_3^-), which is then converted to CO_2 in the cell and used in photosynthesis. Examples are the stoneworts (Characeae), which are characteristic species in calcareous waters, and several species of *Myriophyllum* and

Ceratophyllum (Hutchinson 1975). Given that there are enough plants, they can produce the oxygen required for respiration themselves.

In several aquatic plants only the flowers emerge above water. The flower buds often start to develop underwater and are sent up above the surface rather late in their development. *Callitriche* even flowers underwater.

Some of the aquatic species have their leaves floating at the water surface (e.g. *Nuphar* and *Nymphaea*) but since they have roots and rhizomes in the bottom sediment they have the same problem as the emergent sedges, grasses and herbs; they need aerenchyma to provide the roots with oxygen.

A critical factor influencing the distribution and occurrence of submergents and emergents in marshes is the degree of exposure to wave action and currents. The submergents and floating-leaved plants tend to occur in sheltered locations with little wave action; only tall emergents such as *Phragmites australis* and *Schoenoplectus* spp., and underwater plants that are firmly fixed to the bottom such as *Isoetes* spp., can maintain themselves in exposed sites. However, in such situations the marsh would not have a peat substrate.

Tussock development

Even plants with aerenchyma will be at risk if the water table rises so much that their leaves become inundated. Wetland *Carex* species, for instance, are very tolerant to water saturation in the root zone (they have aerenchyma, and in general are not dependent on mycorrhiza). Development of high tussocks is a way for species to endure a highly variable water table (Fig. 3.2). This can be seen in swamp forests where the underlying soil has low permeability so that water table rises drastically after snow melt or heavy precipitation and then drops gradually. It can also be seen in some types of marsh, and along brooks with variable water flows (see e.g. Fig. 1.2). Examples are the stilt-like tussocks of *Carex cespitosa* and *C. nigra* var. *juncea* in Europe, *C. stricta* in North America, and *Juncus effusus* on both continents.

Many bryophytes and lichens that are themselves unable to form high tussocks avoid flooding by growing on top of *Sphagnum* hummocks, and other species grow on stumps, rotting logs, and the bases of trees, examples being *Ptilidium pulcherrimum* and *Cladonia* spp.

Floating mats and rafts

Another way to cope with water table fluctuations is simply to follow the water surface. Floating mats are common features where fens border lakes. Such *Schwingmoor* or quaking mires often consist of a *Sphagnum* or brown moss cover, but are actually held together and kept afloat by the rhizomes of *Carex rostrata*, *C. lasiocarpa*, *C. elata*, *Menyanthes trifoliata*, and other species. Sometimes parts of the mat break loose, forming floating rafts.

Fig. 3.2   Plant adaptations to flooding: (A) Tussock formation in *Carex nigra* var. *juncea*. (B) *Alnus glutinosa* forming stilts with adventitious shoots.

These may be transported by flooding or currents, and most plants are capable of establishing a root system once the raft has stranded.

Shallow root systems

In many peatlands there is a microtopography with hummocks extending so high above the average water level that they are never inundated. The aerated zone is, for most of the time, a few tens of centimetres deep. Lack of aerenchyma, in combination with a dependence on mycorrhiza, forces many woody species to have a shallow root system. Examples of trees that can develop root systems in the aerated zone are *Pinus sylvestris*, *Picea abies*, *Alnus glutinosa*, and *Betula pubescens* in Europe, and *Picea mariana*, *Larix laricina*, and *Thuja occidentalis* in North America. After the sapling stage, trees require a thicker layer for the roots, and areas with larger trees in a bog indicate that the oxic layer is thicker, as is the case at the margins of raised bogs. Also, tree species that in upland forests have a deep taproot (*Pinus sylvestris* in Europe, *P. contorta* in North America) are forced to spread their roots laterally in peat. As the trunk grows and becomes heavier, the tree base may sink deeper into the peat, at the same time as the surrounding *Sphagnum* continues to grow upwards. The ends of the lateral root system may then even be directed upwards as peat accumulates.

Several ericaceous dwarf shrubs characterize the hummock level. In Scandinavia, the lowest level of *Calluna vulgaris* is flooded only rarely and for short periods. The level correlates almost perfectly with the highest water level, and can therefore be used as a convenient definition of the lower limit of the hummock microhabitat (Du Rietz 1949; Malmer 1962a, b).

Developing adventitious roots and layering

A different way to survive the burial caused by tree weight and *Sphagnum* growth is for the tree to develop adventitious roots on the trunk. Adventitious roots can also develop on low branches of trees and shrubs that become overgrown by mosses. Outside the new rooting point the tip of the branch starts to grow upwards and to form a new main shoot. It will then be living on the new roots, and gradually loses contact with the mother plant. The phenomenon, known as *layering*, is common in *Larix laricina* and *Picea mariana* and leads to a shallow root system and at the same time to clonal expansion of the plant (Fig. 3.3).

Tolerance to flooding

Some plants have the ability to live underwater for some periods of time in anoxic or poorly aerated conditions. Bryophytes, lacking roots, usually stand flooding without difficulty. In places of periodic flooding, such as marsh, meadow marsh, or thicket swamp, woody species like *Acer negundo*,

Fig. 3.3   Development of adventitious roots is a way for *Picea mariana* to ensure that the active root system remains close to the surface in aerated layers. Layering is the predominant method of vegetative regeneration in *P. mariana* in bogs. Branches bending down to the ground are covered with *Sphagnum* or other mosses and develop adventitious roots, and the branch tips turn upwards and become vertical stems. Photo René Doucet.

Alnus glutinosa, Betula nigra, Fraxinus nigra, Myrica gale, Nyssa spp., *Quercus palustris, Salix* spp., and *Taxodium distichum* have the ability to tolerate shorter or longer periods of flooding. Often flooding occurs in winter or early spring, when the trees are less active and do not need as much oxygen. With lack of oxygen toxic levels of ethanol can be produced in plants, but it appears that this happens less quickly in flood-tolerant plants, and non-toxic malate is produced instead of ethanol (Crum 1988).

Plant adaptations to low nutrient availability

Nutrient conservation

A general way in which plants adapt to nutrient-poor habitats is by conserving nutrients. In Grime's (2001) classification of plant strategies, most mire plants are 'Stress-tolerators'. As opposed to 'Competitors' they have a low growth rate, and in contrast to 'Ruderals' they are long-lived and invest little into production of seeds. Where availability is low it is hard to increase uptake. A key feature of a peatland plant is instead to conserve nutrients. This is achieved by several mechanisms.

Long lifespan

Virtually all true mire vascular plants are perennial. This is a most effective way to ensure a large biomass, both below and above ground. In a nutrient-poor environment, a relatively large root biomass is required to obtain enough resources, and this cannot easily be built up within one season. Also, the large above-ground biomass which may be necessary for light capture in wooded mires can be built only by perennials.

Long-lived leaves

Mire plants typically develop leaves over some months in the beginning of the season, and these leaves are active for one growing season. However, there are several species that retain their leaves over one or several winters, such as *Empetrum, Rhododendron, Chamaedaphne, Andromeda, Erica*, and *Kalmia*. For leaves to have long duration and withstand frost in the winter they have to be stiff or leathery and often narrow and rolled-up (to the extent that they may be circular in cross-section). Such constructions are of course less efficient in catching sun energy and taking up CO_2 than are flat, broad leaves. Evergreen leaves also have low N content and therefore tend to have lower photosynthetic rate than deciduous leaves, but on the other hand they can be active for a longer part of the season. Most important in nutrient-poor peatlands is probably that once N is incorporated into the leaf it can stay for a long time and contribute to plant growth.

Somewhat paradoxically, the species with long-lived leaves give the impression of *xeromorphy* – a morphological adaptation to drought by reduced transpiration. The same plant species could be found in dry heaths, and the resemblance in vegetation between a bog hummock and a coniferous forest, with the dominance of ericaceous dwarf shrubs, is remarkable. Curled, leathery, and hairy leaves reduce water losses, and are common xeromorphic traits. These adaptations were noted by some of the earliest mire workers, who emphasized the low pH and nutrient levels, and anoxic conditions, as the main reasons. In bogs there is nothing to indicate that drought tolerance should be important in vascular plants (Small 1972; Chabot and Hicks 1982).

Cladonia lichens also occur on bog hummocks and in dry coniferous forests. For them the situation is different. As they do not have roots and grow exposed on the surface, drought tolerance is important even in a bog. These plants, as well as many bryophytes, cannot avoid desiccation, but they tolerate it and recover easily when re-wetted.

Recycling of nutrients

In species with broad deciduous leaves, seasonal transfer of nutrients to the roots and rhizomes in the autumn, and back to the new leaves in the spring, is crucial. Examples are few in *Sphagnum*-dominated mires (*Rubus chamaemorus* and some *Vaccinium* species), but in rich fens and swamp forests this strategy is more common. *Larix laricina* is a deciduous conifer that is able to translocate nutrients from needles to stem or roots in the autumn before the needles are shed. There are a few peatland plants that wilt completely in the autumn. Examples are *Drosera* spp., *Rhynchospora alba*, and *Utricularia* (especially *U. intermedia*). They condense nutrient for next year's growth in a winter bud, which gives a more rapid regrowth than development from a minute seed.

Many peatland plants (especially Cyperaceae) are clonal – they expand by having above-ground stolons or below-ground rhizomes from which new shoots emerge at intervals. In such plants nutrients can be translocated from old tissue to these new growing points of the clone, and thereby conserved.

Short lifespan on ephemeral substrates

In many other ecosystems dominated by perennials (for instance boreal forests), disturbances by animals open up the plant cover, expose mineral soil, and thus create good conditions for establishment of annual plants. However, in the mire, disturbances will only expose peat which is nutrient poor and prone to desiccation or flooding. Peatlands are therefore unsuitable for short-lived plants which require establishment from seeds or spores each year. Atypical cases are the remarkably well-adapted moss species of *Splachnum* and *Tetraplodon* (Splachnaceae) growing on droppings of large mammals (e.g. moose, *Alces alces*). The moss spores are dispersed from one dropping to another by flies, which also are dependent on the dung for their

Fig. 3.4   *Splachnum luteum* grows on dung patches on peatlands. The yellow umbrella-like base of the spore capsules attracts flies, which disperse the spores to other fresh dung-patches. Photo Tomas Hallingbäck.

breeding. Both the moss and the insect have to conclude their life cycle before the dung patch becomes inhospitable. The moss species attract somewhat different sets of flies (Marino 1991), and each moss is dispersed by 10–17 fly species. The adaptation is remarkable: the mosses have large capsules raised on long seta to attract the insects. In *Splachnum* the section under the capsule is bright red or yellow and in some species umbrella-shaped (Fig. 3.4). It seems also that the capsules produce volatile attractants, and the sticky spores attach to the fly (Bates 2000).

Mycorrhiza

A majority of vascular plant species form mycorrhiza. This is a mutualistic association of roots with a fungal species. The fungal partner receives carbohydrates from the photosynthesis of the plant, and the fungal mycelium (the subterranean network consisting of a large number of threads, hyphae) acts as an enlarged root system which helps the plant to capture nitrogen (N) and phosphorus (P). This association is often crucial for the plant's survival in nutrient-poor environments. Fungal growth requires oxygen, and non-mycorrhizal species are more common in peatlands than in most other ecosystems. In Britain, about 45% of the species in bogs and fens are rarely or never infected by mycorrhizae, compared with 34% in salt marsh, 26% in wet forest, 21% in heathland, 11% in deciduous forest, and 0% in coniferous forest (Peat and Fitter 1993). These statistics do not preclude that mycorrhizae are important, and even necessary in those peatland species that have them (and this could be more than half of the peatland flora).

There are different types of mycorrhizae. In peatland woody plants of Betulaceae (birches), Salicaceae (willows), and Pinaceae (conifers) have ectomycorrhiza. The root cells are surrounded by the fungal mycelium, and the fungi involved are often basidiomycetes.

Many more plant species (as much as 80% of all species) have vesicular–arbuscular mycorrhizae (VAM, also referred to as arbuscular mycorrhizae, AM, or endomycorrhizae). Here the hyphae penetrate into the root cells. The arbuscles are branched structures that increase surface area, and probably assist in the transfer of substances between the host and the fungus. Vesicles are probably storage organs for the fungus (Raven *et al.* 1999). Betulaceae and Salicaceae in peatlands may have both ectomycorrhizae and VAM at the same time (Thormann *et al.* 1999). Even though as many as 80% of all vascular plant species have VAM, the association probably only involves about 200 fungal species globally (zygomycetes). Orchids have their own special mycorrhizal type, and association with the fungus is required even at seed germination. Ericaceae (heather and related species) also have their own type of mycorrhiza. Here the fungus forms a web around the root, and this web releases enzymes that enhance mineralization and thus availability (particularly of P) for the plant root (Raven *et al.* 1999).

Mycorrhizae require oxygen and are therefore disfavoured by flooding. Cantelmo and Ehrenfeld (1999) found that in *Chamaecyparis thyoides* (Atlantic white cedar) the degree of VAM infection is much higher in the elevated parts of the mire hummocks than further down where arbuscles are almost absent. Even so, VAM seem more tolerant to flooding than ectomycorrhizae, and are more common in peatland trees and shrubs.

A general view has been that Cyperaceae, including the cotton grasses and sedges dominant in peatlands, do not have mycorrhizae. More recently, mycorrhizal infections have been found in quite a few species. Still, in a study of 23 *Carex* species (Miller *et al.* 1999), about half of the species had no or only few infections of arbuscular mycorrhiza, and mycorrhizae were especially rare in wet and acid habitats. Thormann *et al.* (1999) screened over 42 common species of bogs, fens, and marshes in Alberta, and although they found fungal contacts with Cyperaceae, no arbuscles were formed and their conclusion was that they are non-mycorrhizal. Conflicting reports on the role of VAM in some species depends on whether the presence of vesicles or arbuscles was used as the criterion, and possibly also on the sampling season. Turner *et al.* (2000) found arbuscles in most sedges, but this was in prairie fens with non-stagnant aerated groundwater. Some other common peatland species that probably lack mycorrhizae are *Equisetum fluviatile*, *Menyanthes trifoliata*, *Rubus chamaemorus*, and the carnivorous *Drosera* (Taylor 1989; Thormann *et al.* 1999), but there are also earlier reports indicating that they may in fact have VAM.

Biological N fixation

One way for plants to increase their N uptake is to form mutualistic associations with N-fixing microorganisms. Some Cyanobacteria are capable of biological N fixation in association with *Sphagnum*, and there are also species living freely among other bryophytes in the mire (Chapter 4). The common N-binders in mineral soils, *Rhizobium* bacteria, are not important in peatlands, since their plant partners, legumes, rarely grow in peat. Trees and shrubs of *Alnus* and *Myrica* have root nodules in which N fixation is performed by mutualistic Actinomycetes of the genus *Frankia*. Actinomycetes are bacteria, but they grow a fungus-like mycelium.

Carnivorous plants

One special adaptation to overcome nutrient deficiency is the trapping of small animals. Several carnivorous plants are found in peatlands, and they have different methods of catching and digesting invertebrates.

Insects are attracted by the scent produced by *Drosera* species (sundews). The leaves are strongly specialized, with glandular tentacles that produce a sticky and honey-smelling substance. The insect is caught by the sticky glands, and the tentacles slowly bend over around their prey which will then be digested by excreted enzymes. More dramatic is the sudden folding of leaves as soon as a prey touches their sensory hairs in the Venus fly trap, *Dionaea muscipula* (occurring in the southeastern USA). In *Pinguicula* species (butterworts) the leaves form a normal-looking basal rosette, but their surface is sticky with glandular hairs, and they present a similar danger to insects.

In the submerged *Utricularia*, small invertebrates (e.g. water fleas) are trapped in bladders. As soon as the prey touches the sensitive bristles at the entrance, the lid springs open. The bladder has a negative pressure, and as it opens the prey is sucked into it. The lid closes, and the bladder resumes its negative pressure (Sitte *et al.* 2002).

Finally, in *Sarracenia* and related pitcher plants the middle parts of their leaves form bowls. Insects, attracted by colour and the scent of nectar, are coaxed to move down inside the pitcher by slippery surfaces and downward-directed stiff hairs. This makes it easy for prey to fall in, and impossible for them to climb out. They probably drown in the water in the bowl, and secreted enzymes and activities from symbiotic bacteria digest the corpses. Quite a diversity of arthropods live in the pitcher and, strangely, avoid being digested. *Sarracenia* occurs in bogs and fens in northern and eastern North America, and is also introduced and naturalized in some European bogs (most abundantly in Ireland).

It is tempting to think of carnivory as a perfect adaptation to an extremely poor environment, but carnivorous plants actually occur in a wide range of habitats. Within *Drosera* and *Pinguicula* there are typical bog species: *Drosera rotundifolia* is common on rather high *Sphagnum* hummocks, *D. anglica* in wetter places, and the northern and alpine Eurasian *Pinguicula villosa* is restricted to high *Sphagnum* hummocks (it is not much bigger than the *Sphagnum fuscum* capitula, and very difficult to detect). Other species, such as *D. intermedia*, are mostly found in fens, and *P. vulgaris* even in richer ones. *Utricularia minor* and *U. intermedia* are found in pools in fens, and do not grow in ombrotrophic mires.

The specialized trapping structures photosynthesize less efficiently than normal leaves, so carnivory has a physiological cost. In open peatlands where light is abundant and nutrients are scarce, carnivory is thus a feasible strategy (Karlsson *et al.* 1996), but it would not be so in shaded, nutrient-rich habitats. With increasing N deposition other plants are favoured, and the carnivores indirectly disfavoured by increased shading and litter cover (Redbo-Torstensson 1994; Gotelli and Ellison 2002).

The contribution to the annual N turnover could be as little as a few per cent, but over 50% in some situations (Karlsson *et al.* 1996). A secondary effect is that capture of prey enables the plants to grow a larger root system, and thus enhance nutrient uptake through the roots (Adamec 2002). In calcareous fens the P from the prey may be more important than the N.

Direct and indirect effects of pH and calcium

The wide range of pH and calcium (Ca) contents in peatlands causes different problems, and a distinction is sometimes made between *calcicole* plants, adapted to conditions of high pH and high Ca concentration, and *calcifuges*, not occurring in such conditions. However, many plants occur at intermediate conditions with weakly acid pH and quite low Ca content.

At low pH, the hydrogen ion may cause direct physiological problems, but may also compete with other cations, making them less easily available for plant uptake (Tyler 1999). The toxicity of aluminium ions (Al^{3+}), released at low pH, is a most serious problem for plants on mineral soils, but much less so in peats where Al is rarely abundant.

At oxic alkaline sites, the problem is instead that manganese (Mn), iron (Fe), and most importantly P form insoluble compounds, and the amounts available for plant uptake are small. It appears that calcicole plants can exude large amounts of oxalic and citric acids from their roots, which increases the solubility of phosphate (Ström *et al.* 1994; Tyler 1999).

In many countries lime has been used to counteract lake acidification. Sometimes the lime is spread on bogs and fens, allowing it to slowly dissolve into downstream lakes. This leads to large changes in the peatland – most *Sphagnum* mosses are killed and may be replaced by naked peat and ruderal mosses. There may also be peat erosion (Hallingbäck 2001).

Shortage of pollinators

Many plant species depend on insect pollinators, and such insects are often rare on peatlands. Bog dwarf shrubs have separated flowering times. For instance, in Ontario the flowering sequence is *Chamaedaphne calyculata*, *Andromeda glaucophylla*, *Kalmia polifolia*, *Rhododendron groenlandicum*, *Vaccinium macrocarpon* (with wide overlap in flowering time only between *Andromeda* and *Kalmia*). The pollinators (e.g. bees) are quite generalist and serve several species, so it may well be that the differentiation in flowering time has evolved to avoid competition for pollinators (Reader 1975).

Adaptations in animals

Soil invertebrates living in peat have to cope with an environment that is often acid, nutrient poor and, worst of all, anoxic at depth, and in wet periods even near the surface. Even so, the peat fauna is species rich and abundant (Chapter 2) although largely restricted to the near surface – as much as 97% of all rotifers appear in the top 5 cm of *Sphagnum* (Bateman and Davis 1980). Ant mounds are common in bogs and fens. They build nests in hummocks, which become better aerated, warmer, and nutrient enriched by the tunnelling activity of the ants (Lesica and Kannowski 1998). Some ant species can also raise the height of the hummocks by several tens of centimetres with plant material (e.g. needles, *Calluna* twigs, and even *Sphagnum* branches). Carabid beetles on open boreal bogs spend most of their time on *Sphagnum* lawns, but these become flooded and frozen in winter, and many species migrate to better-aerated high hummocks with dwarf shrubs where they hibernate (Främbs 1996).

Among the aquatic invertebrates in bog pools, predators are common (Gibbons 1998). Here, the growth of algae is strongly nutrient limited, and one might ask how food chains with little plant production can sustain the predator populations. It appears that in waters with a lot of humic substances in the form of dissolved organic matter (see Chapter 5), the food chain is based not only on photosynthesis, but also on bacteria feeding on these humic substances (Fig. 3.5) which are constantly dissolved from the peat.

Many invertebrates are habitat generalists, but for aquatic invertebrates specializing in peatlands and similar habitats a number of traits have been suggested (Rosenberg and Danks 1987):

- Small size may be advantageous when nutrients are limiting and where water level fluctuates seasonally.
- Very precise microhabitat selection to avoid the most severe conditions.
- Tolerance to environmental extremes and anoxia.

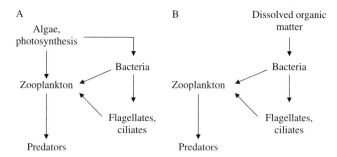

Fig. 3.5   Different bases for food webs in peatland water bodies: (A) Food web based on algal photosynthesis, such as may be common in shallow waters in marsh and nutrient-rich fens. (B) Food web based on input of dissolved organic matter in humic-rich 'brown water' such as in bog pools. Modified from Jones, R. I. (1992). The influence of humic substances on lacustrine planktonic food chains. *Hydrobiologia*, **229**, 73–91. With kind permission of Springer Science and Business Media.

- Suitable life cycle; either very rapid development during suitable period, or slow development with dormant stages that are resistant to adverse conditions.

Some invertebrates have elaborate ways of acquiring oxygen in peatland waters (Speight and Blackith 1983). Molluscs such as *Lymnaea* can live in oxygen-free water, and they climb up to the surface and store air in a mantle cavity. Some annelid worms can use their tail end as a periscope, while the rest of the body is immersed. The large mire spider *Dolomedes* can trap air in the hair covering the body, and this air can be used as the spider plunges into water in a defence reaction. The water spider *Argyroneta* lives almost permanently underwater. Many of these fascination adaptations in animals are not unique to peatlands; for instance, how to acquire oxygen in water is a problem to solve wherever animals occur in water. We refer the reader to the book *The Biology of Lakes and Ponds* (Brönmark and Hansson 2005) for a full treatment on aquatic adaptations in animals.

4 *Sphagnum* – the builder of boreal peatlands

Boreal peatlands would have neither their extent nor their particular features if it were not for the peculiar characteristics of the peat mosses – the genus *Sphagnum*. Peat mosses are very special bryophytes, particularly adapted to acid, cool, waterlogged, and extremely nutrient-poor conditions, and they create these hostile environments themselves. In short, *Sphagnum* mosses are successful in boreal mires because:

- they make their surroundings acid, wet and anoxic
- they tolerate and require only low concentrations of solutes
- they are resistant to decay
- there are a number of species that occur across peatland gradients.

If we boldly assume that half of the northern peatland areas are covered and built by *Sphagnum*, the astonishing result is that this moss genus covers over 1.5 Mkm^2 of ground and has stored some 300 Gt dry mass or 150 Gt of carbon (peatland areas are presented in Chapter 11, and carbon stores in Chapter 12). Clymo and Hayward (1982) suggested that there may be more carbon in *Sphagnum* than in any other plant genus in the world.

Morphology

A *Sphagnum* carpet is typically formed by vertically growing shoots with a density of 1–7 shoots cm^{-2} depending on species. There is an apical meristem in the top of the vertical stem (Fig. 4.1). As the shoot grows upwards the meristem gives rise not only to the new cells in the stem but also to side branches with numerous branch leaves and to leaves attached directly to the stem. In the apex, known as the *capitulum* (pl. capitula), the branches are tightly packed along the stem and do not grow to their full length, which

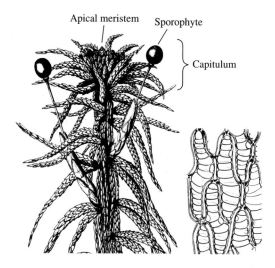

Fig. 4.1   A *Sphagnum palustre* shoot with sporophytes. The top meristem is embedded in the new, not yet fully developed branches. This cluster is called the capitulum. Below the capitulum the clearly distinguishable hanging (pendant) and spreading branches are seen. The close-up of the leaf shows the single layer of cells. The darker, narrow chloro-phyllous cells are alive and perform photosynthesis. The larger hyaline cells with their fibres and pores are dead and normally filled with water. From Schimper (1858).

gives the capitulum its conspicuous appearance. Further down along the stem (between 0.5 and 1.5 cm) the side branches extend to their full length. Here also the elongation growth of the stem takes place and it becomes clear that the branches are attached to the stem in clusters (Fig. 4.2). There is a species-specific number of branches per cluster (3–5 in most species). The branches are of two types – hanging (pendant) branches (more or less hanging down along the stem) and spreading ones. The branch leaves and stem leaves have similar anatomy, but in most species the stem leaves are tongue-shaped and the branch leaves typically longer and more pointed.

Peat mosses are notoriously plastic, and their morphology depends on the wetness and exposure of the habitat. In plants growing in water the leaves tend to be longer, and the difference between stem leaves and branch leaves smaller, than in plants growing well above the water table. In the higher positions the shoots are often smaller and more tightly packed (Rydin 1995). Many *Sphagnum* species have distinguishing red or brown colours, especially conspicuous late in the summer, but these tend to be poorly developed in shade or in very wet conditions. The variable colours and plastic responses make field identification more difficult.

When species like *Sphagnum cuspidatum* and *S. majus* occur in permanently or periodically inundated mire habitats they tend to grow horizontally instead of vertically. A larger part of the shoot is then exposed to sun, but

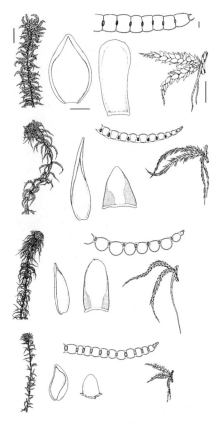

Fig. 4.2　Four *Sphagnum* species with wide geographical distribution representing different sections of the genus. From top: *S. magellanicum* (sect. Sphagnum), *S. cuspidatum* (sect. Cuspidata), *S. fuscum* (sect. Acutifolia), *S. subsecundum* (sect. Subsecunda). For each species the pane shows a shoot (scale bar = 1 cm), a branch leaf (left), a stem leaf (right) (bar = 0.5 mm) a cross-section of a branch leaf (bar = 0.01 mm), and a branch cluster (bar = 0.5 cm). Section Sphagnum (*S. magellanicum*) are distinguishable as very robust, large-leaved plants. Some sections differ in the way the chlorophyllous cells are embedded between the larger hyaline cells, as seen in the cross-section of the branch leaf. In section Cuspidata (*S. cuspidatum*) the chlorophyllous cells are triangular with the base exposed to the convex (outer) side of the leaf, in Acutifolia (*S. fuscum*) they are facing the concave side, and in Subsecunda (*S. subsecundum*) they are almost equally exposed to both sides of the leaf. Reprinted by permission from Crum, H. (1984). *North American flora*. Series II, Part 11. *Sphagnopsida*. Copyright 1984, The New York Botanical Garden, Bronx, New York.

the capitulum is still clearly discernible. In bog pools or lakes these species and others (such as *S. auriculatum*) can also grow submerged.

As the *Sphagnum* shoot grows upwards in the carpet, its lower parts soon become buried. Light penetrates only a few centimetres in the carpet; in a dense hummock perhaps only 1 cm, so photosynthesis is almost restricted to the capitulum. The lower part may still be alive, as witnessed by its ability to create a side shoot that can replace the apex if it is damaged. At 5–15 cm depth, however, the shoot becomes part of the peat and is dead (although still physically connected to the capitulum).

The leaves are only one cell layer thick, so each cell is in direct contact with the mire water. The leaf anatomy in *Sphagnum* is peculiar, with two cell types (see Fig. 4.1). The *chlorophyllous cells*, where all biological processes take place, are green and narrow with only a small part facing the leaf surface. They are embedded between the dead transparent *hyaline cells* which cover the larger part of the leaf.

The species are grouped into 'sections' within the genus *Sphagnum*. The division is largely based on microscopic morphology, especially the cross-sectional shape of the chlorophyllous cells in the leaf (see Fig. 4.2). It is useful to be able to recognize the sections in the field from their gross morphology and general appearance of the whole plants, because there is some overlap between classification in sections and the ecological grouping of species.

Sphagnum life cycle

A detailed treatment of the reproductive biology of *Sphagnum* is given by Cronberg (1993). *Sphagnum* are mosses, and therefore the shoots that we see are haploid, that is they have a single set of chromosomes (in contrast to vascular plants where the dominant life stage is diploid). The haploid shoot (*gametophyte*) carries the sex organs: male gametes are formed in flask-like structures, antheridia, embedded between the branch leaves (Fig. 4.3). Female gametes are located in archegonia in leafy structures on the stem (see Fig. 4.1)

Almost 75% of the species are unisexual (dioecious, carrying either antheridia or archegonia), and the others are bisexual (monoecious, carrying both antheridia and archegonia). For sexual reproduction, the male gametes must be transported in water to an archegonium, and a zygote is formed by fertilization. The zygote then grows to become a sporophyte which is the minute, ephemeral diploid stage of a moss resting on a short stalk (pseudopodium). In the sporophyte, haploid spores are formed during meiosis. *Sphagnum* has a unique mechanism for spore dispersal: as the capsule ripens and dries out, air pressure is built up. In the end the force is high enough to throw the lid off, and the spores are dispersed in an audible explosion. The mechanism ensures that the spores are lifted several centimetres in the air and thereby reach a moving airstream. It also ensures that the spores are dispersed in dry conditions – in damp air or during rain they would be immediately forced to the ground. When the spore germinates

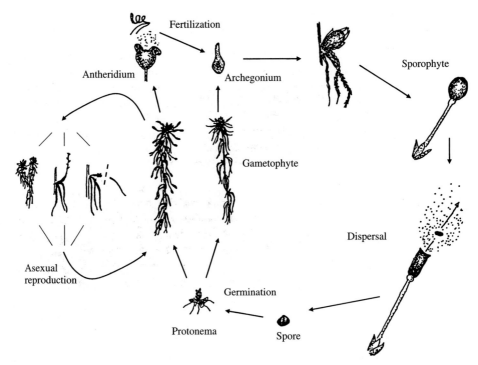

Fig. 4.3   Life cycle of *Sphagnum*. The main plant is the haploid (*n*) gametophyte, while the diploid (2*n*) sporophyte is supported as an organ growing on the gametophyte. In this example the species is dioecious, with male antheridia and female archegonia on different shoots. Drawing by Nils Cronberg.

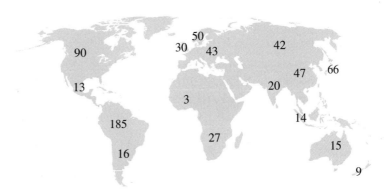

Fig. 4.4   Geographic patterns in species richness of peat mosses. Diversity is high in Eurasia and North America where there are vast peatlands dominated by *Sphagnum*, but there is an astonishing species richness in the New World lowland tropics, where *Sphagnum* is not abundant. Redrawn from Shaw *et al*. (2003).

it forms a *protonema*, which in *Sphagnum* is flat, one cell layer thick and about 1 mm wide. The new gametophyte grows out from the protonema.

Sphagnum can also disperse and expand over the mire surface by vegetative reproduction. A large capitulum can split its meristem into two, and side shoots can also emerge from lower parts of the stem. A method of vegetative reproduction that is probably common in some species is that fragments (capitulum or branches) can form new independent shoots. If the capitulum is detached the original shoot can still survive, as a side shoot emerges from the lower part to form a new main stem.

Diversity of *Sphagnum*

Because of the peculiar morphology of the peat mosses there has not been much disagreement about the taxonomic delimitation of the genus *Sphagnum*, but the global diversity is not completely known. There are about 100 well-known species in temperate and cold climates in North America and Eurasia, and perhaps another 10–15 in such climates in the southern hemisphere. To this a substantial number of poorly known tropical and subtropical species should be added (Fig. 4.4).

Quite a number of subspecies and varieties have been described, but only a few instances of hybrids. Because the morphology varies with the light and wetness of the habitat, it is difficult to distinguish a hybrid from normal variation within the species. Interestingly, there is now strong evidence that several species have originated as allopolyploids. This type of speciation starts with a sporophyte hybrid, when a male gamete from one species fertilizes a female gamete of a different species. Since sporophytes live only for a few weeks, sporophytic hybrids would be difficult to detect even if they were common. The next step is that spores are formed, but the normal reduction in chromosome number at meiosis does not take place. These allopolyploids therefore have 38 chromosomes in the spores and, consequently, in the gametophyte, whereas the basal number is 19 in the parent species. An example is the recently described *S. troendelagicum* which has a very small distribution in Scandinavia. The parental species are both widespread and common: *S. balticum* and *S. tenellum* (Såstad *et al.* 2001).

Sphagnum characteristics

Several biological features of *Sphagnum* are important for their role in peatlands (for reviews see Clymo and Hayward 1982; Andrus 1986; Rydin 1993b). The growth form can explain their water relations and how they affect mire hydrology. The peculiar chemical attributes explain how they acidify the substrate and slow down decay.

Capillarity and water-holding capacity

Vascular plants have stomata, that is pores that can open and close on the leaf surface to control gas exchange and water loss. Like all mosses *Sphagnum* lacks leaf stomata, and therefore cannot actively control water loss. This means that water lost by evaporation must be replaced either by rain or by a capillary movement of water from the peat below. There are no specialized cells in the stem to transport water, which is instead transported in a capillary network formed by spaces between the leaves, and between the stem and the branches.

It appears that species with hanging branches closely surrounding the stem have a particularly efficient capillary network; these branches act as a wick for water movement. Such species (e.g. *S. fuscum*, *S. rubellum*) can grow at a higher position above the water table than species with poorly developed and loosely packed hanging branches (e.g. *S. balticum*, *S. tenellum*) (Hayward and Clymo 1982; Rydin 1985). Even though mires are by definition wet habitats, the growth of *Sphagnum* is often limited by desiccation in the summer, especially in continental areas. The water table will drop during dry periods, and as water is lost by evaporation, the typical hummock species can keep their capitula wet by capillary water transport and maintain a high rate of photosynthesis (Fig. 4.5). Another factor here is that *Sphagnum* populations on the hummock are very dense, and the collective surface area

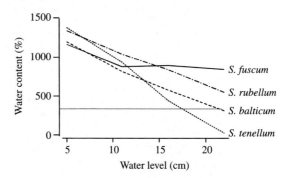

Fig. 4.5   Decrease in water content in the capitula of four *Sphagnum* species as the water table was experimentally lowered from 5 to 22 cm below the surface. The ranking in water content at the end of the experiment (22 cm) is the same as the ranking of how high up on the hummock the species can grow in nature. The horizontal line is the approximate water content needed for *Sphagnum* to maintain 50% of the maximum photosynthetic rate. As the water table drops below 20 cm photosynthesis is seriously reduced in the hollow species *S. balticum* and *S. tenellum*, but not in the hummock-forming *S. fuscum* and *S. rubellum*. Redrawn from Rydin (1985).

exposed for evaporation is rather small. In hollows, the carpets are much looser, and especially when the water table drops the carpet has a tendency to fall apart, with each shoot much more exposed to wind and desiccation. The result is that some species that grow in hollows will dry out rather quickly, even though their capitula are closer to the water table than those of the hummock species. Somewhat paradoxically, we can observe that hollow *Sphagna* are often, and hummock species rarely, limited by desiccation (Rydin 1985).

The upper limit that a *Sphagnum* species can reach on the hummock is thus related to its capillarity (review in Rydin 1993b). Intriguingly, some hollow species can grow rather high up on the hummock, but then only as individual shoots surrounded by hummock species. As an example, such individual shoots of the hollow species *S. balticum* that are completely surrounded by the hummock species *S. fuscum* receive water by lateral transport from the *fuscum* shoots. These *balticum* shoots may be somewhat drier than the surrounding *fuscum* carpet, but wetter than the continuous *balticum* carpet growing closer to the water table (Fig. 4.6; Rydin 1985). In general, hollow species have higher growth rate than hummock species

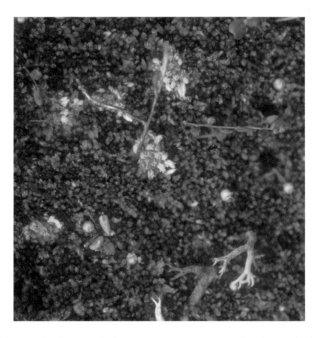

Fig. 4.6   Isolated shoots of *Sphagnum balticum* growing in a matrix of *S. fuscum* high up on a hummock. After a long dry period these *S. balticum* shoots are somewhat drier (as indicated by the whitish branch tips) than the surrounding *S. fuscum* carpet, but wetter than the continuous *S. balticum* carpet growing closer to the water table. The round light-coloured objects are spore capsules (not fully ripe).

when water is freely available (Rydin 1993b). We could then imagine that *S. balticum* should expand during wet periods, but expanded patches of *S. balticum* at the hummock level will not survive when drought sets in. A few vigorous individuals of *S. balticum* will maintain themselves in the hummock, protected by the surrounding moist *S. fuscum*.

In addition to capillary water transport ability, the peat mosses have a unique capacity to store water. The dead hyaline cells are usually saturated, but most of the water is stored between the leaves. If a bunch of *Sphagnum* shoots is put in a cylinder and submerged, and then allowed to drain freely for 20 minutes, the water content can be as much as 15–20 times the dry mass.

The consequences of the capillarity and water storage of *Sphagnum* are enormous for the ecosystem. With stagnant and continuously high water table, anoxic conditions prevail. Decomposition is hampered, and the peat layer gradually builds up. As peat thickness increases over time, the mire surface rises above the surrounding mineral soil surface. The mire will then create its own groundwater table, higher up than that of the surroundings. This is the fundamental process of bog formation (*ombrotrophication*) which isolates the plants from minerotrophic influence.

Chemical attributes

The concentration of essential elements and nutrients is low in *Sphagnum* peatlands in general and in bogs in particular (Chapter 8). *Sphagnum* mosses have special adaptations to cope with low levels of solutes. As the *Sphagnum* plant grows it continuously creates cation exchange sites. The active substances on these sites are uronic acids (galacturonic acid and 5-keto-D-mannuronic acid, 5KMA) (Painter 1995, 1998). The uronic acids appear in the cell wall as a pectin-like polymer, sometimes referred to as sphagnan (Børsheim *et al.* 2001). These acids exchange hydrogen ions for cations in the mire water at the carboxyl group.

Sphagnum also contains a number of phenolic compounds (historically referred to as 'sphagnol', although this term has been abandoned since it represented a mixture of different phenolics). The most common phenolic substance is sphagnum acid (*p*-hydroxy-*β*-carboxymethyl-cinnamic acid). A polyphenolic substance is sphagnorubin, which gives the typical colour to *S. magellanicum*. Verhoeven and Liefveld (1997) give an overview of these and other secondary metabolites in *Sphagnum*, together with their chemical structure. It appears that the uronic acids are responsible for cation exchange at low pH, as in bogs or *Sphagnum*-dominated fens, while the phenolic acids are active at higher pH (Richter and Dainty 1989). Note, though, that the phenolics are present in much lower concentrations than the uronic acids.

These organic compounds make *Sphagnum* litter very resistant to decomposition. Ecologists have often focused on the anoxic conditions as responsible for the preservation of peat and of objects preserved in peat. Børsheim *et al.* (2001) strongly argued that 5KMA is the important compound that prevents decay, and it acts by binding nitrogen (N) and making it unavailable for microorganisms. According to Clymo and Hayward (1982), 10–30% of the dry mass of *Sphagnum* is uronic acids, and their concentration is higher in typical hummock species, and also higher in shoots growing higher up from the water table. For this reason hollow species are less resistant to decay than hummock species (Johnson and Damman 1991; Hogg 1993), and in combination with differences in capillarity this helps us understand why only some species are hummock formers. 5KMA also acts by tanning objects such as the famous bog bodies (Chapter 6) and archaeological artefacts found buried in the peat (Painter 1995).

We saw above how *Sphagnum* could create ombrotrophic conditions as the peat accumulation alters mire hydrology. The organic acids produced by *Sphagnum* can explain another marked ecosystem shift. Mires with high pH (say ≥7) will gradually undergo a natural acidification, for instance by gradual leaching of Ca. As *Sphagnum* mosses colonize, the acidification accelerates and a decline from pH 6 to pH 5 can occur over a short period (Kuhry *et al.* 1993). This leads to a rapid shift from rich fen to poor fen (see further in Chapter 7) and the strongest change in vegetation that we can observe in boreal mires. The *Sphagnum* species responsible for this transition are not those that finally come to dominate the poor fen and bog. The latter species, for example, *S. fuscum* and *S. capillifolium*, cannot directly invade rich fens; it appears that the combination of high pH and high Ca concentration is fatal (Clymo and Hayward 1982). Examples of species that can grow in rich fens are *S. warnstorfii*, *S. teres*, and *S. contortum*, and a typical species for the rich to poor transition is *S. subsecundum*.

The cation exchange is one explanation why *Sphagnum* can grow in extremely poor habitats. Another factor is the ability to conserve nutrients. As the lower parts of the shoots are incorporated into peat, the plant faces the risk of losing essential nutrients and minerals. By tracer techniques (^{14}C, ^{32}P) it has been shown that *Sphagnum* can translocate metabolites to the growing capitulum from further down. This transport occurs internally and is dependent on the plant's being alive (Rydin and Clymo 1989). This is somewhat surprising, since *Sphagnum* mosses lack specialized conductive tissue. It is made possible since the cell ends in the stem are connected by small perforations (plasmodesmata) through which the transport occurs. Nitrogen is accumulated in new biomass, and it is likely that it is translocated internally in the same way (Aldous 2002).

Sphagnum as an environmental indicator

With the close link between *Sphagnum* and its environment, it is not surprising that the different species are useful indicators of habitat conditions. Presented with a list of *Sphagnum* species, one can say quite a lot about pH, Ca concentration, shading, water level, and oceanicity in the locality from which the list was assembled. Figure 4.7 demonstrates the affinities for a number of species to different habitat conditions, and there are several publications on *Sphagnum* as indicators (e.g. Andrus 1986; Daniels and Eddy 1990; McQueen 1990; Sims and Baldwin 1996). When using *Sphagnum* as environmental indicators one must be aware of the pitfalls with indicators as discussed in Chapter 2. Remember also that peat mosses create their own habitat, so the tables do not necessarily show in what sort of conditions the different species grow best, but rather a combination of the kind of habitats that they can tolerate and the type of conditions they create. The external influence of mineral soil water becomes less and less influential as one progresses from rich fen to poor fen to bog.

Sphagnum mosses can also indicate environmental impacts. We have noted their sensitivity to liming (Chapter 3), and below the impact of N is discussed (see also Chapter 9). In the Pennine area of the UK, large areas of *Sphagnum* were killed as an effect of industrial sulfur dioxide (SO_2) as early as the eighteenth century (Lee and Studholme 1992), and from this peat erosion followed on the blanket bogs.

Biological interactions in *Sphagnum*

Where *Sphagnum* grows it commonly completely covers the ground, by itself or together with other bryophytes, and also closely cohabits with a number of vascular plants (such as sedges and dwarf shrubs). Within a radius of 2 cm it is possible to find as many as five species of *Sphagnum* (Rydin 1986), which further suggests that there are ample opportunities for competition, both between *Sphagnum* shoots and with other plants.

Competition in *Sphagnum* carpets

Sphagnum grows in a single layer, and the interaction between individual shoots can be viewed as competition for space: the occupation of an area by one shoot precludes its occupation by another (Rydin 1986; 1993a; 1993b). The monolayer is maintained and kept very smooth by forces acting on individuals. Shoots growing more rapidly in height than their neighbours will be exposed to desiccation, which leads to diminished growth (Hayward and Clymo 1983). Conversely, reduced light levels stimulate the growth in length and this can save a shoot from being overtopped.

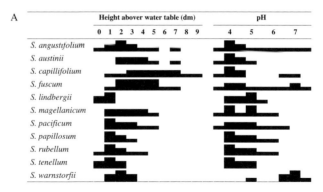

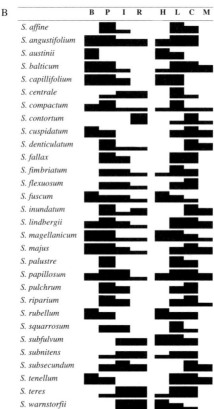

Fig. 4.7   Ecological ranges of *Sphagnum* species: (A) Range of common *Sphagnum* species along water level and pH gradients in western Canada. Three bar heights indicate the frequency of the species in mire spots with certain conditions: 1, present in <10% of sites; 2, 11–40%; 3, >40%. Based on graphs in Gignac (1992). (B) Occurrence of species along the bog to rich fen and hummock to mud-bottom gradients in Sweden. A higher bar indicates habitats where the species is commonly found. The habitats are Bog, Poor fen, Intermediate fen, Rich fen, Hummock, Lawn, Carpet and Mud-bottom. Rare species and species largely confined to swamp forests are not included. Redrawn from Rydin *et al.* (1999).

It follows that shoots with lower growth in mass must remain slender to keep their position in the moss surface and that therefore they will cover a smaller part of the surface area. In the extreme case, such shoots will be buried. Since *Sphagnum* cannot grow in several layers (as trees and shrubs can) competition can result in changes in mire area occupied by the competing species.

Interactions with vascular plants

Quite a number of vascular plant species, the typical inhabitants of bog and poor fen, depend on *Sphagnum* as their growth substrate. At the same time they compete with *Sphagnum* for light, space, and nutrients.

The mechanisms for competition between *Sphagnum* and vascular plants are rather peculiar (Fig. 4.8). First, mosses do not have roots, so they cannot compete for nutrients in the peat beneath the surface. Second, in bogs by definition, and to a large extent also in other *Sphagnum*-dominated fens, the nutrient input is not from weathering of rocks, but from precipitation. These peculiarities make competition highly asymmetric. For light competition, the advantage is of course to the vascular plants that overtop the *Sphagnum* carpet. But for nutrients, the photosynthetic capitulum is also the part of the moss which takes up nutrients, and the odd situation is that the *Sphagnum* mat will have first access to the nutrients as the rainwater percolates through their capitula. Nutrients in the rainwater are effectively

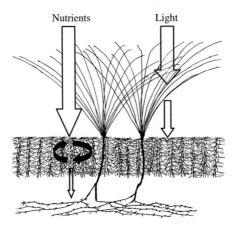

Nutrients   Light

Fig. 4.8   Conceptual model of the asymmetric competition between *Sphagnum* and vascular plants in an ombrotrophic mire. Light is reduced by the canopy of the vascular plant, and only *Sphagnum* is affected by the competition. Nutrients in precipitation are caught by *Sphagnum* and recirculated in its photosynthetic layer. The roots of the vascular plants are reached by nutrients that percolate through *Sphagnum* and are released by mineralization. A wider arrow indicates a larger amount of light or nutrients.

trapped by *Sphagnum* before they reach the root zone of the vascular plants. As *Sphagnum* also tends to relocate nutrients, the vascular plant roots are to a large extent constrained to nutrients passing through *Sphagnum* or mineralized in the dying zone of *Sphagnum* (Svensson 1995). With increasing amounts of N in the precipitation, the *Sphagnum* carpet will not be able to capture all of it. Increasing amounts will then reach the vascular plant roots and these will no longer be hampered by competition from *Sphagnum* (see also Chapter 9).

A serious threat for vascular plants is the risk of being buried in the moss if the upward growth cannot keep pace with that of *Sphagnum*. The risk is obvious for very small plants such as *Drosera rotundifolia*. Each autumn *Drosera* forms an overwintering bud containing the leaf primordia for next year's basal rosette (Fig. 4.9). The summer growth must enable *Drosera* to place this bud just beneath the *Sphagnum* capitula. To some extent it is easy for a small plant to follow *Sphagnum*, since the whole plant with its roots

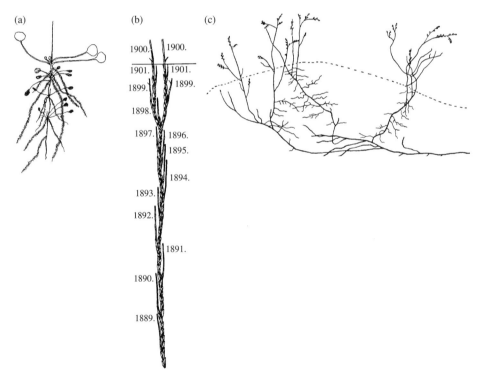

Fig. 4.9   The growth of vascular plants in a *Sphagnum* carpet: (a) Rosettes *of Drosera rotundifolia* formed annually at the *Sphagnum* surface. (b) Growth segments *of Scirpus cespitosus* indicating the annual growth of surrounding *Sphagnum*. (c) *Chamaedaphne calyculata* reinforcing the peat structure and facilitating the hummock formation. Drawings (a) and (c) from Metsävainio (1931); (b) from Weber (1902).

is lifted by the moss carpet. The situation is more dangerous for a tree seedling: as its root system develops it becomes anchored in the peat, and to develop into a sapling the height growth must be faster than that of *Sphagnum*. Seeds may germinate well in *Sphagnum* hollows, but survival is low because the seedlings are overgrown by *Sphagnum* and because of anoxic conditions in the root zone (Gunnarsson and Rydin 1998; Ohlson *et al.* 2001). Seedling survival is higher in the hummock where the annual height growth of *Sphagnum* is lower.

The fact that vascular plants with a strictly vertical growth form keep pace with *Sphagnum* can be used to measure the growth of *Sphagnum* retrospectively. For instance, *Scirpus cespitosus* forms a winter bud close to the moss surface each autumn, and annual moss growth can then be measured as much as 20 years back in time by the length of the yearly segments of *Scirpus*. (Fig. 4.9). This method was used by Backéus (1988) to relate annual *Sphagnum* growth to annual weather conditions.

Dwarf shrubs with branching above ground and formation of adventitious roots beneath the *Sphagnum* cover (e.g. *Betula nana*, *Calluna vulgaris*, *Rhododendron* spp., and *Empetrum* spp.) act as a supporting structure for *Sphagnum* (Fig. 4.9), and it may well be that the presence of these vascular plants is a prerequisite for the formation of hummocks (Malmer *et al.* 1994). Note that the reverse is also true – well-developed individuals of these species on bogs can grow only on *Sphagnum* hummocks.

Herbivory and other interactions

Sphagnum provides an environment for a number of invertebrates and microorganisms (Chapter 2), but since these mosses have very low N concentration, and high contents of organic acids, they do not seem to be eaten by any herbivores.

Lyophyllum palustre (= *Thephrocybe palustris*) is a common parasitic fungus which may cause necrosis in *Sphagnum* (Limpens *et al.* 2003). This fungus utilizes ammonia and amino acids, and there are indications that the increased N content in *Sphagnum* following from N deposition may enhance both growth and fructification and lead to more severe effects on *Sphagnum*. Overgrowth by algae is a potential hazard for *Sphagnum* in really wet situations, but it is unclear how common this is.

On the positive side, N-fixing cyanobacteria (notably *Nostoc*) live within the hyaline cells and in some habitats make a considerable contribution to the N economy of the mosses. It appears that they enter the hyaline cell through the pores. By their ability to live intracellularly, it is clear that the cyanobacteria can endure rather acid habitats (e.g. *S. lindbergii* at pH just above 4). There are also heterotrophic bacteria in the hyaline cell, and their relationship with the cyanobacteria seems mutualistic. The bacteria utilize fixed N and their respiration stimulates N fixation by increasing CO_2 levels and reducing O_2 levels (Granhall and von Hofsten 1976). Bog communities with the lowest pH (<4.0) seem to be too acid for this association, and above pH 5 N fixation could be much higher, but then it would be carried out by epiphytic cyanobacteria (Granhall and Selander 1973).

Dispersal and colonization

As *Sphagnum* often forms a dense cover, it leaves little space for colonization of new individuals. For most of the time a peatland is characterized by persistence of the *Sphagnum* species combined with some lateral clonal expansion. Thus, one could easily get the impression that dispersal and colonization in *Sphagnum* is of little ecological importance. However, there are occasions when establishment of new shoots must take place. Within the mire there are numerous small patches where *Sphagnum* dies because of trampling, fungal growth, bird droppings, or shading by dwarf shrubs or graminoids, and the species are able to re-cover these patches. On a larger scale we can follow *Sphagnum* colonizations on land exposed after glacial retreats, on new islands or shores formed in regions with land uplift, on bare peat exposed after peat extraction, and so on.

Vegetative dispersal and establishment

The best evidence for the capacity for vegetative dispersal comes from peatland restoration experiments. Left to themselves, virtually no recolonization takes place in sites left after industrial-scale peat extraction even after decades, and the method used to establish a new *Sphagnum* cover involves spreading of fragments of healthy mosses from a nearby site (see Chapter 13). In natural ecosystems, colonization by fragments seems most likely in patches with bare peat that are permanently wet, or in strongly shaded depressions, for instance in mud-bottoms or depressions in swamp forests. Foster (1984) gave examples of *S. lindbergii* encroaching on bare peat, and similar ruderal characteristics have been attributed to *S. tenellum*, *S. molle*, and *S. compactum* (Heikkilä and Lindholm 1988; Økland 1990b; Slack 1990).

Spore dispersal and establishment

Most *Sphagnum* species produce spores bountifully. The number of capsules that give ripe spores varies according to wetness, and in some years there may be almost none. Dry weather is harmful at the stage of the initiation of sex organs (probably late summer in most species), fertilization (often in the early spring), and sporophyte maturation (Sundberg 2002).

As with growth, a hummock former like *S. fuscum* with good capillarity is able to produce spores even in dry years when other species fail. At the dispersal stage, dry and sunny conditions are required, and wet weather can prevent spore dispersal.

The number of spores produced in a good year is astonishing. Each capsule can hold from about 20 000 spores (*S. tenellum*) to over 200 000 (*S. squarrosum*) (Sundberg and Rydin 1998), and the total output in a mire can be over 15 million spores per square metre (Sundberg 2002).

The ecological significance of these spores has long been questioned, since spores have failed to germinate in experiments on natural peat and peat water because of the low phosphate concentrations. However, experiments have now shown that even trivial nutrient sources, such as leachate from decaying birch leaves, are sufficient to promote germination (Sundberg and Rydin 2002). More substantial phosphate sources may be moose dung or bird droppings, but in either case the protonemata probably require some protection from desiccation by, for example, litter or vascular plants.

Spores buried in the peat can remain viable for decades (Clymo and Duckett 1986; Sundberg and Rydin 2000). These may be a source for re-establishment after small disturbances, even though wind-dispersed spores are probably much more important. Several instances of long-distance dispersal in *Sphagnum* point to the ecological importance of the *Sphagnum* spore. A classical case is *S. lindbergii*. In Sweden it has a northern distribution, but with scattered occurrences in southern Sweden in peat pits (Sjörs 1949). Soro *et al.* (1999) noted appearances of the rare species *S. aongstroemii* and *S. molle* in abandoned hand-dug peat pits in eastern Sweden, scores of kilometres from their nearest known localities. The presence of several other species in the peat pits (notably *S. fimbriatum*, *S. riparium*, and *S. compactum*) suggests dispersal in the order of kilometres.

Dynamics and persistence in *Sphagnum* assemblages

Fine-scale dynamics over 1–10 yr

Dynamics in *Sphagnum* can be studied on three time scales:

- Experiments and permanent plots followed by a researcher rarely go beyond a decade in duration, and the scale of interest has often been cm^2 to dm^2.
- Retrospective studies using old aerial photos or previous detailed investigations can reveal changes over several decades. Here we are mostly concerned with population and vegetation changes at the community scale.
- For the century–millennium time scale, we resort to stratigraphic data.

The *Sphagnum* shoots form a carpet which expands over the mire surface by lateral growth and vegetative reproduction. A clonal front may expand at a rate of 1 cm yr^{-1} (Rydin 1993a), so there is potential for complete species turnover at the dm^2 scale within a decade.

A number of transplant experiment and permanently marked *Sphagnum* patches were followed for several years, in some cases as many as 11 years in Sweden (Rydin 1993a). The experiment contained several dominant bog species in the following order of occurrence from hollow to hummock: *S. cuspidatum, S. tenellum, S. balticum, S. rubellum,* and *S. fuscum.* As predicted from their poor ability to keep wet, when transplanted to high hummock positions the hollow species died because they dried out. Both hollow and hummock species survived well at lower positions along the gradient, but, surprisingly, the hummock species were not outcompeted by the hollow species. Changes over the years were instead rather erratic. Most interesting was perhaps that no cases of competitive exclusion were observed despite the fact that the transplanted *Sphagnum* patches were only 6–8 cm in diameter. The difficulty is then to explain why the hummock species cannot establish and maintain themselves closer to the water table. It could be that occasional very high water levels can be detrimental to the hummock species, perhaps in combination with expansion of algae during periods of flooding.

In southern Norway, Nordbakken (2001) analysed almost 7000 plots of 4 × 4 cm in 1991 and again after 5 years. All *Sphagnum* species (*S. cuspidatum, S. tenellum, S. balticum, S. rubellum, S. fuscum*) decreased in plot frequency, and the area with naked peat or covered by plant litter increased. In quite a few instances *S. tenellum* gained area from *S. fuscum* and *S. rubellum,* The study was made in a well-developed raised bog where such directional changes are unexpected, and most likely the results were caused by particular weather during the study period.

Long-term dynamics in *Sphagnum*

The species composition in peat cores can be used to reveal long-term changes in *Sphagnum* assemblages (see also Chapters 6 and 7). Since *Sphagnum* species indicate different degrees of wetness of the habitat, such changes can be linked to the past climates. A well-known example is the disappearance of *S. austinii* (referred to as *S. imbricatum* in the earlier accounts of these changes) in the British Isles during the period 1400–1800 AD, probably caused by a wetter climate (Mauquoy and Barber 1999).

Peat cores can also be used to study competition and coexistence. Rydin and Barber (2001) made use of stratigraphic data from 10 × 10 cm excavated columns from England where ^{14}C-dating was used to establish a depth–time relationship. A salient result was that a *Sphagnum* clone can

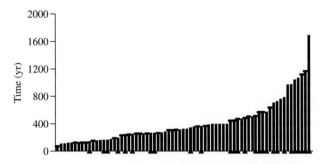

Fig. 4.10   Duration of single-species occurrences of *Sphagnum* (*n* = 54) in peat profiles in northern England. Based on data from 10 × 10 cm peat profiles in Barber (1981; see further Rydin and Barber 2001). A bar across the top of a vertical bar indicates that these occurrences reached to the very top of the profiles, that is, they were interrupted by the sampling. A triangle at the base indicate that they could be traced all the way to the lower end of the peat profile.

persist within the plot for centuries (Fig. 4.10). As old respiring tissue is gradually buried in peat the growing capitula are 'eternally young', and the shoots we see on the mire surface today may well be the same genetic individual that was established when the mire was formed thousands of years ago.

5 Peat and organic soil

Peat is an organic material composed mainly of dead plant matter in various stages of decomposition, accumulated under conditions of excessive moisture. Although most of the peat matrix is dead, peat contains a great variety of living organisms, in particular the microorganisms that act to decompose the plant remains. In addition, many diverse invertebrates live in the surface layers of peat, and living roots of some vascular plants are able to penetrate below the groundwater level and provide the microorganisms with new carbon sources in the form of dying roots and root exudates.

The colour of peat varies from almost white (just below the living surface mosses) to yellow, orange, brown, or black. The colour was even used as classification in traditional peat-cutting in Britain: 'black peat' is the highly humified bottom layer with swamp or fen peat, above it comes the moderately humified 'grey peat' of bog species, and on top the 'white peat' consisting of almost intact *Sphagnum* mosses (Berry *et al.* 1996). The texture and decomposition vary from very fibrous and poorly decomposed to weakly fibrous and highly decomposed. Because of the slow rates of solution and diffusion of oxygen in water, waterlogged conditions are associated with low levels of oxygen and reducing conditions. The waterlogged organic material acts as an insulator, and tends to keep the peat cooler than surrounding mineral soils. These conditions engender slow rates of decomposition of the organic materials, and hence organic matter accumulates over time.

In geological terms, peat is coal at its primary stage. Peat is changed only slightly from wood, having a lower proportion of volatiles and a slightly higher heat content. If peat continued to age and compact in its natural environment, lignite, charcoal, anthracite, and graphite would successively be formed (Bélanger *et al.* 1988).

Sedentation versus sedimentation

As outlined in Chapter 1, peat is formed in place, that is, as sedentary material, in contrast with aquatic sedimentary deposits. Quite different plant materials may be involved in the process of peat formation: woody parts, leaves, rhizomes, roots, and bryophytes (notably *Sphagnum* peat mosses). Much of the peat material originates below ground. Other terms for sedentary and sedimentary are *autochthonous* (forming in place) and *allochthonous* (transported from elsewhere before or during deposition).

Aquatic (limnic) sediments are plant fragments, organisms, precipitates, or minerals that settle to the bottom of pools, ponds, and lakes. This concept emphasizes the allochthonous genetic origin rather than composition. *Gyttja* is formed on lake bottoms as the deposited remains of animals and plants are transformed by microorganisms and mixed to varying degrees with inorganic sediments. When exposed and dried, gyttja becomes light-coloured. *Dy* accumulates as a gel-like, colloidal precipitate which, when exposed, maintains a dark colour. Both gyttja and dy give way under pressure when wet, but when drained they contract more and more into a dense mass that hardens on air drying, showing vertical cracks and fissures. Other kinds of aquatic sediments are marl which is dominantly a calcium (Ca) carbonate deposit, and diatomaceous earth, which is a sediment rich in siliceous diatom cases.

Some of the aquatic sediments contain high percentages of organic matter, but the term aquatic (limnic) peat should not be used unless the deposit has formed as a sedentary deposit. Submergents and floating plants, such as *Ceratophyllum demersum*, *Myriophyllum* spp., *Potamogeton* spp., and *Nuphar*, have aquatic parts that die and settle to the bottom (sedimentary), but their roots can form peat (sedentary). Limnic peat can form from the root mats of emergents in open-water marsh in species such as *Schoenoplectus* spp., *Phragmites australis*, and *Equisetum fluviatile*. Normally peats accumulate in these conditions as thin layers.

Bog iron (Fe), often known as *iron ochre*, forms in springs and spring rivulets on or close to mineral soil, and is subsequently covered over by peat (Troedsson and Nykvist 1973). Under reducing conditions below ground, the water contains Fe^{2+} in solution, but on reaching the surface the Fe is oxidized to Fe^{3+} and then becomes insoluble. The oxidation is at least partly caused by Fe bacteria which build cases and an orange gelatinous deposit, for example *Leptothrix* species (Chapter 2). Iron may also occur as an oil-like surface film in springs and waters upwelling from anoxic layers. This material is not organic or biological. It forms even in sterile inorganic media containing Fe, and is probably some kind of Fe^{3+} oxy-hydroxide (D.B. Johnson, pers. comm.).

Organic versus mineral matter content

Although peat is formed by organic matter, it also contains various minerals as bound components in the organic matter and as free minerals in the peat matrix. One of the characteristics that has been used to attempt to define peat is the total organic matter compared to mineral matter content, since this is strongly related to the overall nutrient regime of the peat. Two techniques used to estimate the mineral matter are *wet ashing* and *dry ashing*, and both methods result in a measure for inorganic matter called *ash content*.

With wet ashing (digestion) the ash is the residue after dissolving away the organic matter using strong acids (e.g. HNO_3 + $HClO_4$, 10:1; for others see Landva *et al.* 1983a). With dry ashing the ash is the residue remaining after burning. For dry ashing, the weight lost is called *loss on ignition* (LOI), which is the same as *organic matter content* (OM). Both ash and organic matter content are expressed as a percentage of the dry mass of the original sample. Different temperatures have been used for dry ashing, but the most common one is 550 °C (Andrejko *et al.* 1983).

Organic-rich materials have been classified into various categories – peat, muck, organic soil, and organic-rich mineral soil. Some of the proposed cut levels for organic matter contents in these categories are summarized in Fig. 5.1. There is no general agreement on how to define peat using organic matter content: the minimum percentage of OM required has ranged from as high as 80% to as low as 20%. In several classifications, muck (cf. mud) is defined as dark, well-decomposed, organic-rich material (peaty or sedimentary), with moderate to high mineral matter content, which has accumulated under conditions of imperfect drainage (Stanek and Worley 1983; Pakarinen 1984; Paavilainen and Päivänen 1995). Some muck layers contain quantities of microscopic siliceous material from diatoms and sponge spicules. In drained agricultural fields they can dry out and be blown about (Dachnowski-Stokes 1926), causing irritation to the eyes and skin of workers, giving rise to the term 'itchy' muck.

The various classifications in Fig. 5.1 may be based on only two classes – peat versus non-peat – or on several classes, e.g. peat, carbonaceous sediment or muck (mud), and mineral material. A classification from Russia (Mankinen and Gelfer 1982, Fig. 5.1), uses botanical content, acidity, degree of decomposition, heating value, and ash content. Peat is defined as >50% organic matter, and divided into six subtypes with increasing OM. These categories are used to define various uses; for example, peats can be used in power stations, public heating systems, fertilizers, and soil conditioning as long as the OM is greater than 77%.

Organic soil is the term preferred by soil classifiers, and they often avoid the use of the term peat (see Organic soils later in this chapter). The breakpoint

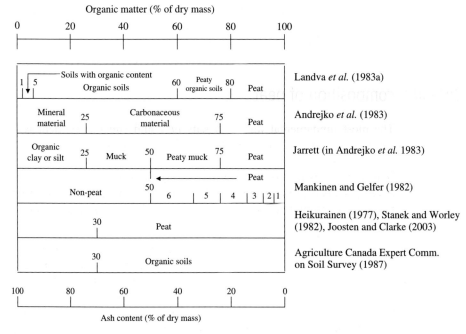

Fig. 5.1   Some levels of organic matter content (loss-on-ignition) used to delimit various categories of peat and organic soil.

>30% OM content, used by several authors to define peat, is the same as the one used in the Canadian system for organic soil (Agriculture Canada Expert Committee on Soil Survey 1987). Therefore, in the interests of simplification and rapprochement, using the 30% OM criterion for peat would make it equivalent to the definition for organic soil. However, what ecologists think of as typical peat – the thick accumulations in fens, bogs, and swamps – normally has at least 80–90% OM.

It is a basic principle of classifications that they are *purposive*, that is, they have a particular user group in mind. Soil classifiers aim to provide a general-purpose classification of organic-rich soils for a wide variety of uses, such as soil mapping, agriculture, forestry, and extraction; thus, they view 'organic soil' as a broad unit with a minimum organic matter content of >30% OM. Other classifiers, however, have recognized a need for more specialized classifications aimed at specific user groups. Peat harvesters are one specific group, interested in the more organic-rich materials mainly for horticultural or energy uses. This group would tend to restrict the definition of peat to to having OM >50% or even as high as >80%, putting materials with less than these levels into other categories such as muck, carbonaceous material, or organic-rich mineral soils.

Further classifications of peat are based on botanical composition and the degree to which the material has been modified by decomposition (see 'Fibrosity and humification', below).

Botanical composition of peat

The most fundamental way to classify peat and aquatic sediment is by composition, that is, the kinds of botanical, organic, or mineral materials that contributed to its formation. In the case of peat, the main composition is determined by the dominant plant species, whereas in the case of aquatic sediments the composition varies from coarse organic particles to gel-like flocculants, precipitates such as marl, or various amounts of mineral particles, clay, silt, and sand.

Botanical composition is the fundamental property for determining the nature of peat (e.g. Lévesque *et al.* 1980; Kivinen and Pakarinen 1981; Bohlin *et al.* 1989; Bohlin 1993). Not only the composition but also the relative amounts of the main plant species or species groups will determine the physical and chemical properties of the peat. A classification of peat is presented in Table 5.1 (Kivinen and Pakarinen 1981), and this reflects the peat that is laid down beneath the dominant kinds of vegetation. Three basic peat types are recognized – *Sphagnum*, sedge, and woody peats – and various subtypes are recognized based on abundant species or species groups. Different regions and different parts of the world will have different kinds and proportions of peat types, based on the kinds of dominant vegetation types which determine the peat that is laid down. For example, lowland tropical peat swamps are dominated by woody peat, predominantly the stems and trunks of large trees and understorey woody shrubs, and have virtually no *Sphagnum*.

Table 5.1 A classification of peat based upon composition and dominance of plant groups (Kivinen and Pakarinen 1981). In the subtypes, the second plant mentioned is the dominant material

Basic type	Subtype
Sphagnum peats – S	*Sphagnum* peat
	Sedge–*Sphagnum* peat
Sedge peats – C	Sedge peat (*Carex* peat is most common)
	Sphagnum–sedge peat
	Eutrophic *Sphagnum*–sedge peat
	Woody sedge peat
	Brown moss–sedge peat
Woody peats (wood content > 50%):	*Sphagnum*–woody peat
L – lignid (large tree wood)	Sedge–woody peat
N – nanolignid (shrub wood)	

- *Sphagnum peat* is mainly composed of three sections of the genus *Sphagnum*: Acutifolia, Sphagnum, and Cuspidata. Each section and component species forms specific peat types that vary depending on their morphology, the peatland type in which they formed (see Chapter 4), and their degree of decomposition. Sections Acutifolia and Sphagnum are probably the most important for peat formation worldwide (Malmer and Wallén 2004).
- *Sedge peat* is the most heterogeneous group. Many genera and species with the graminoid life form are included here in addition to species of *Carex* (e.g. *C. rostrata, C. lasiocarpa, C. limosa, C. diandra*). For example, sedge peat includes *Cyperus, Dulichium, Equisetum (E. fluviatile, E. sylvaticum), Eriophorum (E. angustifolium, E. vaginatum* (Europe), *E. spissum* (North America)), grasses (several genera, e.g. *Calamagrostis, Molinia, Phragmites*), rushes (several genera, *Cladium, Eleocharis, Juncus), Scheuchzeria palustris, Schoenoplectus, Trichophorum, Scirpus, Triglochin*, and a multitude of wetland herbs (e.g. *Menyanthes, Potentilla*).
- *Woody peat* has two main categories, the large stems and branches of trees and tall shrubs (*lignid*), and the small stems and branches of the Ericaeae, Myricaceae, Rosaceae, and other low shrubs (*nanolignid*).

The dominant plants forming peat are in turn determined by the main ecological factors, and these include moisture and nutrient regime. These factors affect the characteristics of the peat. For example, the surface peats of ombrotrophic bog have the lowest ash content, fen and swamp somewhat higher, and marshes that form under the influence of lake or stream waters

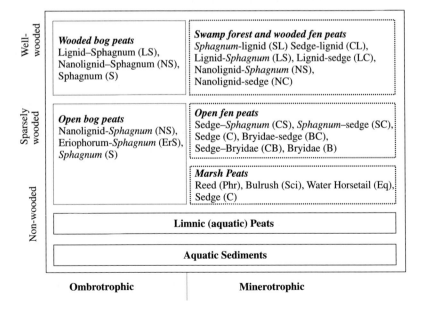

Fig. 5.2   The relationship of the main classes, bog, fen, marsh and swamp, to main botanical peat types. Modified from Pakarinen (1984).

have the highest ash contents. Figure 5.2 is a simple diagram showing how the main botanical peat types fit into the four main classes of peatlands, and how these relate to the main gradients of wetness and pH–Ca. This ordering of peat types corresponds to the ordering of wetland classes presented earlier (Fig. 1.1), and it makes it possible to distinguish for instance lawn, open hummock, and wooded hummock peats. One can also identify poor, intermediate, and rich fen peats. It is at these finer levels that differences in peat type become most interesting, since they show how the site has developed.

Sampling the peat profile

Three commonly used peat samplers are the Russian (Macaulay), Livingston, and Hiller side-opening samplers. These have extension rods that allow for full sampling of deep profiles. Landva *et al.* (1983b) maintain that these cannot be used to collect perfectly 'undisturbed' samples. The Hiller sampler cannot collect volume samples, but the Russian sampler can. There are many other samplers that can be used to take volumetric samples of short or medium-length cores, blocks, or monoliths (e.g. Day *et al.* 1979; Landva *et al.* 1983b; Waardenaar 1987; Jeglum *et al.* 1992). The small-diameter Russian ('minimized Macaulay') sampler is judged the best all-around sampler for one-man operations in unfrozen, undisturbed peats. The smallest version is 3.5–5 cm in diameter; a larger version may be 8–10 cm. There are also piston corers for collecting aquatic sediments and mineral samples from peatlands and lakes (Wright *et al.* 1984). Some samplers are shown in Fig. 5.3.

Normally, a peat profile is described from the surface of the moss or litter layer downwards. Important measures to record are depth of peat (over aquatic sedimentary, mineral material, or bedrock), depth and kind of aquatic sediment if present, and kind of mineral material beneath the peatland. Transitions between horizons of different peat types and bands of different colour should be noted, and whether the transitions are gradual or distinct. Other important descriptors are layers or lenses of water or soft mushy peat, logs and woody peat, mineral matter (wind or water deposition, volcanic ash, soot), and charcoal as a fossil indicator of fire.

The variable depths of transitions between different peat horizons present a problem of how to sample for further analyses – by horizon or by depth intervals. In the former method samples are collected from visually discerned homogeneous layers of different thicknesses, 2–5 or more, depending on the complexity of the core and detail of the study. In the latter, fixed sampling intervals are used, for example 50, 20, 10, 5, 2, or even 0.5 cm intervals, depending on the detail of the study and time and resources available for analysis. In both methods subsamples from each horizon or interval can be thoroughly mixed, or a subsample taken from the middle of the horizon or interval.

Fig. 5.3   Peat samplers: (A) the Russian (Macaulay) side-opening sampler which collects a half
cylinder (4 cm diameter), 50 cm at each extraction; extension rods allow sampling to
several metres. (B) A cylindric shallow peat sampler, 11.5 cm in diameter at the cutting
edge; the slits in the cylinder are 5 cm apart to allow cutting layers of known depth and
volume. (C) A box sampler; dimensions of the square cross section are 8 × 8 cm; the
cutting tool has a curved spring steel leading edge, which follows a groove at the bottom and undercuts the sample to allow extraction without loosing sample.

Peat properties

A large number of peat properties can be measured. Choice of factors to
be measured depends on the purpose for the data, for example, soil survey,
peat inventory, ecological survey, terrain stability, and so on. Numerous
manual and guides exist describing both field inventory and laboratory
methods for sampling and testing (e.g. Day *et al.* 1979; Keys 1983;
Agriculture Canada Expert Committee on Soil Survey 1987; Riley 1989;
Riley and Michaud 1994).

Physical properties of peat

Fibrosity and humification

Together with botanical composition, the most important peat descriptor
is the degree to which it has been modified by decomposition, that is, how
much of the original fibrous structure of the vegetation remains and how
much of the peat consists of humified material in which the original plant
organs are no longer recognizable. *Humification* is a concept that describes
how decomposed the peat is. High fibre content is associated with low
humification, and vice versa. Classification systems for peat always use
fibrosity and humification at a high level. For example, Canada uses three

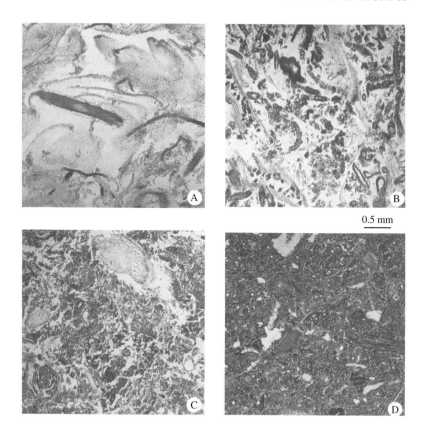

0.5 mm

Fig. 5.4   Thin sections of peat, showing several levels of decomposition: (A) fibric peat – *Sphagnum*, very little decomposition, only a few soil fauna droppings, leaves, branches, and stems easily seen. (B) fibric peat – sedge material, mostly roots (rootlets), rhizomes, or stems of sedge, some epidermic tissues with peculiar yellowish brown colour, lack of orientation of material; (C) mesic peat – accumulation of disintegrated materials, fine substances resulting from microbial and soil fauna activities, which resulted in granular structure; wood fragments, cross-sections of rootlets and some meristematic elements (bud scales); (D) humic peat – wood fragments (cross sections) embedded in organic plasma, and densely layered leaves of *Larix laricina*; humic substances indicated an advanced degree of decomposition evidenced by the dark brown colour. From Lévesque *et al*. (1980). Originally published by Agriculture Canada, reproduced with the permission of the Minister of Public Works and Government Services Canada, 2005.

basic classes – fibric, mesic, and humic peats – and the USA uses three similar concepts although with differences in terminology – fibric, hemic, and sapric peats, respectively. Both sequences correspond to the level of decomposition. To illustrate the range of fibrosity from fibric to mesic to humic, we present some thin sections of peats in Fig. 5.4.

The von Post humification method is one of the most commonly used, simplest, and best for the field characterization of peat. In this test you squeeze a sample of organic material in your hand and observe the colour of the solution that is expressed between the fingers, the nature of the fibres, and the proportion of the original sample that remains in the hand. Ten classes are defined, which can be grouped to correspond with the three Canadian classes (Table 5.2).

Janssens (1990) selected two key measures to characterize fibrosity and humification: fine particle content (<0.3 mm) and light absorbance of a pyrophosphate extract. Malterer *et al.* (1992) tested 11 measures, and found a high degree of correlation amongst them, and also found that the von Post field method is a good measure of of peat decomposition.

Table 5.2 The von Post system of humification, H1–H10 (von Post 1924; given in Agriculture Canada Expert Committee on Soil Survey 1987). This is combined with the three-part fibre content system used in Canada (Lemasters *et al.* 1983), and a similar one, but with different terminology, used in the USA

Class of fibre content, Canada (USA)	von Post humification	
Fibric (Of) = Fibric (USA) Weakly decomposed, botanical origin readily identified, 40% or more of rubbed fibre (by volume)	H1	Undecomposed: plant structures unaltered; yields only clear, colourless water when squeezed
	H2	Mostly undecomposed: plant structures distinct; yields yellow-brown water, still almost clear
	H3	Very weakly decomposed: plant structures distinct; yields somewhat turbid, brown water; no peat substance passes between the fingers. Residue is not mushy (pasty)
	H4	Weakly decomposed: plant structures distinct; yields very muddy, dark, and very turbid water; no peat substance passes between the fingers, and no distinct ridges after squeezing. Residue slightly mushy (pasty)
Mesic (Om) = Hemic (USA) Medium decomposed, botanical origin indistinct but still recognizable, material partly altered both physically and chemically into amorphous humus, 10–40% rubbed fibre	H5	Moderately decomposed: plant structures clear but becoming indistinct; yields much turbid brown water, some peat (about 1/10) escapes between the fingers, and distinct ridges remain after squeezing. Residue very mushy (pasty)
	H6	Strongly decomposed: plant structures somewhat indistinct, but clearer in the squeezed residue than in the undisturbed peat; about 1/3 of the peat escapes between the fingers. Residue strongly mushy
Humic (Oh) = Sapric (USA) Strongly decomposed, botanical origin very indistinct, in advanced stage of decomposition, dominated by amorphous humus. Less than 10% of rubbed fibre	H7	Strongly decomposed: plant structures indistinct but recognizable; about 1/2 the peat escapes between the fingers
	H8	Very strongly decomposed: plant structure very indistinct; about 2/3 of the peat passes between the fingers, residue of resistant root fibres and wood
	H9	Almost completely decomposed: plant structures very indistinct almost unrecognizable; nearly all the peat escapes between the fingers
	H10	Completely decomposed: plant structures unrecognizable; all the peat escapes between the fingers

Density

Bulk density (BD) is dry mass per unit volume (g cm^{-3}). To establish dry mass, the sample is dried at 105 °C to constant weight. When measuring bulk density it is crucial that the peat is not compacted during the sampling. Bulk density varies from 0.02 g cm^{-3} in the surface layers to 0.1–0.26 in lower layers (Boelter 1969), and shows a positive and approximately linear relationship with von Post humification (Päivänen 1973). *Specific density* (SD, also called particle density) is the dry mass of solid packed matter (with no pore spaces). The SD for peat varies from 1.3 to 1.6 g cm^{-3} (Okruszko 1993).

After drainage (or drought) the peat surface will sink. The reasons for this *subsidence* are an initial shrinkage caused by loss of water and collapse of pore spaces, and increased humification. This leads to increased bulk density, up to 0.4 g cm^{-3} in highly consolidated and decomposed, drained peats (Okruszko 1993). (See Chapter 13 for more details on subsidence.)

Porosity

There are three main phases in peats – air, water, and solid material (Fig. 5.5). *Total pore space (TPS)*, also called *total porosity*, is the total volume minus

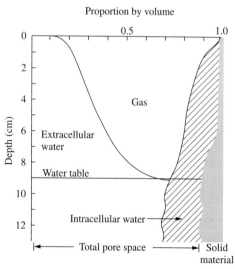

Fig. 5.5   Air, water, and solid material, and their volume relationships in a peat core under *Sphagnum capillifolium*. The gas volume decreases from about 85% near the surface to zero just below the water table (which is at 9 cm below the surface in this example). The total pore space is the sum of volumes occupied by extracellular (capillary) water, gas, and intracellular water. The proportion occupied by solid peat increases with depth as the material becomes increasingly compressed and humified. Simplified after Hayward, P. M. and Clymo, R. S. (1982). Profiles of water content and pore size in Sphagnum and peat, and their relation to peat bog ecology. *Proceedings of the Royal Society of London B Biological Sciences*, **215**, 299–325; with permission from the Royal Society, London.

the volume of the solid material. Total pore space is calculated as TPS = [1 −(BD/SD)] × 100. For example, if BD = 0.2 and SD = 1.4 the total pore space is 86%.

TPS ranges between 78% and 93% (Ilnicki and Zeitz 2003). There is high porosity in fibric peat, but decreasing porosity in mesic and humic peat. Therefore, porosity decreases with increasing humification, bulk density, and degree of compaction. Furthermore, with increasing humification and bulk density, the quantity of large pores decreases and that of small ones increases (see Chapter 8).

If one knows the TPS and the moisture content, one can calculate the percentage of the pore space that is water-filled and air-filled. If we oven dry 1000 cm^3 in the above example and lose 700 g water, this means that the water-filled pore space is 70%, and hence the air-filled pore space = 86 − 70 = 16%.

For roots of vascular plants to function well they require some minimum soil air space. Research suggests that the minimum air space for roots of Scots pine (*Pinus sylvestris*) is 10% (Paavilainen and Päivänen 1995), and for grasslands established on drained fens it is 6–8% (Okruszko 1993).

In the water-filled portion of the pore space, some of the water is intra-cellular while the rest is in the extracellular spaces between the peat fragments and particles (Fig. 5.5). Hayward and Clymo (1982) developed an apparatus that used absorbance of soft gamma radiation to produce water-content profiles in *Sphagnum* and underlying peat, and to infer water-fillable spaces of different sizes. The intracellular component of water is relatively high in *Sphagnum* peat owing to the hyaline cells. How pore size distribution changes with depth is important to the interpretation of water flow in peatlands (see hydraulic conductivity in Chapter 8).

Water content

Water content can be determined by taking a sample of peat, drying at 105 °C to obtain water content in g (= cm^3), and expressing water as a percentage of the original mass or volume. Three ways of expressing per-centage moisture content are:

- water content/dry mass of sample
- water content/original wet mass of sample
- water content/volume of sample.

As an example, a volume sample of fresh peat of 1000 cm^3 weighs 1050 g. We dry this sample to constant weight at 105 °C to remove the water, and obtain 950 g of weight loss. The water content is then: on a dry mass basis 950 g/100 g dry peat = 950%; on a sample mass basis 950 g/1050 g water plus peat = 90.5%; and on a volume basis 950 cm^3/1000 cm^3 = 95%. It is therefore clear that statements about water content are dependent on the method of expressing the water.

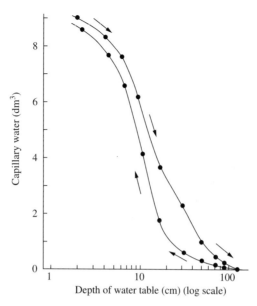

Fig. 5.6   Capillary water of a 16 cm deep, 30 cm diameter core of *Sphagnum capillifolium*. The water table was lowered to 100 cm below the surface and then raised again (following the arrows). Depth scale is logarithmic. Simplified after Hayward, P. M. and Clymo, R. S. (1982). Profiles of water content and pore size in Sphagnum and peat, and their relation to peat bog ecology. *Proceedings of the Royal Society of London B Biological Sciences*, **215**, 299–325; with permission from the Royal Society, London.

Field measurements of water contents in unsaturated surface layers, and their variations over time, are uncommon. However, moisture content in peat can be assessed with a more sophisticated indirect method, *time domain reflectometry (TDR)*, which permits the determination of soil water storage and dynamics in peatlands (Lapen *et al.* 2000; Kellner and Lundin 2001).

Drying peat tends to hold on to water, whereas very dry peats tend to resist rewetting. Hence the moisture curves for drying and wetting form a loop, which is termed *hysteresis* (Fig. 5.6). Hayward and Clymo (1982) give a reasonable theoretical explanation of these curves, based on how water moves through a complex system consisting of 'caverns' linked by narrower 'bottlenecks'. They suggest that the emptying (drying) curve is controlled by the bottlenecks and the filling (wetting) curve by the caverns.

Calorific value

When peat is used as a fuel, a relevant measure is calorific value describing the energy content (Asplund 1996). After drying and collecting in the field, milled peat usually has a slightly higher average water content (about 50% dry mass basis) than sod peat (about 40%), and thus slightly lower heating values (10 and 12 MJ kg^{-1}, respectively). Tolonen (1983) showed that the lowest calorific values are in the least humified, fibrous peat. For *Sphagnum* peat, there is a good correlation between von Post humification and calorific value.

Electrochemical and chemical properties of peat

pH of peat

The pH of peat can be measured by directly inserting a combination electrode into the moist peat, or by adding distilled water at a volume ratio of 1:1 or 1:2 fresh peat to water. The diluting agent can also be a standard concentrations of a salt solution such as 0.01 mol L^{-1} $CaCl_2$ or 1 mol L^{-1} KCl (Day *et al.* 1979). For field survey purposes, modern pocket-size pH meters give good accuracy, and for simple field observations pH indicator paper may suffice. As mentioned in Chapter 1, there is a strong correlation between pH and the vegetational variation in peatlands. Since pH is frequently measured in water rather than in peat, we discuss this further in Chapter 8.

Oxygen content and redox potential

The oxygen content of the peat is broadly correlated with the water table depth. If oxygen is present chemical reactions move towards oxidized states, whereas when soil oxygen becomes limited the reactions are driven towards reduced states. Many of the reactions, both oxidations and reductions, are usually microbially mediated (e.g. McBride 1994; Stumm and Morgan 1996). When wetland soils are depleted of oxygen the reduced forms of several ions and compounds become more mobile, such as manganese (Mn^{2+}), ferrous iron (Fe^{2+}), hydrogen sulfide (H_2S), and methane (CH_4).

Redox potential (E_h) is a quantitative measure of electron availability and indicates intensity of oxidation or reduction in both chemical and biological systems. It can be measured in the peat with platinum electrodes and a portable voltmeter (e.g. Faulkner *et al.* 1989; McBride 1994; Stumm and Morgan 1996). The more oxidizing the environment is, the higher the voltage. A specially designed multilevel redox probe has been described (Day *et al.* 1979). Redox conditions can also be expressed as pϵ, which is the negative $\log_{10}$ of electron activity.

The oxidized and reduced forms of Fe have different colours – reddish and grey or black, respectively – and the reduction of ferric Fe (Fe^{3+}) to the ferrous form (Fe^{2+}) can be diagnostic for reduced conditions in wetland soils. A striking example of this is where pyrite (FeS) occurs in the gyttja along a shore. When the gyttja is below the water surface, it is black due to the presence of reduced ferrous sulfide (FeS). When the water drops below the surface, the FeS oxidizes to a thin skin of reddish-brown ferric sulfate, $Fe_2(SO_4)_3$, but there is a rapid transition, within millimetres, to the black reduced Fe^{2+} below. Faulkner *et al.* (1989) described a simple method using α,α-bipyridine (dipyridyl) as a colorimetric indicator of Fe^{2+}.

If the oxygen (O_2) concentration is high enough, organisms will use O_2 as an electron acceptor in their metabolism. As O_2 falls to micromolar concentrations, other electron acceptors can be used. In the sequence of increasing reduction intensity reduction of O_2 is followed by reduction of nitrate (NO_3^-) to ammonium (NH_4^+), and this can proceed before the complete removal of O_2 (Stumm and Morgan 1996). Manganic manganese (Mn^{3+}) closely follows nitrate in the reduction sequence, and will begin reducing before NO_3^- completely disappears. Next in the reduction sequence is Fe^{3+} to Fe^{2+}, which does not overlap with NO_3^-, and which does not occur in the presence of O_2 (Fe-reducing bacteria are obligate anaerobes). The approximate level of redox potential where the reduction of Fe^{3+} begins (at pH 7) is $E_h = 100$ mV (Fig. 5.7).

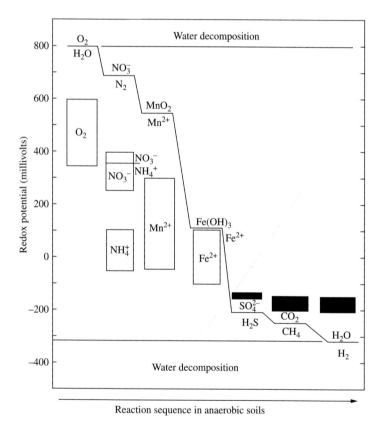

Fig. 5.7   The redox sequence in soil solutions at pH 7. Theoretical potentials are indicated, assuming equal activities of reduced and oxidized species unless otherwise noted. Measured ranges of soil potentials over which the indicated species react (change concentration) during the reduction sequence are specified by boxes, open for reduction, black for initial appearance of the reduced form. Modified from McBride (1994).

Simpler methods have been suggested to enable qualitative measurements of redox potential at a large number of points. Belyea (1999) showed that PVC electrical insulation tape becomes discoloured when buried in reducing levels of the peat. Sticks with such tape attached to them can be inserted into the peat, and when they are recovered a faint discoloration indicates the highest water level during the time the sticks have been buried, and the depth of full discoloration indicates the lowest water level. Although the method does not give the exact redox potential, it can be used to measure both temporal and spatial variation in water levels and reducing conditions.

Gases in peat

In addition to oxygen, other gases in peat have become the focus of much interest in recent peatland work, with particular relevance to the release of greenhouse gases, CH_4, carbon dioxide (CO_2), and nitrous oxide (N_2O). This subject is dealt with in Chapter 12. Various sulfurous gases (H_2S, SO_2, CH_3SCH_3, COS) are also of interest. These gases, similar to oxygen, are dynamic and rapidly changing in response to water level fluctuations, and to aerobic and anaerobic metabolic processes of microorganisms.

Essential elements

Among the essential inorganic nutrients there are nine macronutrients which can be ranked in order of decreasing concentration in plants: carbon (C), oxygen (O), hydrogen (H), nitrogen (N), potassium (K), calcium (Ca), phosphorus (P), magnesium (Mg), and sulfur (S). The important micronutrients are Fe, chlorine (Cl), copper (Cu), Mn, zinc (Zn), molybdenum (Mo), and boron (B) and some plants also require sodium (Na), silicon (Si), and cobalt (Co), and possibly nickel (Ni).

Analyses of the peat can include the main macronutrients, especially the main nutrients N, P, and K, and the main divalent cations, Ca and Mg. Other nutrients may also be analysed, depending on the purpose. Total analyses and extractable analyses will give different results. *Total analyses* of peat are carried out using the same methods as for plant tissues, which includes complete dissolution with strong acids. *Extractable analyses* use weaker reagents, and are intended to represent amounts that are available for uptake by plants. Extractable NH_4^+ and NO_3^- can be analysed after extraction with KCl; extractable P is analysed using such extractants as Bray's solution (Kalra and Maynard 1991). For the remaining cations scientists have used water, potassium chloride, ammonium chloride, or ammonium acetate as extractants (Nömmik 1974; Starr and Westman 1978; Day *et al.* 1979; Wiklander and Nömmik 1987). In studies of nutrient store and balance, the totals are usually measured. However, measuring extractables gives a better reflection of nutrient supply, in particular for N and P, which can have a high proportion of their totals tied up in organic

Table 5.3 Total elemental analysis and other properties of peat from a boreal bog, a fen, and a marginal swamp forest in a peatland complex near Cochrane, northeastern Ontario (National Wetlands Working Group 1988)

Depth (cm)	von Post humification	Peat type	Ash (%)	Total elements (%)					
				Ca	Mg	Fe	N	P	K
Raised bog with concentric patterns									
0–50	2	*Sphagnum–Carex*	1.8	0.12	0.02	0.03	1.35	0.06	0.02
50–100	4	*Carex–Sphagnum*	1.1	0.20	0.03	0.04	0.81	0.03	0.01
150–200	6	*Carex–Sphagnum*	1.5	0.36	0.01	0.04	0.98	0.002	0.01
310–350	2	Brown moss	2.2	0.61	0.21	0.09	1.77	0.04	0.01
Basin fen									
0–20	2	*Sphagnum–Carex*	3.3	0.24	0.04	0.14	1.48	0.07	0.06
70–100	7	–	9.0	0.84	0.04	0.33	2.59	0.06	0.08
135–175	5–6	–	6.0	1.11	0.04	0.30	2.36	0.04	0.06
Peat margin swamp									
0–50	3	*Sphagnum–wood–Pleurozium*	7.9	2.16	0.13	0.22	1.63	0.04	0.04
201–215	4	*Sphagnum–wood–Pleurozium*	7.2	2.42	0.41	1.31	1.13	0.05	0.78

form in the peat and hence not available for plant growth (see further Chapter 9).

The elementary composition of peat is strongly related to ecosystem type, peat type, and nutrient richness–trophic gradient. Some chemical analyses for a boreal bog, fen, and swamp, collected from a raised, concentric bog, basin fen, and marginal swamp forest complex near Cochrane in northeastern Ontario, Canada are shown in Table 5.3. Several points about elementary composition of peats can be made from this study. There are generally increasing values of nutrients from bog to fen to swamp, and nutrients tend to be lowest at the surface, increasing with depth. Values for P and K tend to have variable patterns. In the bog and fen they are usually highest at the surface, although this is variable. In the swamp forest, the values increase with depth. The tendency for P and K to be highest at the surface in bog and fen is also seen when one examines short surface cores of 40–50 cm depth with depth intervals of 10 cm. This may be owing to nutrient cycling and conservation in the living 'skin' at the surface of mires, or partially owing to leaching of P and K out of subsurface layers as humification progresses.

Cation exchange capacity

Cation exchange capacity (CEC) is a measure which describes the ability of a soil to hold positively charged ions (cations). The major ionized elements are H, K, Na, Ca, Mg, and Al, which are adsorbed on the surface soil particles when they have negative charges. On decomposition, organic matter produces organic acids, lignin, and many other products which can exhibit exchange properties. In frequently used methods to quantify CEC, all the adsorbed cations in peat are replaced by a common ion, such as barium or potassium, and then the amount of that ion is determined.

Peat has high CEC, and therefore high buffering capacity. This buffering capacity increases with peat humification owing to increasing numbers of exchange sites with smaller particles. Because CEC is influenced by botanical or depositional origin, and by humification, it is not as strongly correlated with richness of the site or main lines of vegetational variation as some other measures. As outlined in Chapter 4, the production of uronic acids and other substances in *Sphagnum* strongly contributes to a high CEC, especially in hummock-inhabiting species.

Base saturation

Percentage base saturation is the sum of the extractable values for main base cations, e.g. Ca, Mg, K, and Mn, as a percentage of total CEC. Base saturation is often more strongly related to the poor–rich botanical gradient than CEC (e.g. Westman 1987) because it reflects the proportion of the base cations on all exchange sites, not including the hydrogen ions. Because base saturation is usually dominated by exchangeable Ca, it is usually highly correlated with the pH–Ca species richness main line of variation. Total exchangeable bases can be determined with a single procedure, extraction with NH_4Cl and EDTA titration (Nömmik 1974); this method mainly extracts Ca, Mg, and Mn.

Toxic elements and compounds

Several micronutrients and non-essential elements can reach concentrations in soils that are toxic to plants, animals, or microorganisms. Some of the most toxic to higher animals are mercury (Hg), lead (Pb), and cadmium (Cd), and those that can be phytotoxic include Cd, Cu, and Ni (McBride 1994). Toxic levels of H_2S can develop in reducing conditions where S is abundant; and toxic levels of Mn^{2+} are most likely in waterlogged, acid mineral soils with low humus content.

Organic composition

Peat is composed of an enormously complex mixture of organic compounds, derived mainly from vegetational remains. These organic compounds consist of carbohydrates (cellulose, hemicellulose, sugars), nitrogenous compounds (proteins, amino acids), polyphenols (lignins, humic acids), and lipids (waxes, resins, steroids, terpenes). In addition, small amounts of nucleic acids, pigments, alkaloids, vitamins, and other organic substances are present, along with inorganic materials. The composition varies depending on the species groups, and even species, from which the peat is derived. Thus, there are differences among *Sphagnum*, sedge, and woody peats, and among species within these groups. Peat composition is also influenced by the peat acidity, alkalinity, mineral content, redox conditions, and degree of humification. When a plant dies the sugars, amino acids, and other water-soluble components of its living tissues are rapidly released to the surroundings and quickly metabolized by microorganisms. Starches, although insoluble, are quickly digested to sugars and consumed as energy in the dying plant or in the organisms of decay. Thus, sugars, amino acids, and starches may be present in low concentrations. Hemicelluloses, celluloses, pectins, gums, and waxes are attacked more slowly and tend to persist in varying amounts. Also the bacteria and fungi of decay die and add their organic matter to the peat, although this may be rather minor in comparison to the higher plant components (e.g. Fuchsman 1983).

Humic substances

Humic substances – humic acids, fulvic acids, and humin – are dark-coloured, acidic compounds with molecular weights ranging from hundreds to thousands, which play a key role in determining peat characteristics. They contribute to the high CEC of peat, to its water-retentive characteristics, to the hydrologic properties of a peatland, and to the carbon balance of peatlands. The origin of humic substances, which occur in the peat but not in the living plants from which the peat is derived, is key to explaining humification, and how this may relate to industrial utilization and chemical conversions of peat (e.g. Fuchsman 1980, 1983; Mathur and Farnham 1985). The exact mechanisms of their formation are not well known; there are probably both microbially mediated and abiotic processes. The final forms of complex humic substances are relatively resistant to further microbial degradation, and tend to form the bulk of the highly decomposed peat. Humic substances are probably a major component of dissolved organic matter (DOM) and particulate organic matter (POM), further discussed in Chapter 8.

Other organic compounds

The organic composition of the peat is dependent on the botanical composition, and even on the specific organs (e.g. leaves, roots, and rhizomes) within species and on the degree of decomposition. Table 5.4 summarizes the various organic compounds in terms of fen, transitional, and bog peat (Efremova *et al.* 1997). The trends across this sequence are for decreasing percentages of humic acids, and increasing percentages of cellulose, lipids, and insoluble residues. There is less variation in the amounts of fulvic acid and hemicellulose.

The reserves of different organic compounds have been calculated for Russian peatlands (Efremova *et al.* 1997). Of the total carbon, the carbon of humic substances amounts to 37% (with humic acids dominating in fen peats, and fulvic acids more important in bog peats). The carbon of insoluble

Table 5.4 Organic compounds in different types of peat (percentage of the total carbon). From Efremova *et al.* (1997). The reserves and forms of carbon compounds in bog ecosystems of Russia. *Eurasian Soil Science*, **30**, 1318–1325, with kind permission of Springer Science and Business Media.

Compound or index	Low-moor peat (cf. fen)	Transitional peat	High-moor peat (cf. bog)
Humic substances	47.2	35.8	26.3
Humic acids	30.5	19.2	11.0
Fulvic acids	16.7	16.6	15.3
Humic : fulvic acids	1–2	0.8–1.5	<0.5–1.2
Polysaccharides	14.2	19.5	21.4
Hemicellulose	8.2	9.0	9.1
Cellulose	6.0	10.5	12.3
Lipids	5.8	7.7	8.1
Insoluble residue (lignin and humin)	32.8	37.0	44.2
Humic substances : polysaccharides	3.3	1.8	1.2

transitional products of humification accounts for 38%, carbohydrates for 18%, and the carbon of lipids for 7%. Water-soluble carbon compounds constitute only about 0.11% of the total carbon. Despite this low concentration, their role in the functioning of peatland ecosystems is extremely important. These compounds exert a strong influence on the redox potential of peatland waters, which dictates the regime of oxygen supply for plants and microorganisms; they supply a source of organic matter for decomposition and production of CH_4 by microorganisms; and they actively participate in the migration of elements and dictate the quality of water in streams that drain peatlands.

Interrelationships of peat properties

Botanical composition, physical properties, and chemical properties of peat (see Chapter 8) are highly interrelated. A detailed survey and analysis of peat properties was carried out for 14 mires in northern Sweden (Bohlin *et al.* 1989; Bergner *et al.* 1990). Sixty-one attributes were analysed for botanical composition (16 attributes), basic peat properties (19, including physical characteristics and minerals), monosaccharides (7), amino acids (16), amino sugars (2), and bitumen (1). Principal component analysis (PCA) was then used to group the peat samples according to similarities and differences in the measured attributes.

The analysis revealed that the main variation among the peat samples was related to factors that have generally been considered as the most important ones, i.e. botanical composition and degree of decomposition. The analysis yielded eight peat groups. The main division in the data set (the first principle component, accounting for 45% of the variation) separated the peat samples mainly on the basis of their botanical composition. The most distinct groups

Table 5.5 Selected attributes for eight main peat groups derived from principal component analysis. Data from northern Sweden (Bergner *et al.* 1990). Bitumen in this study was the fraction that is soluble in acetone, consisting of waxes, fats, etc. All percentages are of dry mass

Peat type	S	S	S	Mixed	C	BC	C	C
von Post humification	2–4	5–6	7–8	4–7	2–3	4	5–6	7–8
Composition (%)								
Sphagnum	89	70	47	24	5	9	6	2
Tree and shrub wood	5	21	36	15	1	1	1	3
Eriophorum vaginatum	2	8	14	8	0	0	3	0
Other mosses (incl. brown mosses)	0	0	0	1	3	18	1	5
Carex	0	0	0	29	89	66	86	78
pH	4.3	4.4	4.5	4.9	5.7	5.5	5.6	5.2
Ash (%)	1.4	1.7	1.6	3.8	4.6	3.9	8.2	13.7
N (%)	0.9	0.9	1.2	1.6	2.8	2.6	2.5	3.0
C/N ratio	61.3	64.3	50.2	36.0	20.4	22.1	22.0	17.9
P (%)	0.02	0.02	0.03	0.06	0.11	0.07	0.09	0.10
K (%)	0.016	0.026	0.017	0.020	0.021	0.020	0.018	0.025
Ca (%)	0.20	0.29	0.29	0.49	0.50	0.48	0.35	0.31
Mg (%)	0.068	0.082	0.070	0.064	0.077	0.072	0.047	0.027
Fe (%)	0.15	0.12	0.12	0.56	1.14	1.02	0.42	1.07
Al (%)	0.07	0.04	0.05	0.19	0.46	0.15	0.47	0.77
S (%)	0.10	0.10	0.15	0.53	0.20	0.25	0.47	0.43
Organic compounds (%)								
Carbohydrates	31	27	23	17	19	20	15	11
Lignin	36	44	53	55	51	53	57	59
Amino acids	3	3	4	6	11	9	9	11
Uronic acids	13	10	6	7	6	7	6	6
Bitumen	5	6	8	7	3	3	5	4
Energy (MJ kg^{-1})	20.9	22.2	23.6	23.7	23.5	23.0	23.9	24.5

S, *Sphagnum*; C, *Carex*; BC, brown moss–*Carex*.

were the *Sphagnum* and *Carex* peats, while the mixed peats had intermediate attributes (Table 5.5). *Carex* peats are formed in fens, and often have higher pH, ash content, and concentrations of the major elements than *Sphagnum* peats, due partly to the higher protein content in the original *Carex* plants and partly to higher transformation into microbial biomass in *Carex* communities.

The second principle component (accounting for 16% of the variation), described variation related to degree of decomposition and the nanolignid content. When all the peat types were studied together, changes during decomposition were characterized by (1) decreases in hemicellulose compounds, (2) increase of small particles and decrease of large ones, (3) increase in bitumen and woody material, and (4) increase in von Post humification value.

The chemical and physical changes during decomposition differed for *Sphagnum* and *Carex* peats, and indicated that *Sphagnum* plants are less chemically transformed in the peat than the *Carex*. *Sphagnum* peats tend to have higher carbohydrates, uronic acids, and bitumen, whereas the *Carex* peats had higher lignin, amino acids, and sugars (Table 5.5). The more easily degraded *Carex* may have cell walls that are more physically resistant owing to a higher lignin content, whereas the uronic acid in *Sphagnum* generally inhibits the rates of decomposition (see Chapter 4). Bohlin *et al.* (1989) concluded that analysis of monosaccharide constituents – rhamnose, fucose, arabinose, oylose, mannose, galactose, and glucose – is particularly useful in characterizing peats, because it provides information on both botanical composition and degree of humification.

Organic soils (histosols)

Histosol and *organic soil* are alternative terms for organic-rich soil profiles. In early work on peatlands in the USA, it was common to designate peat profiles based on the name of the locality where the unit was first described, for example, the 'Carbondale' peat profile (e.g. Lucas 1982). With this approach it was possible to observe similarities and differences by comparing a given profile with established peat profile types. This was appropriate for soil mapping, but it is a complex system that cannot be simplified into fewer main types. More recent work on classification has given rise to simpler systems based on fibrosity and humification of the peat.

A useful reference is Driessen and Dudal (1991) which provides an overview of peat soils in the northern hemisphere and the tropics. A more recent publication by Parent and Ilnicki (2003) focuses largely on organic soils in Canada, Poland, and Russia. For a more general overview of wetland soils we refer to Richardson and Vepraskas (2001).

The Canadian classification of organic soils

The Canadian system (Agriculture Canada Expert Committee on Soil Survey 1987) provides a logical system for describing organic profiles. This classification is based on recognizing horizons of fibric, mesic, and humic peat. It also treats the variability of depths of organic matter, and depths to contacts with rock, mineral, aquatic sediments, and water layers. The depth of organic matter required to place a soil in the organic order is generally 40 cm. There are four main organic soil groups: fibrisol, mesisol, humisol, and folisol. The first three groups originate in waterlogged conditions, but the folisols originate on upland. The system employs depth tiers: the *surface tier* is from 0 to 40 cm deep, the *middle tier* from 40 to 120 cm, and the *bottom tier* deeper than 120 cm. The middle tier is the layer used to name the soil group.

- *Fibrisols* have a dominantly fibric middle tier (von Post 1–4). They occur extensively in the whole boreal zone, particularly in deposits dominated by *Sphagnum* mosses. These soils are commonly found in ombrotrophic or weakly minerotrophic types.
- *Mesisols* are at a stage of decomposition intermediate between fibrisols and humisols, with a dominantly mesic middle tier (von Post 5–6). These soils are commonly found in minerotrophic types, but some can be in early stages of bog development.
- *Humisols* are at the most advanced stage of decomposition of the organic soils groups. They have a dominantly humic middle tier (von Post 7–10), and there are few recognizable fibres. These soils are commonly found in marshes, fens, and swamps.
- *Folisoils* are composed of upland organic materials, generally of forest origin, that are thicker than 40 cm (60 cm or more if the organic materials consist mainly of fibric materials, or a bulk density of less than 0.1 g cm^{-3}), well-drained, and are never saturated with water for more than a few days.

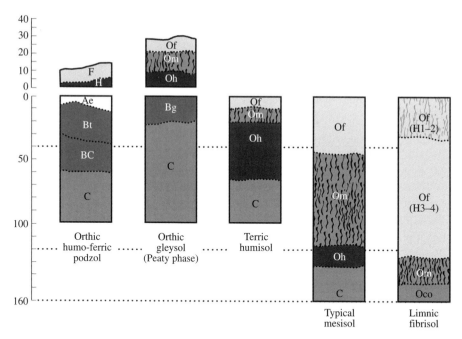

Fig. 5.8   A conceptual sequence of profiles from mineral soils adjacent to peatland, podzol, and gleysol, grading into organic soils – humisol, mesisol, and fibrisol. F = upland fermentation layer; H = upland humus; Ae = eluviated, grey-white podzol layer; B = depositional layer; BC = depositional B mixed with parent C; C = parent material; Oco = organic carbonate sediment; Oh = humic peat; Om = mesic peat; Of = fibric peat; H1–2 and H3–4 = von Post humification ratings. Simplified from Agriculture Canada Expert Committee on Soil Survey (1987). Originally published by Agriculture Canada, reproduced with the permission of the Minister of Public Works and Government Services Canada, 2005.

Peatlands show gradations at their margins into mineral soil uplands, and where the peat becomes less than 40 cm, a completely arbitrary depth in the soil classification, it grades into an adjacent mineral soil type. Some examples of profiles showing a sequence from mineral soils into the three great groups of organic soils are presented in Fig. 5.8.

Relationships of organic soils to other soil orders

Organic soils are regarded as having 'wet' moisture regimes. The other soil orders, such as gleysols, podzols, brunisols, and luvisols (Agriculture Canada Expert Committee on Soil Survey 1987) occur upslope from the organic soils, and usually are rated as having 'moist', 'fresh', or 'dry' moisture regimes.

Often the soils just upslope from the organic soils are in the gleysolic order, or the gleyed subgroups of other upland soil orders. *Gleyed* soils develop where the water table is close to the surface and anoxic or poorly aerated. They are characterized by a gley layer, with a grey to blue-grey colour caused by the reduced state of iron, Fe^{2+}. Often at the top of the gley layer is a zone with mottles, orange to brown spots of oxidized material in the gleyed, reduced soil matrix, reflecting a fluctuating water level with alternating periods of aeration and saturation. A gleyed site may be rather stable, but there are also sites that are becoming drier (e.g. by isostatic land uplift), or becoming wetter by paludification and developing towards organic soils.

The organic soils also grade into the *cryosolic* soils. These are formed in mineral or organic material that have permafrost at certain depths and can be mixed by freeze–thaw processes. The cryosolic soils are most commonly found in arctic and subartic regions.

6 The peat archives

In a peat profile there is a fossilized record of changes over time in the vegetation, pollen, spores, animals (from microscopic ones to the giant elk), and archaeological remains that have been deposited in place, as well as pollen, spores, and particles brought in by wind and water. These remains are collectively termed the *peat archives* (Godwin 1981), and they provide a wealth of information and evidence pertaining to the development of the peatland and environmental changes through time.

The Quaternary Period includes the Holocene Epoch, the time period after the last ice age including the last 10 000 years, and the Pleistocene Epoch or 'Ice Ages' from 10 000 to about 1.6 million BP (before the present). In this chapter we deal for the most part with the Holocene, although we occasionally delve a little further back into the Pleistocene. We consider the microscopic study of peat, the dating and chronology of peat profiles, climate and environmental changes indicated by the peat archives, and archaeological remains found in peatlands and wetlands. The subject has a vast and diverse literature (e.g. Birks and Birks 1980; Davis 1983; Porter 1983; Wright 1983; Berglund 1986; Barber 1993; Berglund *et al.* 1996; Smol and Last 2001–2005; Charman and Chambers 2004)

Peat fossils

A peat profile contains two main kinds of fossil remains. *Macrofossils* are the remains and fragments of plants – leaves, roots, rhizomes, seeds, fruits, twigs, larger pieces of wood, and some animals such as insects. There are also small macrofossils, such as moss leaves, that have to be studied under the microscope. *Microfossils* consist of the pollen grains, spores, and other unicellular or very small multicellular organisms entrapped in place, or transported by wind or water onto the surface of the peat. If one is interested primarily in the succession and development of peatland plant communities, macrofossil analysis will provide the most detailed information about the plants that contributed to the peat development. The pollen and spore microfossils derived from plants and animals provide more information about the development of the particular peatland being sampled, but in addition provide a good picture of the surrounding upland and regional vegetation.

Macrofossil and microfossil analysis are complementary and together give a more complete picture of peatland development than either approach separately. Birks and Birks (1980) listed six questions that can be asked:

- What taxa are present?
- What were the relative abundances of the taxa in the past?
- What plant communities were present?
- What space did each community occupy?
- What time period did each community occupy?
- What were the other factors operating in the ecosystem in that time and place?

For both pollen and macrofossil sampling, peat corers such as the side-opening Russian sampler can be used (see Fig. 5.3). The core can be divided vertically in two halves – one for pollen and one for macrofossils – and may be cross-cut into depth intervals as described in Chapter 5. Macrofossil and pollen studies are often combined with detailed descriptions of the recognizable peat horizons, and of any distinct bands or narrow horizons that are different in colour or humification. The peat horizons may be further characterized for pH, bulk density, organic matter content, and particle-size analysis (Chapter 5). These analyses assist in determining the types of communities and conditions at the time each layer was deposited.

Macrofossils

Various techniques have been used to prepare macrofossil samples. These include extraction with KOH (Bohlin *et al.* 1989), acetic acid anhydride ('acetolysis'), or with a non-foaming wetting agent (Janssens 1983), and sieving and separation into size fractions (e.g. Dinel *et al.* 1983). For example, Bohlin *et al.* (1989) thoroughly mixed the peat from a uniform layer, then took 1 cm^3 of peat and boiled it in 10% KOH for 1 minute. After washing with water on a 0.045 mm sieve, the plant fragments were mounted on microscope slides. A useful reference for the study of macrofossils is Birks (2001).

Some representative peat macrofossils are shown in Fig. 6.1. The number of fragments of defined species or classes are counted by systematic point

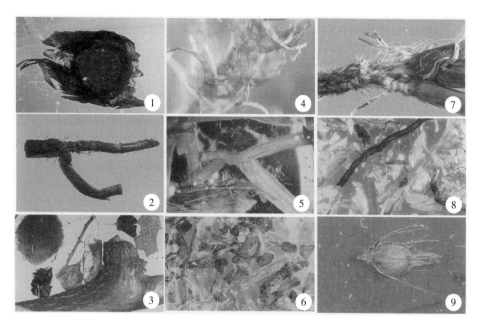

Fig. 6.1   Some representative macrofossils in peat. 1, Leaf junction scars on Pteridophyta (cf. *Equisetum*); 2, *Carex stricta* rhizome; 3, Fragments of stems, bark, roots, and Ericaceae leaves; fine roots of herbs; aggregates composed of macrofossil elements and fine materials; 4, Stem and leaves of *Drepanocladus* (brown moss), leaf sickle-shaped; 5, Fraction >2000 μm with fragments of rhizomes and roots of Cyperaceae; 6, Fraction 450–1000 μm with root fragments and tissue remains of herbs, wood fragments, needles, bark of shrubs and *Larix laricina*, and a few carbonized wood pieces; 7, *Eriophorum vaginatum* var *spissum*; 8, Fraction 1000–2000 μm with roots of shrubs, roots and tissue remains of herbs, and leaves and branches of *Sphagnum*; 9, *Rhynchospora alba* fruit. From Lévesque *et al.* (1988). Originally published by Agriculture Canada, reproduced with the permission of the Minister of Public Works and Government Services Canada, 2005.

count methods. The percentages of each kind of species or group are charted according to depth in a macrofossil diagram (Fig. 6.2). There are limits to the numbers of species that can be consistently identified in macrofossil analysis. It is also possible to obtain information consistently for a limited number of species or species groups, which will still reveal the main peat types and allow one to reconstruct the peatland development. It is possible to identify quite a number of plant groups and draw conclusions about the habitat conditions (Table 6.1). *Sphagnum* species are particularly important in characterizing peat and habitat types, as shown in Chapter 4. For example, in southern Sweden Svensson (1988) identified 11 different species or sections of *Sphagnum*; and in Minnesota, Janssens *et al.* (1992) identified 12 species.

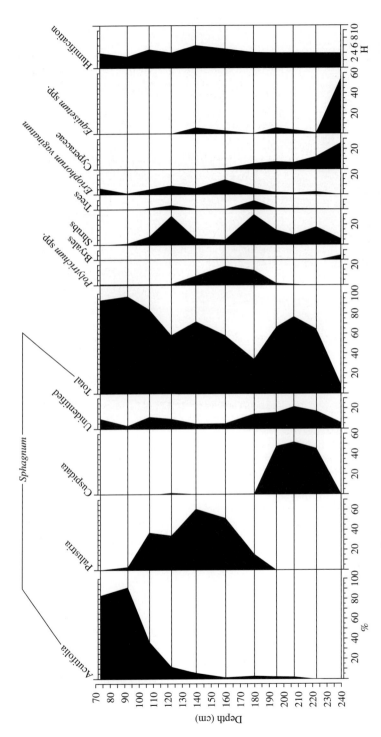

Fig. 6.2. Macrofossil profiles from Stor-Åmyran, a mixed mire complex in Sweden, boreal zone. The profile was collected on a weakly elevated eccentric bog with hummocks arranged in strings. Peat development started about 2400 years BP, and the underlying sediment was gyttja clay with rhizomes of *Equisetum fluviatile*. After some years of sedimentation, a peat dominated by *Sphagnum* Sect. Cuspidata along with *Carex* developed (2400–2000 BP). Dwarf shrubs and brown mosses developed at the end of this period. At about 2000 BP the *Carex* species vanished and the plant composition indicated transition to ombrotrophic bog. The indicators to appear were *Sphagnum* Sect. *Sphagnum* (= Palustria) *Eriophorum vaginatum*, and *Polytrichum* species, followed by *Sphagnum* Sect. Acutifolia, especially *S. fuscum*. Profile from Bohlin (1993), with author's permission. Further explanations from Klarqvist (2001).

Table 6.1. Generalized categories of macrofossils that can be used to characterize peats

Macrofossil category	Indication
Sphagnum	Particularly important to characterize peat and habitat types, using data on their relation to pH, shading, and water level as described in Chapter 4
Brown mosses Ericaceous dwarf shrubs (nanolignids)	Indicators for moderately rich to rich fens (higher pH and Ca) Indicate hummock level of acid peats
Coarse wood (lignids)	Indicates aerated peats of swamp, fen, or bog. A stump horizon indicates an earlier period of drier conditions (e.g. Fig. 6.5)
Eriophorum	*Eriophorum vaginatum* (Europe) and *E. spissum* (N Am), can dominate ombrotrophic or near ombrotrophic peats, often along with *Scirpus cespitosus*
Cyperaceae (e.g. *Carex* spp.)	High abundance indicates wetter conditions, open water, carpet, or lawn level. Some species indicate richer peatlands (*C. pseudocyperus*, *C. vesicaria*, and *Cladium mariscus*). Others can cover the whole range of minerotrophy (e.g. *Carex lasiocarpa*, *C. rostrata*)
Equisetum, Phragmites australis, Typha	Indicate open water marsh or marsh or fen at water's edge
Poaceae	Indicate marsh or meadow marsh or moderately rich to rich minerotrophic fens
Herbs	Most common in marshes, moderately rich to rich fens, wooded fens, and swamps

Microfossils

As the peat surface and water table grow upwards, pollen grains and other microscopic organisms pass below the water table into the anoxic layers of the peat. Other single-celled or colonial organisms are included, such as the algae *Pediastrum, Botryococcus,* and *Scenedesmus*; diatoms; and testate amoebae. Fungal hyphae can be found in certain layers, and bryophyte, fern, and fungal spores are also identifiable (e.g. see papers in Berglund 1986). To the list we may add small aquatic fauna, such as Cladocera, chironomids, and ostracods (Barber *et al.* 1999).

Pollen analysis was developed in the early decades of the twentieth century. It is the basis for much of our knowledge about the postglacial climate and vegetation history, and an interesting summary of the early days was given by the leading pioneer, Lennart von Post (1946). Pollen analysis is one of the commonest palaeoecological techniques. Pollen grains show a great deal of variation in size, external coat, pores, furrows, ridges, and wings, allowing plant families, genera, and even species to be identified. For a general description of pollen analysis, its application and pitfalls, and description of microfossils reference may be made to several textbooks and review papers (e.g. Birks and Birks 1980; Berglund 1986; Faegri and Iversen 1989;

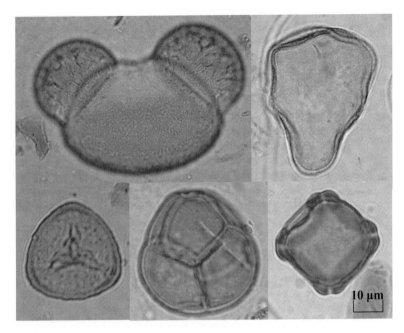

Fig. 6.3   Some representative pollen grains and spores. Top from left: *Pinus*, Cyperaceae, Bottom from left: *Sphagnum* spore, Ericaceae, *Alnus*. Photos by Henrik von Stedingk.

Moore *et al.* 1991; Bennett and Willis 2001; Punt *et al.* 2003). Some representative microfossils are shown in Fig. 6.3.

A common technique is to take a small sample (about 0.5 cm³) from each sampling level, loosen it up by soaking in alcohol, and boil it in 10% KOH for 1 minute. This may be followed by addition of 10% HCl and acetolysis at 95 °C for 10 minutes (Janssen 1968). The objective is to remove the cellulose, lignin, and humic substances and retain the highly resistant pollen grains. The suspension may be stained with gentian violet or safranin to differentiate pollen from other organic matter. A small amount of the suspension is mounted in glycerin or silicone oil, and examined under the microscope. Using a technique to scan the slide systematically, 300–500 (sometimes 800) pollen grains or spores are identified. The percentages of each kind of pollen/spore are charted according to depth in a pollen diagram.

It is essential to distinguish between *local pollen*, representing the specific peatland being studied, and *regional pollen*, representing the surrounding upland vegetation species which have long-distance, wind-dispersed pollen (e.g. Janssen 1968, 1992; Janssens *et al.* 1992). The local pollen can imply, for instance, changes in wetness of the peatland and development from minerotrophy to ombrotrophy. The upland pollen provides evidence for changes in forest vegetation, or increases in grasses, composites, and weedy pioneer species, indicating advance of anthropogenic activity. In small,

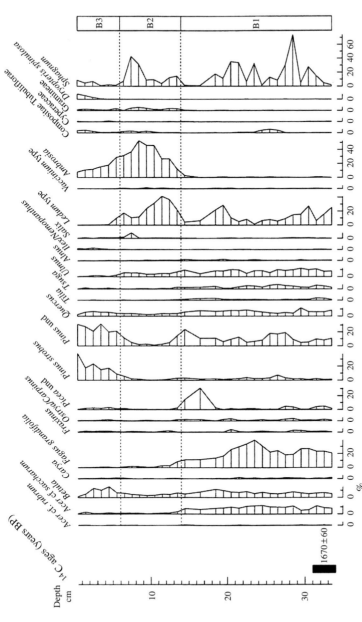

Fig. 6.4. Microfossil diagram (pollen and spores) from a coniferous swamp forest in southern Ontario reflecting vegetation changes both in the peatland and in the region. Three zones were described, B1–B3. The high values of *Ledum* (*Rhododendron*) pollen in zones B1 and B2 indicate that the site was a low shrub bog. Thereafter it developed into a conifer swamp (high values of *Pinus* pollen) with the fern *Dryopteris spinulosa*. The regional pollen composition in B1 indicates a mixed deciduous forest dominated by *Acer* and *Fagus*, and with some *Ulmus*, *Quercus*, and *Betula*. This was typical for the region for several thousands years. The decline in these trees at the top of B1 marks the clearance by European settlers, and the following agricultural expansion is revealed by the peak in *Ambrosia* pollen in zone B2. Note that the abundance of each pollen type is given as a percentage of the total. This means that a large increase in one group can mask subtle changes in other groups – the expansion of pines in the peatland in zone B3, for example, probably masked some changes in the upland vegetation. Simplified from Bunting *et al.* (1998).

shallow peatlands in forested terrain short-core methods have been used with considerable success in recording successional developments (e.g. Hulme 1994; Segerström *et al.* 1994), tree species migrations (Björkman 1997), and local forest disturbance and dynamics history (Bradshaw and Hannon 1992; Björkman 1999; Hellberg *et al.* 2003). Figure 6.4 gives an example of a pollen analysis for pre-European settlement conditions and human disturbance of a coniferous swamp in southern Ontario (Bunting *et al.* 1998). Here it is possible to see some local pollen types, such as *Ledum* (*Rhododendron*) and *Vaccinium*, and also regional pollen, for example *Ambrosia*, which signals the European settlement.

The modern pollen rain can be quantified by exposing collecting surfaces, e.g. glass slides with sticky surfaces, to the air during pollen seasons. By comparing modern pollen rains with vegetational structure and composition of the surrounding vegetation, one may arrive at certain correlations and corrections to adjust the raw data (Andersen 1973; Webb and McAndrews 1976). However, such corrections usually are not made, and these differences in abundances and distribution of pollen must be taken into account during the interpretation. For example, Janssen (1992) noted that when tamarack (*Larix laricina*) grows in a low stunted form (in poor swamp or fen) it produces little pollen, whereas when it grows taller and more densely in richer wooded fens or swamp forests it produces much more. Since the pollen is heavy and most of it falls only a short distance from the tree, abundance of tamarack pollen suggests denser and taller trees, and also richer conditions. For peatland plants other than trees, it appears that most pollen comes from the nearest few metres, so local pollen from a single peat core reflects predominantly small-scale vegetation dynamics rather than the development of the whole peatland (Bunting 2003).

Macrofossil and pollen analyses are based on two different data types, and are normally presented separately. These analyses are highly complementary, and together provide for a more complete interpretation of the succession of peat communities at a specific site in the peatland, as well as interpretations about past regional floras, vegetation, and environment.

The problem of dating profiles

In all studies of peat profiles and peatland development the age of the peat layers is important, and there are a number of methods that can be used to age peat layers.

Radiocarbon dating

The most common method for peat dating is based on the fact that the ^{14}C carbon isotope is present in the air in a small ratio to ordinary ^{12}C carbon.

As plants take up carbon dioxide (CO_2) for photosynthesis they acquire ^{14}C in the same proportion, and this is incorporated into the organic compounds that are laid down as peat. Since ^{14}C decays over time with a half-life of 5568 years, it is possible to calculate the age of organic material from its ^{14}C content. The dates are reported as years before the present (BP), where 'present' is defined as 1950.

Selection of the specific profile depths for radiocarbon dating is the most important step. Usually the dating levels are chosen where there are sharp rises or drops in several of the pollen profiles, indicating a change between zones. Other profile features for which dating may be desirable are abrupt boundaries between distinct layers. The bottom of aquatic sediment (if present) and of the peat are particularly important in determining when the aquatic sediment and peat started to form. It is also valuable in profile interpretation to determine the level and age at which bog formation was initiated over fen or swamp.

Before the radiocarbon dates are used, they may be calibrated into calendar years (Reimer *et al.* 2004), to allow comparison with dates obtained by other techniques. The calendar-year calibration corrects for long-term changes in atmospheric radiocarbon content. However, some workers choose not to calibrate ^{14}C dates because the dates quoted in earlier research were often not calibrated (e.g. Korhola 1992). When referring to dated peat profiles it is important to note whether they were calibrated or not. For instance, the start of the Holocene in Europe is defined as 10 000 ^{14}C years BP (Mangerud *et al.* 1974; Godwin 1981), which is equivalent to 11 268–11 553 calibrated years BP.

There are problems with sampling and with accuracy of traditional radio-carbon values. One must have a large enough sample for normal analysis; a 5–10 g dry mass sample is necessary, although smaller samples can be used with more modern methods (see below). Another problem is that of possible contamination of the original layer of peat by movement into it by 'younger' or 'older' carbon. Dilution of older carbon with younger carbon may take place by downwards percolation of particulate or dissolved organic from above, by downward transport of pollen grains, or by transport of younger carbon to lower layers by root exudations or dying rhizomes and roots. This is quite important in fen peats that contain a large percentage of much younger rhizomes and roots penetrating down from more recent vegetation (Sjörs 1991). The possibility also exists that older carbon in the form of methane (CH_4) or CO_2 may move upwards from lower layers, and become biologically fixed in the younger layers. A detailed study (Nilsson *et al.* 2001a) revealed that, depending on the sampled depth and peat type, the difference in calibrated ^{14}C age within specific 2 cm thick samples, varied between 365 and about 1000 years. There is also the possibility in calcareous fens that old carbon from $CaCO_3$ may influence the ^{14}C dating.

Both the sample size and peat contamination problems may be avoided by collecting small fragments of plant remains and obtaining ^{14}C dates with accelerator mass spectrometry (AMS) (e.g. Hedges and Gowlett 1986; Olsson 1986). This can be applied to very small fragments (less than 1 mg dry weight) of *Sphagnum* stems, *Carex* leaves, or wood, but the disadvantage is the somewhat higher cost per sample.

The precision of the ^{14}C method becomes low for very young ages, and [14]C *wiggle-matching* has been used to increase the resolution. In this method ^{14}C AMS dating is made at close intervals, a few centimetres between each sample, in the profile. Over time the ^{14}C content in air has varied, and the 'wiggles' in this fluctuation have been exactly dated by comparing with ^{14}C content of tree-ring sequences of exactly known age. Now the wiggles in the peat can be matched with the calibration pattern from tree-rings and this gives good dating over the recent periods when ^{14}C itself is highly unreliable (Kilian *et al.* 2000; Blaauw *et al.* 2003).

Other dating methods

Numerous other methods of dating peat deposits have been used, and for details we refer to the excellent review by Turetsky *et al.* (2004). Since ^{14}C is uncertain for the recent decades, the radioactive lead isotope ^{210}Pb, with its convenient half-life (22.3 yr), may be used for dating these younger layers (e.g. Oldfield *et al.* 1979; El-Daoushy *et al.* 1982). [210]Pb is gradually incorporated from the atmosphere when plants grow, and it is possible to date over the last decades to centuries by measuring how much of it remains at a certain depth. However, ^{210}Pb may move in the peat, which creates some uncertainties (Belyea and Warner 1994).

There are several techniques that use specific time markers in the peat. Well-defined *recurrence surfaces* (see below) that have been dated by ^{14}C can be quite useful for defining certain periods of peat accumulation (Malmer and Wallén 2004). One technique is that of characteristic pollen limits, especially finding the layer where a species migrated into an area, as indicated by the first (or first significant) occurrence of the pollen. Commonly used in Fennoscandia is the *Picea* limit (see Fig. 7.5), though it is not synchronous, but other limits are those of *Alnus, Corylus, Fagus, Tilia,* and *Quercus* (Svensson 1988; Korhola 1992; Björkman 1997, 1999). In North America *Ambrosia* pollen has been associated with the agriculture introduced by the European settlers (Fig. 6.4). A simple biological time marker was used by Ohlson and Dahlberg (1991) who excavated small pines from *Sphagnum* bogs. The position of the original growing point of the pine, the root–stem transition, can be aged by counting the annual tree rings of the pine. All these methods have uncertainties and even pitfalls.

There are several recent events from which depositions of various kinds have occurred, and these can be used if it is known when the fall-out occurred or started, and if the deposit is immobile in the peat. Examples are layers of volcanic ash (tephra), wildfire ash (Alm *et al.* 1992), pollutions of heavy metals, organic pollutants (PCB, DDT), and spheroidal carbonaceous particles ('soot balls' from combustion of fossil fuel). Fallout of radioactive caesium ^{137}Cs (half-life 30.2 yr) from nuclear tests in the 1950s and early 1960s and from the Chernobyl nuclear power plant disaster in 1986 has also been used, but Cs appears to be more mobile than Pb in peat.

Independently dated archaeological findings in or beneath peat can be used, such as remnants of villages and middens of wetland dwellers, stone walls and huts, wooden trackways (Coles and Coles 1989), and remnants from agricultural fields such as cultivated grains found at the juncture of the peat and mineral soil (e.g. Segerström *et al.* 1994).

It is not always possible to obtain datings, or enough of them to fully characterize a profile, so one way to obtain some idea of age is to compare the macrofossil and pollen layers of the undated profile with other profiles from nearby that have been ^{14}C dated. For example, this was done in the study of the Myrtle Lake Peatland in Minnesota where comparisons of a raised bog profile were made with a lake profile with dates spanning an 11 000 year sequence (Janssen 1992).

Peatlands as archives of changes in climate and vegetation

The early work on pollen analysis focused on interpreting the vegetation and climate from the peat and pollen records of the Holocene. Since modern understanding of Holocene climate comes from other sources (e.g. Greenland ice cores), some of this work is now mainly of historical interest. Currently, it is more common to use bryophyte macrofossil composition to infer the wetness of the peat surface, and then interpret these changes as indicators of the wetness of the climate and how that affects the peatland vegetation (Barber *et al.* 1998, see also Chapter 7).

Blytt–Sernander scheme of climate change

The main scheme of postglacial climatic change is the Blytt–Sernander scheme, named after the Norwegian Axel Blytt and the Swede Rutger Sernander. It was developed in the late 1800s and early 1900s (von Post 1946). The scheme recognized four postglacial periods of climate represented by different layers observed in raised bogs. These were, from bottom to surface, the Boreal, Atlantic, Subboreal, and Subatlantic periods. The Boreal and Subboreal had darker, well-humified peats often associated with layers of wood, the so-called *stump layers*, and were interpreted as being formed during drier continental climatic periods. The Atlantic and Subatlantic had lighter, poorly humified peats, without wood, interpreted

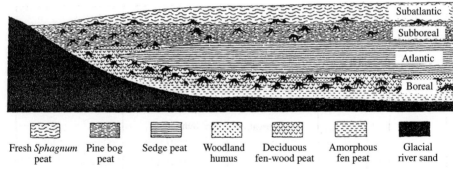

| Fresh *Sphagnum* peat | Pine bog peat | Sedge peat | Woodland humus | Deciduous fen-wood peat | Amorphous fen peat | Glacial river sand |

Fig. 6.5   von Post's visualization of the four major climate periods in the Blytt–Sernander sequence, as represented in the schematic profile through Lerbäck Bog in Närke, Sweden (von Post 1946). The bottom two layers are fen peat, and the upper two layers are ombrotrophic bog peat. Drier periods are represented by the two woodland peat layers, moister phases by sedge peat between the stump layers and by *Sphagnum* peat highest up in the series.

as being formed during more maritime, humid climatic periods. A conceptual diagram of these layers in a Swedish raised bog is presented in Fig. 6.5. For the late glacial and earliest postglacial, other terms have been added later (see below).

Early work on peat profiles in Scandinavia and Germany had identified transitions between dark, highly humified peat to light, poorly humified peat. C.A. Weber, the German pioneer in peat bog investigations, had named the sharp transitions *Grenzhorizont* ('boundary horizon') (Weber 1902). The peat transitions have also been called recurrence surfaces. Numerous recurrence surfaces have subsequently been described and dated to 800, 1600, 2600, 3200, 4300, 4800, and 5700 BP (Frenzel 1983), indicating that the climate has fluctuated repeatedly from warm–dry to cool–wet during the Holocene. Early work suggested that the recurrence surfaces were synchronous over large areas of Europe, but this has been contradicted by other investigations. Therefore it is not clear if these surfaces were always caused by climate change, or if they sometimes represent transitions from minerotrophic to ombrotrophic, or an autogenic process related to the hydrologic development of the bog (Frenzel 1983). Also, wildfire may cause abrupt changes during dry periods.

The Blytt–Sernander scheme has been applied to other areas of Europe, North America, and elsewhere, and has come to be used generally as a framework for describing the main climatic periods of the Holocene. Deevey and Flint (1957) and Mangerud *et al.* (1974) have provided syntheses of pollen chronozones for North America and northern Europe, respectively, and many others have given similar syntheses for different areas, for example, recently for Italy (Ravazzi 2003). These are all based on the early scheme of Blytt and Sernander, and a composite scheme is presented in Fig. 6.6.

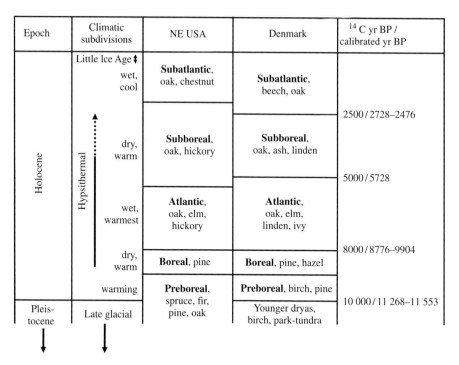

Epoch	Climatic subdivisions	NE USA	Denmark	^{14}C yr BP / calibrated yr BP
Holocene	Little Ice Age ⭥ wet, cool	**Subatlantic,** oak, chestnut	**Subatlantic,** beech, oak	
	dry, warm	**Subboreal,** oak, hickory	**Subboreal,** oak, ash, linden	2500 / 2728–2476
	wet, warmest	**Atlantic,** oak, elm, hickory	**Atlantic,** oak, elm, linden, ivy	5000 / 5728
	dry, warm	**Boreal**, pine	**Boreal**, pine, hazel	8000 / 8776–9904
	warming	**Preboreal,** spruce, fir, pine, oak	**Preboreal**, birch, pine	10 000 / 11 268–11 553
Pleis- tocene	Late glacial		Younger dryas, birch, park-tundra	

Fig. 6.6   Postglacial periods, with subdivisions and times before present. Based on Deevey and Flint (1957), Mangerud *et al.* (1974), and Ravazzi (2003).

A huge amount of pollen information has accumulated, giving more detailed perspectives on zonal and regional variations. The dates for the chronozones vary somewhat, depending on the interpretation of boundaries and the times that various climatic changes occurred in different geographic regions. However, the general timing of events was about the same in North America and Europe (e.g. Davis 1983). The following are brief sketches of the climate and vegetation periods of the Holocene with time given as uncalibrated ^{14}C years BP (see Fig. 6.6. for calibrated calendar years).

Preboreal

The Holocene begins with the Preboreal period, at 10 000 to 9500–9000 BP. This period was much warmer and moister than the final stage of the Pleistocene (in Europe the Younger Dryas). In Europe there were already thick stands of pine and birch, and a few plants indicating warm climate. The preboreal in the northeastern USA was interpreted differently from the European, and was thought to begin much earlier, around 12 500 BP, and characterized by spruce, fir, pine, and oak (Deevey and Flint 1957). On the other hand, the ice remnants disappeared later in Canada than in northwest Europe.

Climates are thought to have warmed by several degrees Celsius over a short time period (less than a century) beginning around 10 000 BP. In the British Isles, there was a rapid and very substantial warming that had a profound impact on species and ecosystems, including turnover of a large proportion of the flora and extinction of about 90% of the beetle fauna (Huntley *et al.* 1997). In North America there was also substantial warming, although it was of lesser magnitude and occurred at varying times; for instance, it was completed before 10 000 BP in some areas. In Scandinavia, the treeline (especially of pine) reached higher up on the mountains than the present one, but the altitudes at that time were lower.

Boreal period

This period lasted from 9000 to 8000 BP, and was warm and dry. In Europe birch decreased in abundance while pine increased and hazel became abundant. During this period plants migrated northwards. For example, the water chestnut, *Trapa natans*, now missing from the Swedish flora, was one of the commoner plants of the lakes of southern Sweden. Postglacial warmth resulted in drying of peatlands, which then became partly covered with forests, chiefly hazel, birch and pine. In northeastern USA the zone is similar, and characterized by pine.

The Boreal period is the beginning of the *warm period*, also referred to as the *thermal optimum* or *hypsithermal*. During this period there were warmer and cooler intervals, and several oscillations of atmospheric moisture. Thus, the periods Boreal, Atlantic, and Subboreal are contained within the hypsithermal.

Atlantic period

This moister period lasted from 8000 to 5000 BP and was the peak of the hypsithermal. A mixed deciduous forest of *Quercus, Ulmus, Alnus, Corylus,* and *Tilia* partly replaced *Pinus* in Europe, and *Quercus, Ulmus* and *Carya* dominated in the northeastern USA. *Sphagnum* bogs began to appear more frequently.

Subboreal period

The Subboreal lasted from 5000 to 2600 BP, and represents a rather dry period of continued warmth but with slow cooling. Several thermophilic species started to decline. Permafrost zones have been mapped for west-central Canada (Zoltai 1995). At 6000 BP permafrost distribution zones were 300–500 km further north than today, corresponding to a mean annual temperature that was about 5 °C warmer than at present. Macrofossils indicate that, except in the far north, permafrost development began about 4000 BP by the onset of a cooler and moister climate.

Subatlantic period

One of the best known recurrence surfaces was the Grenzhorizont (numbered III) at about 2600 BP (Granlund 1932; Deevey and Flint 1957). As the boundary between the Subboreal and Subatlantic, it was interpreted by Sernander as the end of the hypsithermal. It is recognized as a recurrence surface with woody, dark to brown, well-humified peats below and light-coloured, poorly humified *Sphagnum* above. The climate changed to cooler and wetter conditions in which *Sphagnum* growth became widespread, soils became flooded or waterlogged, and fens could expand. There was steadily increasing human impact in Europe, with an increase in meadows and farms replacing the forests. The abrupt change to the Subatlantic shows more losses of thermophilic species, but the pollen assemblages still include species that existed in the hypsithermal. Apparently some elements of the dominant vegetation that was established during the hypsithermal could still maintain themselves. In Scandinavia (earlier in Finland) *Picea* forests expanded, also southwards, and *Fagus* and *Carpinus* invaded the south.

Medieval Warming and Little Ice Age

The Medieval Warming occurred about 900–1200 AD, and the Little Ice Age cooling about 1300–1850 AD (Fagan 2000). During the Little Ice Age glaciers advanced in the mountains, but ice accumulation rates on the Greenland ice sheet were generally lower. Work by Barber *et al.* (1999) on a large number of proxy records of climatic changes in the UK supports the idea of a Little Ice Age, and shows particularly cool, wet decades between 1700 and 1850. Historical events and famine records confirm this, especially from about 1650 on.

Geographic shifts in vegetation

One of the universally repeating patterns revealed in pollen profiles and pollen assemblages for the Holocene is the geographical shifting of vegetation zones represented by different pollen assemblages. These displacements occurred in response both to postglacial changes in temperature, usually north–south movements, and to moisture. A general reference for these shifts is Huntley and Birks (1983). Today, much of the understanding of these processes come from lake sediments rather than peat. Peat is more directly useful for local changes, including succession, to which we will return in Chapter 7.

An example of migration of pollen assemblage zones related to changes in moisture regime is from the work of McAndrews (1966), who studied pollen assemblages along a 106 km transect of peatlands through the main vegetation types in Minnesota. The transect progressed from mixed pine – hardwood forest in the east, to deciduous forest, to prairie in the west. From 12 000 to 11 000 BP, the whole transect was occupied by a *Picea–Populus* pollen assemblage, for which no modern equivalent has been found. At about 11 000 BP, the prairie spread from the west into the transect area, and further east a *Pinus banksiana–P. resinosa–Pteridium* assemblage replaced the *Picea–Populus* assemblage. At about 8500 BP, prairie or oak savannah expanded throughout the transect, represented by Poaceae–*Artemisia* and *Quercus*–Poaceae–*Artemisia* assemblages. A dry, warm period is recorded throughout the midwestern USA at this time. At about 4000 BP, savannah was replaced by deciduous forest on the fine-textured soils, represented by a *Quercus–Ostrya* pollen assemblage zone, as a response to a cooler, wetter climate. At about 2000 BP *Pinus strobus* immigrated into the deciduous forest from the east and became dominant on the coarser soils. It was followed by *Pinus banksiana–P. resinosa* at about 1000 BP in the east, where the mixed pine – hardwood forest of today became established. The invasion of *Picea* into Fennoscandia from the east and of *Fagus* into Germany, Denmark, and southern Sweden are other well-known cases.

Pleistocene peatlands

The great majority of studies of pollen deal with the Holocene, since most of the extant northern hemisphere peat was formed on terrain opened after the retreat of the Pleistocene glaciers. However, there are older peatlands that had developed on terrain that was not glaciated during the Pleistocene, and there are many subtropical and tropical peatlands that date back to the Pleistocene. Nakaikemi Wetland in south-west Honshu, Japan, is a deposit of peat with interbedding of mineral layers up to 45 m deep formed at least 50 000 years BP (Saito 2004). The deepest peat/lignite layer in the world is probably the Phillipi Peatland in Greece, reputed to be 190 m deep (Kalaitzidis and Christanis 2002), and largely from the Pleistocene. There are other ancient peat beds that were laid down during interglacial or interstadial periods, and subsequently covered by glacial till and lacustrine deposits during later ice readvance and retreat; for example, the Missinaibi beds in Ontario, Canada, dating to 38 000 and 53 000 BP (Terasmae and Hughes 1960).

Environmental research

Rapid progress in technical and natural sciences has opened new possibilities for the use of peatland palaeoecology in the field of environmental research. Fire is an ecological factor that has received increasing attention among palaeoccologists in recent decades. The analysis of fire history and forest succession from peat layers can be an important approach to understanding local and regional ecology. Ombrotrophic peat layers are superior to lake sediments as fire archives, because the airborne microscopic charred particles deposited on the peatland surface remain in place and do not migrate. Thus, a precise fire chronology can be read from peat cores when the consecutively cut sections are thin enough. When combined with pollen analysis, counts of charred particles give information on fire events, both natural and owing to human activity. A good summary is given in Patterson *et al.* (1987), and it has been recommended that counting of charcoal particles should be included as routine practice in pollen analysis (Tolonen 1991).

Since the surface peats of ombrotrophic bogs get their water supply only from the atmosphere, they are useful for mapping areal distribution of airborne fallout such as radioactive pollutants, heavy metals, organochloride compounds, and polycyclic aromatic hydrocarbon compounds. There are numerous examples of the successful use of surface peat and *Sphagnum* for monitoring local or regional heavy metal pollution. Similarly, the surface peats of bogs show good retention of magnetic iron (Fe) particles originating from industrial and urban pollution, and studies have shown good correlations between consumption of coal and the magnetic Fe deposition values for dated peat profiles (Oldfield *et al.* 1978; Tolonen and Oldfield 1986).

In prehistoric peats lead deposition apparently corresponds to the aerosol influence of dust and smoke, and this indicates that older peat is potentially a good historical heavy metal archive. In the surfaces of some bog profiles near the forest–grassland border in west-central Canada there are elevated levels of ash content, calcium (Ca), magnesium (Mg), Fe, sulfur (S), and phosphorus (P) (Zoltai *et al.* 1988). It is tempting to ascribe these higher levels to windborne dust resulting from recent agricultural activity in the grain-growing provinces of Canada.

Wetland archaeology

Ancient people often lived on the shores of seas, lakes, or wetlands. The advantages were many: water for humans and cattle; transport by dugouts and canoes; food from wetland plants, fish, wildlife, shellfish, and so on; and protection of dwellings by building on islands, peninsulas, and even constructed mounds and platforms. The 1800s witnessed a rapid development of the archaeology of wetlands with finds of ancient wetland settlements – pile-dwellings, crannogs (artificial islands supporting ancient dwellings), and terps (mounds of north European coasts). Many ancient wetland dwelling sites have been found in the British Isles, northern and central

Europe, Japan, the United States, and Canada. However, many of these wetlands were probably not peatlands. An excellent overview of wetland archaeology is given in the book *People of the wetlands* (Coles and Coles 1989). An important British archaeological site in peat-like material is Star Carr (Mellars and Dark 1998).

The wetland settlements have yielded many archaeological artefacts dating back as far as the early Holocene, and as recently as a few hundred years ago. Many of the artefacts date from 800 BC to 400 AD, a period of transition from the Bronze Age to the Iron Age when there was a period of agricultural settlement, which was also congruent with Roman invasions and colonization (Godwin 1981). These artefacts include canoes, fish spears, floats, nets, axes, clubs, grindstones, cooking utensils, swords, daggers, coins, and horns. Remains of ancient plank and brush trackways across peatlands, and wooden disc wheels, indicate that a multitude of roadways once traversed the quaking bogs, and where rivers and lakes barred the way, rafts, logs, bridges and dugouts were used (Coles and Coles 1989).

Among the most fascinating of the remains found in peatlands are the preserved human bodies, the most famous of which are the Tollund Man of Denmark and Lindow Man of England. Some of these bog bodies show clear evidence of execution or sacrificial murder, but others may originate from accidental drowning or ritual burial. As many as 2000 bodies have been discovered, and they are especially common in north-west Europe, notably in Germany (about 600) and Denmark (more than 400), but also in Holland, Finland, Norway, Sweden, England, Ireland, and other countries. The majority of these can be dated from the period between 100 BC and AD 100, but isolated finds are more than 5000 years old. Far fewer sites have been uncovered in North America, but in Florida there are ritual burial sites 7000–8000 years old, containing perhaps as many as 1000 bodies that were wrapped and staked to the bottom of a pool in a peaty site. The bodies have invoked much scientific interest as to the circumstances by which they got there, the archaeological significance, and the mechanisms by which they were preserved in the peat (e.g. Glob 1998; Turner and Scaife 1995). A curious feature of the bog bodies is that they have become dark brown or black; the preservation involves a process similar to that for leather tanning. This process involves a series of very complicated reactions which are understood only in outline, known collectively as the Maillard reaction (Painter 1995), and caused by the uronic acid (5KMA) produced by *Sphagnum* (Chapter 4).

7 Peatland succession and development

Change over time is a fundamental dimension in ecosystems, and can be studied better in peatlands than in most other ecosystems because changes are preserved in the peat laid down beneath successive vegetational communities. *Succession* is the term usually applied to short-term directional changes occurring over periods of a few years, decades, or up to several hundred years. *Development* is the term applied to long-term changes, over several hundred years, or millennia in the case of postglacial development of peatland. In this chapter we first present an overview of succession, and follow this with processes of peat formation and development.

Peatland succession

Succession is one of the earliest concepts in the field of plant ecology, dating back at least as far as the mid nineteenth century in Europe. Early seminal work on wetland and peatland succession and development was done by Blytt, Cajander, Sernander, and Weber in Europe, and Cowles and Clements in North America. This was soon followed by detailed peatland inventories and ecological studies by Auer (1930), Granlund (1932), Malmström (1923, 1931), Osvald (1925), and von Post (1937) in Europe, and Dachnowski (1912, Dachnowski-Stokes 1926) in North America. A later landmark paper is that by Walker (1970).

Succession was initially defined as a directional change over time in species composition and community structure, but nowadays the concept is broadened to include concomitant changes in the animal community, soils, and other habitat factors. Normally a line of succession leads to a community in which overall species composition changes only gradually. This stage may be dynamic, but the directional changes characteristic of succession are replaced by short-term fluctuations.

Succession is often divided into primary succession and secondary succession. In *primary succession*, by definition, organisms colonize a previously vegetation-free substrate, that is, a substrate with no biological legacy. Examples are land colonized after glacial retreat or postglacial land uplift. *Secondary succession* occurs on sites where there has previously been some vegetation that has been disturbed. In this case there is some residual influence on the site by the previous vegetation, and some plants, seeds, and organic matter in the soil. Some of the processes initiating secondary succession in peatlands are fires, flooding (high rainfall, beaver damming), drought, and human activities (draining, damming, peat harvest, and forest harvest).

Succession is driven by allogenic and autogenic factors. Allogenic factors are external to the sites, such as fire, climate, wind, and erosional deposition, as well as human activities such as draining and fertilization. Autogenic factors are those resulting from internal influences and changes caused by the interactions of plants, animals, and substrates. Successions change in the relative influences of these factors. Early in a succession allogenic factors may predominate, for example in marshes. As the succession proceeds, allogenic influences may decrease while autogenic factors often increase in importance, for example in fens and swamp forests. When one reaches the ombrotrophic bog stage, autogenic factors predominate, although there are still allogenic influences such as climate and atmospheric pollution.

Successional pathways

Modern succession theory stresses the individualistic behaviour of the species involved – features such as dispersal ability, shade tolerance, competitive ability, and life span explain why some species appear early and others late in the succession (Huston and Smith 1987). On a larger scale we can still recognize a *successional pathway* – a sequence of plant communities developing during or after a particular kind of disturbance, on a particular substrate, and with a particular complement of residual and invading species.

In the case of peatlands, successions normally progress towards decreasing base saturation and increasing acidity. This long-term trend is towards ombrotrophy, and ombrotrophic conditions may be regarded as the final stage of peatland development. Even when ombrotrophication occurs, ombrotrophic conditions are almost never reached in all parts of a peatland, because mineral soil water is continually supplied at the margins of the peatland, or there are pools or flushes of mineral soil water occurring within the ombrotrophic parts. Over time there may be increasing dominance of a peatland area by ombrotrophic ecosystems, but with persistence of inputs of mineral soil water a more or less dynamic balance will be reached between the parts that remain minerotrophic and those that are ombrotrophic.

Successions are considered as *progressive* or *retrogressive*. In earlier writings progressive was often regarded as moving towards more diversity, but with regard to peatlands species richness decreases during most successions. Hence, it is better to think of progressive successions as the 'normal' pathway as peat builds up, whereas retrogressive successions recede to an earlier stage, usually following peat surface destruction. These are often caused by allogenic factors such as changed climate, flooding, fire, grazing, browsing, or human impacts. Flooding is one of the most frequent disturbances that can set the vegetation back to earlier stages. For examples, a bog may revert back to fen; a swamp forest may become wetter and revert to a marsh; and a wooded bog stage may revert to a wetter open bog as a result of a wetter climate (for example at a recurrence surface). A peat core may contain both kinds of sequences, for example, wetter to drier (progressive), and then a change back to wetter community types (retrogressive). These changes may be accompanied by pH and nutrient changes.

Sometimes the spatial sequences observable along a transect in the field are interpreted to represent successional changes over time. Often these sequences represent wet to dry communities, or rich to poor, or a combination of wetter – richer to poorer – drier, because one of the prevailing tendencies in peatlands is for sites to become progressively poorer. However, one must be very cautious about applying this 'space for time' logic – the spatial sequences *may* indicate successional trends, but also *may not*. For example, a spatial sequence of open fen in the centre of a basin to wooded fen to swamp forest at the peatland margin need not be the sequence that is found in a peat core in the basin centre.

Figure 7.1 shows a model of spatial sequences that can be observed in peatlands in northwestern Ontario, Canada. Some potential temporal sequences that progress from wetter-richer to drier-poorer are:

• Aquatic plants (aquatic sediment) → emergent plants (limnic peat) → marsh (marsh peat) → thicket swamp (sedge-lignid peat) → swamp forest (lignid peat)
• Aquatic plants (aquatic sediment) → emergent reeds and rushes (limnic peat) → marsh (marsh peat) → open fen (sedge peat) → wooded fen (sedge-lignid peat) → swamp forest (lignid peat)
• Aquatic plants (aquatic sediment) → open fen (sedge peat) → open bog (*Sphagnum* peat) → wooded bog (*Sphagnum*–lignid peat).

These sequences, which are not the only ones, are portrayed with all the stages, but often sequences in the field are only partial ones. For example, the initial stage of succession may be missing, and begin with emergents or

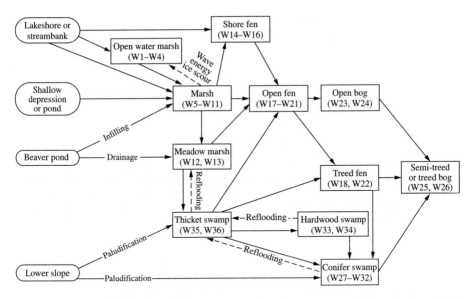

Fig. 7.1   Some potential successional pathways in northwestern Ontario wetlands, with the wetland types in the wetland ecosystem classification for northwestern Ontario. Solid lines represent progressive successions, dashed lines represent retrogressive successions. Redrawn from Harris *et al.* (1996).

open fen or even moist forest. In addition, field sequences do not always end with bog; they may end with relatively stable open fen, wooded fen, or swamp forest. This may happen where there is strong allogenic influence, for instance a long-term relatively constant flow-through of water with the same nutrient contents, such that the nutrient and moisture regimes are more or less stable. Figure 7.1 also depicts retrogression by flooding, wave action and ice scour.

Processes of peat formation

Traditionally, peatland ecologists have recognized three main kinds of peatland formation processes: infilling (terrestrialization), primary peat formation, and paludification (Fig. 7.2).

Infilling

Infilling is the process whereby peat develops on the margins and in the shallow waters of ponds, lakes, or slow-flowing rivers. An often used term is *terrestrialization* (Weber 1902), but the resulting ecosystem can be quite far from 'terrestrial', and therefore we prefer the term infilling. Peat can form beneath the emergent reeds, rushes, and sedges from roots,

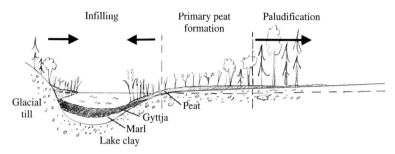

Fig. 7.2   A conceptual diagram showing a lake and the three main processes of peat formation. Infilling (terrestrialization) is development of peat into open water with aquatic sedimentary material below the peat. Primary peat formation is development of peat directly on wet mineral soil. Paludification is the upslope growth of peat over previously 'drier' conditions, usually with an upland humus at the base of the profile.

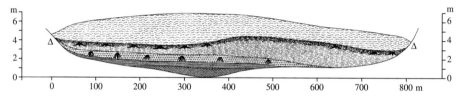

Fig. 7.3   An ombrotrophic raised bog developing through infilling. Store Mosse, boreo-nemoral zone, Sweden (from Lundqvist 1955). The horizons from lowest to highest are a gyttja layer, followed by *Carex* peat, and then a wooded fen peat with *Alnus*. Ombrotrophication occurs as the wooded fen is succeeded by a layer of *Sphagnum* peat, probably ombrotrophic. This was again followed by another layer of woody peat of *Pinus* in the bog, and a final layer of lighter poorly decomposed *Sphagnum*.

rhizomes, and leaf bases. Peat can also expand as floating mats of marsh and fen species out over the water. Sometimes portions of the mat may break loose, as a result of damming and wave action, and become free-floating rafts.

The direction of the succession is usually toward more closed vegetation and formation of peat mats, less wet conditions, and increasing organic matter and decreasing ash content in the mats. The development may continue to fill the basin completely with a fen or swamp forest (as indicated by a layer with woody remains), or it may progress to a bog (Fig. 7.3). The final stage will depend on local hydrological situation, and the influence of climate on water balance and on net peat accumulation. Some basins in west-central Canada close to the forest – prairie border seem to have achieved stability with open marsh or fen communities. On the other hand, in more humid regions, basins may still have rising water and continuous peat accumulation, and they maintain themselves as wet, slowly growing peatlands, with either open or wet wooded communities (e.g. Verry and Timmons 1982; Verry and Urban 1993). At the mire margin substances derived from the peat itself lead to reduced porosity of the soil and thereby a higher water table.

Early work in the USA often described the infilling sequence as ending in broad-leaved, mesophytic forest with moist soil conditions (e.g. Dachnowski 1912). However, much of this was speculative, and some of the observed cases could have been caused by drainage. For example, drained swamp forests can progress to upland-like forest types with typical upland under-storey plants, but still with deep peat beneath (Chapter 13).

A unique type of succession and basin infilling is caused by beavers (Adams 1995). Beaver dams create ponds in streams and drainage ways. These ponds are filled, by successive generations of beaver colonies, with the remains of food stores, branches and old beaver houses. Surrounding trees are killed, either by the beavers or by waterlogging. Beaver meadows are the marsh meadows or open fens that form when the beavers have abandoned a beaver pond. Normally, since these meadows are in or close to stream courses, they may develop only to poor fen or swamp, and never reach a bog stage.

A detailed examination of the development of a floating mat was studied in the Harvard Pond, Massachusetts (Swan and Gill 1970). The pond was originally a black spruce forest on deep peat, which was cut and dammed to form a lake. The origins, spread, and consolidation of a floating mat were explained by the growth characteristics of *Chamaedaphne calyculata*, which originated on cut stumps and spread out over open water, forming a lattice of floating stems for support of *Sphagnum cuspidatum* growth, followed by other 'drier' *Sphagnum* species. Other species may also be involved (see Chapter 3).

A special feature of succession is the occurrence of floating logs and stumps which act as support for communities of vascular plants and mosses. These woody rafts occur in swamps of eastern North America which have high fluctuation of water levels, often rising over the ground surface and floating up large woody debris (e.g. Dennis and Batson 1974).

Paludification

Paludification is a term meaning the development of peatland over previously less wet mineral ground. These sites do not have an aquatic sedimentary deposit at the base of the profile, but instead often have woody peat with stumps and logs at the bottom, indicating a previous forested condition (e.g. Heinselman 1963, 1970, 1975). These layers have been known at least since the 1700s (Granlund 1932) and even then were interpreted as a development of peatland over previously drier, forest-covered land. This is the classic paludification theory, and it is true that paludification can cover previously drier forested upland, progressing from quite dry forest through stages of swamp forest and wooded fen, and ending in open peatland. Mor humus layers have been found at the base of blanket bogs in Wales (Taylor and Smith 1972) and on sands in Sweden. It seems that the larger part of the earth's area of peatland has been formed by paludification. Paludification may occur not only in northern conifer forests, but also in grasslands, cool temperate heathlands (e.g. UK, Ireland), arctic or alpine tundra, and even bare rock areas.

Not all woody layers are found at the bottom of peatlands and indicate paludification. Woody peat layers that occur higher up in the profile (Fig. 7.3) indicate a swamp forest or wooded fen stage occurring on the peatland during an earlier drier climate (recurrence surfaces, see Chapter 6). They may also be a developmental stage above an earlier open mire.

Paludification by mire growth and water level rise

Paludification by upslope movement of peat is caused by concurrent rise of peat and water table with encroachment on to the adjacent upland. As the water table rises, the mineral soils adjacent to the peatlands become waterlogged. Upslope paludification has been described with transects of peat profiles and dating of the bottom peats (e.g. Korhola 1992). It may be speeded up by clearcutting or forest fires, which reduce tree evapotranspiration and interception (Chapter 8), thereby causing water table rises (Vompersky and Sirin 1997), and by beaver damming causing flooding of mineral ground (Adams 1995).

Paludification by pedogenic processes

Paludification can also be initiated by pedogenic processes (soil development) that promote decreasing soil permeability, causing wet and anoxic near-surface conditions which lead to *Sphagnum* invasion. These processes occur commonly in northern conifer forests that are located upslope from peatlands. In such cases it is difficult to separate the pedogenic process from the rise in water table owing to the growth of the adjacent peatland. In other cases the initiation of the process is not next to a peatland, but rather on flat or gently sloping ground, such as terraces, outwash plains, or gentle slopes of high hills or mountains, which have abundant water supply and poor drainage. Several pedogenic processes have been suggested:

- *Pan formation*: This involves the formation of one or several thin layers (commonly < 5 mm thick) that are hard, impervious, and dark reddish brown to black. The mineral particles may be cemented by oxides and organic complexes of iron (Fe), aluminium (Al), and manganese (Mn) (Agriculture Canada Expert Committee on Soil Survey 1987). The pan creates a barrier to water, and at some point the conditions allow the succession of peatland vegetation on to the surface which once supported well-drained forest. This process has been described by a number of workers (Ugolini and Mann 1979; Noble *et al.* 1984). Damman (1965) suggested that the pan may be the result of the development of a water-saturated surface horizon rather than the cause of it, but Wells and Pollett (1983) note that 'It is difficult to rationalize a sequence of events in freely draining soils in which iron pan formation is preceded by surface peat formation.' Indeed, it may be that the two processes occur simultaneously.

- *Development of a thick, water-retaining organic mat*: Northern conifer forests and maritime heathlands are commonly associated with a feather-moss, mor carpet which increases in depth over time. This carpet is acidic, more highly decomposed at the base, and tends to impede drainage. Increasing depth of the surface organic mat and increasing humification of its organic matter both serve to increase its water holding and reduce its oxygen content, and this in turn favours the establishment of Ericales shrubs and *Sphagnum*. This may cause roots to be restricted to the uppermost layers of humus, thus reducing transpiration of the plants, further increasing the water content, which in turn promotes the establishment of peat-forming species, especially *Sphagnum* (e.g. Neiland 1971; Taylor and Smith 1972; Noble *et al.* 1984; Zobel 1990).
- *Self-sealing*: This processes is one in which a water-filled depression is sealed by a layer of fine, organic material over and in the underlying mineral soil. It is far from being fully understood, and probably several other pedogenetic processes may play a role, for example, podzolization in humid areas with creation of Fe layers ('Ortstein'). The process is described by Timmermann (2000, 2003), Succow and Joosten (2001), and Gaudig (2002). It was described much earlier as 'bottom tightening'.
- *Permafrost or frozen soil*: In areas with discontinuous permafrost such as in parts of Alaska, the carpet of feathermosses and mor mat acts to insulate the soil, resulting in permanently frozen soil forming near to the soil surface, hindering drainage, resulting in increased soil wetness, and thus enabling *Sphagnum* to invade (Viereck 1970).
- *Nutrient depletion*: Reduction in nutrients by leaching or immobilization, especially of nitrogen (N), may reduce the vigour of trees. The decrease in tree vigour will cause canopy cover to decrease, increase throughfall and reduce evapotranspiration, resulting in more moisture in the forest floor and favouring growth of *Sphagnum* (Heilman 1966, 1968; Viereck 1970).

Paludification by anthropogenic causes

Natural or human-related forest fires (Heinselman 1975; Patterson *et al.* 1987; Alm *et al.* 1992) and forest clearing (Smith and Taylor 1989; Moore 1993; Vompersky and Sirin 1997) may cause paludification, since both of these remove the forest canopy, decrease evapotranspiration, increase the precipitation reaching the ground, increase the wetness of the surface organic layer, promote acidity, and favour Ericales and *Sphagnum* invasion.

Since forest clearing and grazing have had widespread influence on creation of heathland in Ireland and the UK, it has been suggested that paludification of heathlands and development of blanket bogs has been caused or speeded up by human activities over hundreds or thousands of years. However, Hulme (1994) maintains that

There is no conclusive evidence, however, to support either an anthropogenic origin or a solely climatic origin of blanket mire. The pollen records do not indicate the nature of the relationship between prehistoric man's activity and blanket mire inception; they record sequences rather than establish the cause and effect relationship. . . . Further evidence is required, therefore, to demonstrate whether forest clearance and blanket mire inception in the British uplands were purely coincidental or whether there was a direct cause and effect relationship, or, indeed, whether the effect of forest clearance was to hasten rather than initiate the inception of blanket mire.

An anthropogenic cause for quaking mire formation has been hypothesized by Warner *et al.* (1989). They speculated that the deforestation for agriculture in southwestern Ontario caused the kettle-hole mires to flood in the spring. Pioneering peatland species were thus placed at a competitive advantage, which then led to the establishment of quaking mats of vegetation over open water bodies.

Primary peat formation

Primary peat formation is the process whereby peat is formed directly on freshly exposed, wet mineral soil (Fig. 7.2) without a previous drier state as in paludification. Thus, there is no (or only very shallow) open standing water, and no prior deposition of aquatic sediments. Examples of where this may take place include:

- land emerging from the sea owing to crustal uplift, the main examples in the world being along the Hudson – James Bay lowlands in Canada, on both sides of the Gulf of Bothnia in Sweden and Finland, and along the White Sea in northwestern Russia
- glacial forelands exposed as glaciers melt
- fresh volcanic deposits, dune slacks, alluvial plains, deltas, and other deposits
- open pit mines, gravel pits or quarries, mine spoils, human-created mineral surfaces.

Water tracks on hilly and mountainous terrain can supply water continuously enough at the surface that the surface is saturated and can support peat. The water is provided from the catchment, or from headwater wetlands (e.g. Sirin *et al.* 1998). These soligenous tracks do not normally begin with aquatic plants, but rather by direct colonization of peat-forming plants. Hence, this could be regarded as a type of primary peat formation on sloping ground following land uplift or glacial retreat.

Clearly, there is a relationship of primary peat formation to both infilling and paludification (Fig. 7.2). Primary peat formation occurs on poorly drained flats which include both shallow basins and slightly sloping topography, and which are supplied by enough water to maintain a wet, saturated condition at or close to the surface. If shallow water is present, it is not deep enough or present long enough to permit deposition of aquatic sediments like gyttja and marl. With primary peat formation the waterlogged condition was present from the first exposure of the ground, whereas paludified areas were not waterlogged initially.

The three processes acting together

An example of a peatland showing all three processes – infilling, primary peat formation, and upslope paludification – is taken from the work of Korhola (1992) in Munasuo Mire, southern Finland (Figs. 7.4 and 7.5). This peatland has arisen in a land uplift area that was previously beneath the Baltic sea. The oldest part of the peatland formed around a central lake about 4300–4200 BP directly on wet ground by primary peat formation, presumably with no or little open water. The central area was a lake that was overgrown from the margins about 4000–3600 BP. The dates obtained suggested that at the same time as the lake was filling in towards the centre, the mire was rapidly extending outwards by paludification. The basal horizon at the margins of the mire was a tall sedge peat with large amounts of lignid peat (*Betula* spp.). Lateral growth of peatlands is caused by the paludification process, that is, the expansion of peatland over previously drier mineral soil. For Munasuo Mire, represented in Figs 7.4 and 7.5, the rates of lateral extension estimated from the basal peat radiocarbon dates are given in Table 7.1. The distances between transect profiles ranged from 50 to 400 m. The rates of lateral extension were calculated as 0.7–3.5 m yr^{-1} by infilling and 0.02–8 m yr^{-1} by paludification.

Physiographic settings promoting peat formation

Above we have listed several settings for peat formation. In this section we add several distinct physiographic settings that can promote infilling, paludification, primary peat formation, or a combination of these.

Dune slack and beach ridge swales

Along the shores of lakes and seas, wave and/or wind action may create dune or beach ridges alternating with linear swales, or dune slacks, which are wet enough to promote peatland formation. The actual processes of peat formation in the swales and slacks may include infilling, primary peat formation, and paludification.

Rock pools and small depressions

Primary peat formation takes place in rock pools as they become isolated from the saltwater influence because of land uplift, and in depressions

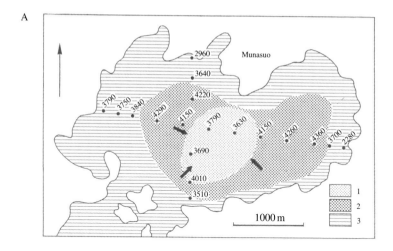

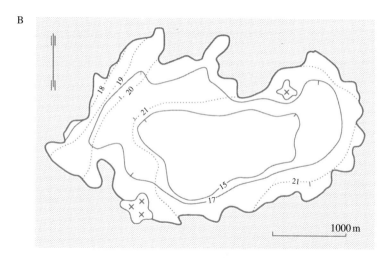

Fig. 7.4   Vertical views of Munasuo Mire, southern Finland. (A) Uncalibrated radiocarbon dates for the basal peat. 1, centre of mire formation, commencing around 4000–3600 BP; 2, littoral zone of the ancient lake, probably primary mire formation around 4300–4000 BP; 3, area of lateral extension, formed by paludification of the mineral soil in the interval 4300–2280 BP. The stippled area is that over which gyttja is found at the base of the mire. (B) Contours (interval 2 m) for the surface (broken lines) and base (solid lines) of Munasuo, based on measurements made at 74 sites. From Korhola (1992).

exposed after glacial retreat. These depressions collect water from the surrounding rock surface. Technically they are therefore minerogenous, but where the rocks are acid and not easily weathered (e.g. granites) they develop into miniature bogs with *Sphagnum* and other ombrotrophic species. Even though they are often only a few metres across and isolated from other peatlands by several kilometres they can contain many of the

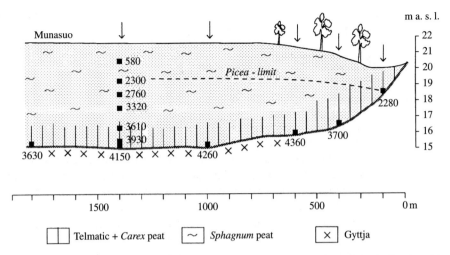

Fig. 7.5 Cross-section of the eastern edge of Munasuo Mire, southern Finland, showing dates obtained for the basal peat and the vertical profile. Ages at the centre, overlying gyttja, are 3630–4360 BP, then falling to 2280 BP closer to the margin where upslope paludification has occurred. At the margin, paludification was over heath forest. From Korhola (1992).

Table 7.1 Rates of lateral extension (m yr^{-1}) estimated from the basal peat dates for Munasuo Mire and distances between the sampling sites. From Table 3 in Korhola (1992)

Type of mire formation	Time interval (years BP)	Distance (m)	Estimated horizontal growth rate (m yr^{-1})
Infilling	4290–4150	350	2.50
	4150–3790	350	0.97
	4360–4260	350	3.51
	4260–4150	350	3.18
	4150–3630	350	0.67
	4010–3690	350	1.09
Paludification	3840–3790	400	8.01
	3790–present	100	0.03
	3700–2280	200	0.14
	2280–present	50	0.02
	3640–2960	250	0.37
	2960–present	50	0.02
	3510–present	100	0.03

typical bog species (e.g. *S. capillifolium*, *Eriophorum vaginatum*, *Rubus chamaemorus*, several ericaceous dwarf shrubs, and *Drosera rotundifolia*).

Discharge/spring

This is a type of peatland formed owing to discharge of groundwater often from the side of a hill or mountain, where the water comes to the surface and wets the surface continuously enough to promote peat formation, especially downslope. Most northern peatlands in hilly terrain are of this kind, and more or less sloping. They may develop ombrotrophy in places where the external water is absent, such as between water tracks.

Ombrotrophication

Ombrotrophication describes the process of development of a peat body from minerotrophic to ombrotrophic, by the upward development of peat surface and its eventual isolation from mineral soil water. An example is the cross section of an ombrotrophic bog in Fig. 7.3. In peat profiles, the change from fen to bog is sometimes sharp, but in other cases, as on slopes and on very shallow peats with little external water, the transition can be indistinct and gradual. In the profiles the loss of minerotrophic indicators (Chapter 2) are used to indicate transition to ombrotrophic peat. For example, if one sees a distinct change from a *Carex* or *Carex – Sphagnum* peat to a *Sphagnum – Eriophorum vaginatum* or pure *Sphagnum* peat, one may be fairly certain this is the boundary between fen and bog.

Bog development is generally thought to be stimulated by climatic shifts (Barber 1981). At the Myrtle Lake Peatland in Minnesota, the raised bogs started to form about 3000 years ago (Janssen 1968, 1992). Janssens *et al.* (1992) suggested a time of between 3250 and 2750 BP at the Red Lake Peatland, some 200 km west of Myrtle Lake. This suggests that bog development corresponded to the shift from warm, dry climate to a cooler, moister period. However, bog development may have started at different times in different locations. In the Finnish Munasuo Mire, the boundary between the *Carex* and *Sphagnum* peat, which occurred around 3500 BP, was interpreted as where ombrotrophication occurred (Fig. 7.5). For Punassuo peatland in southern Finland, the bog phase started slightly before 3000–3200 BP (Korhola 1992). These dates are similar to the dates for Minnesota bog initiation.

However, there is not always a correlation with a distinct time of climatic shift, and individual mires or locations within a mire may become ombrotrophic at different times depending on their particular hydrology and vegetation type (Barber 1981). For example, Svensson (1988) noted several shifts of vegetation in Store Mosse Mire in southern Sweden, from fen to ombrotrophic vegetation of three types, characterized by *Sphagnum fuscum*, *S. fuscum* + *S. rubellum*, and *S. magellanicum*, respectively. There were locations that changed from highly humified *rubellum – fuscum* peat to a slightly humified *magellanicum* peat indicating a strong rise in the water level of the bog. Some changes were driven by complicated local changes in hydrology, so that the development was not always in the direction of ombrotrophication. For example, some marginal parts of the *rubellum – fuscum* bog retrogressed into poor fen when mineral soil water flooded the bog. This change took place in 1000–1200 BP, contemporary with changes in the vegetation around the mire. These sites later developed back again to a *magellanicum* bog.

Tolonen (1987) reported a wide range of dates for transition from fen to bog for raised bogs in the Lammi area of southern Finland – 1900, 3000, 3535, 3780, 4000, and 5620 BP. As a general rule the maximum age of the ombrotrophic stage within European peatlands decreases northwards (Tolonen 1987, and references therein).

Ombrotrophication is undoubtedly still occurring in most if not all regions where peatlands are found. There is a connection between ombrotrophication and peatland formation in general. Peatland formation processes continue as long as the water supply is high enough to maintain saturated ground surfaces. As the peat is added incrementally year by year, parts of the peatlands will become isolated enough, by slight rising above the mineral water surface, to develop to ombrotrophic bog. The ombrotrophication process will be hastened where there is high precipitation combined with low evapotranspiration. If climate changes towards cooler and wetter, one may expect speeding up of peatland formation and ombrotrophication, and vice versa. Therefore, in future scenarios both temperature and humidity changes will be crucial for bog development.

The 'normal' path of ombrotrophication is, as described above, through the accumulation of peat in a wet climate. On very permeable surfaces, such as sand plains, the peat could be ombrotrophic almost from the beginning, a kind of 'dry paludification'. The separation from mineral soil water could also follow from a lowering of the water table, in which case the initial bog stage is relatively dry (Hughes and Barber 2004). For central Swedish raised bogs Almquist-Jacobson and Foster (1995) found that bog initiation occurred during periods that were relatively dry with declining summer temperatures.

Detailed sequences of peatland development

Peat profiles are used to study peatland development. They are most easily studied in wide vertical cuts, such as ditch faces created by ditching equipment or manually digging (e.g. Walker and Walker 1961). Unfortunately, such faces are not easy to dig, and when they are present as peat diggings or ditches they provide only random samples and do not represent the whole peatland. Hence, for a complete study of a peatland one must take samples with traditional peat corers in systematic transects or selected locations in the peatland.

Multiple pathways – Sarobetsu Peatland

Bogs can arise from more than one point of origin, and the points then fuse together to form one bog body. Detailed grids of peat profiles have been taken in the Sarobetsu Peatland in northern Hokkaido, Japan (Umeda *et al.* 1986). The Sarobetsu Peatland has developed in an old river-drained basin, and has a central raised bog around a central mineral island, surrounded by transition peatlands, and then low-lying fens, marshes, and swamps. However, the profiles showed that the bog did not form in the way that might be expected, from marsh to sedge fen to bog. Rather, there was a large variety of sequences of sediments and peats. The ancient river channel was indicated by raised mineral levees in some profiles. Clay and peaty clay occurred at the bottom, but also as thin layers within the peat indicating flooding or changes in river channels. Reed peat normally occurred at the bottom or margin of the peatlands. However, reed peat, sedge peat, and *Sphagnum* peat can all occur directly over the mineral material, in middle positions, or at the top of the profile. There were retrogressions, with reed peat developing over sedge peat, and sedge peat over *Sphagnum* peat. The transects revealed that bog had actually arisen from three separate *Sphagnum* peat bodies which then combined into one. On the surface of the present combined bog was a soak line which traced the path of the original river, to be precise it traced over one of the mineral levees along the original river. In other parts of the peatland which are presently minerotrophic, one finds buried masses of *Sphagnum* peat, presumably bog or poor fen, that have become covered with sedge or even reed peat. Therefore, as the peatland developed the areas of *Sphagnum*, sedge, and reed dominance shifted, influenced by the relative amounts of influx of minerogenous waters, and some of the *Sphagnum* bog areas remained bog while other bog areas were taken over by minerotrophic vegetation.

Inferring wetness and pH from species in the peat – Red Lake

When the peat profile is described in a macrofossil diagram in such a way that changes in dominant botanical constituents can be seen, it is possible to infer changes in wetness and pH that accompanied these changes. Janssens (1983; 1990; Janssens *et al.* 1992) developed a method to show the changes in degree of wetness as indicated by height above mean water table (HMWT) and pH, using the fossil-moss assemblage. This employs a weighted average technique, in which the present-day ecological preferences of species are assigned an index number along a numerical scale of HMWT and pH. Then the fossil-moss assemblage occurring in a particular layer is used to infer HMWT and pH for that layer.

As an example, Fig. 7.6A shows profile 8112 taken in an open bog community in the Red Lake Peatlands, Minnesota (Gorham and Janssens 1992). The figure presents the changes in the *Sphagnum* and other moss composition, and the pH and HMWT inferred from this composition. Brown moss species were abundant in the bottom layers, then intermediate, minerotrophic *Sphagna* where followed by ombrotrophically occurring *Sphagna*. It is seen that the pH starts high at the bottom, and drops rapidly to low values between 2200 and 1800 BP, remaining low to the top of the profile. At the same time the HMWT drops, corresponding to the development upwards of the raised bog above the water table. Hence,

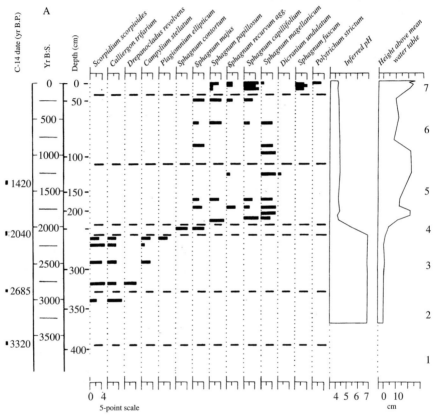

Fig. 7.6 Peatland development as interpreted from macrofossils. pH and height above mean water table (HMWT) are inferred from the species composition. Red Lake Peatlands, Minnesota. The cores are constructed to show comparable ages with ¹⁴C dating. From Gorham and Janssens (1992). (A) Profile 8112 in an open bog community in a large bog. This profile was interpreted to represent the following peatland communities from bottom to top: 1, rich-fen sedge meadow; 2, sedge meadow and *Alnus* carr; 3, brown-moss rich fen; 4, *Sphagnum* intermediate fen (with rapid drop in pH); 5, wooded *Sphagnum* poor fen or bog; 6, *Sphagnum* poor fen; 7, open bog community. (B) Profile 8104 from a fen water track. The communities are: 1, fine sandy silt, no analyses done; 2, minerotrophic aquatic community; 3, sedge meadow; 4, *Sphagnum*-dominated intermediate and poor fen; 5, brown-moss, rich-fen flark; 6, *Sphagnum contortum* rich flark; 7, *Sphagnum centrale* and *S. papillosum* lawn, poor minerotrophic; 8, wooded *Sphagnum* and brown-moss rich fen; and 9, *Sphagnum* poor fen, dominated by section Cuspidata but still with minerotrophic indicators.

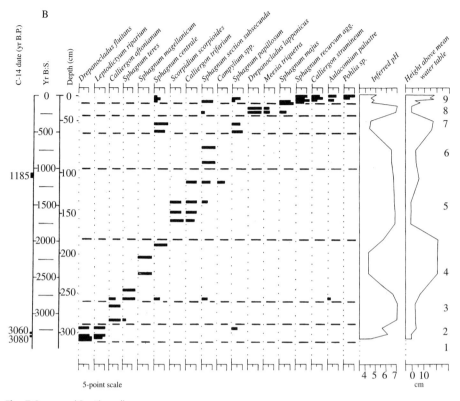

Fig. 7.6   (*Continued*)

at this site the raised bog was established about 1800 years ago. Another profile, 8104, was located in a site close to the boundary between a fen water track and a bog, and in this site there have been alterations through time between bog and fen conditions. As indicated by the pH graph, there have been six reversals in pH through time, from low to high and vice versa (Fig. 7.6B). Janssens explains this series of reversals as being caused by changes in the fen water track from broad to narrow by constriction between growing bog islands, and late development of string patterns.

Cyclic regeneration – a myth?

In 1910 Rutger Sernander presented his thoughts about how bog surfaces develop from hollows to hummocks and back to hollows (von Post and Sernander 1910). This idea of cyclic regeneration of bogs was based on field observations but there were no detailed stratigraphic studies to prove it. The 'theory' was quickly accepted and passed on by many influential peatland scientists in the first half of the twentieth century (Godwin and Conway 1939; Osvald 1923, 1949; Kulczyński 1949). In the words of Backéus (1990b), 'The theory of cyclic regeneration on bogs is a splendid example of how a theory can survive a long time, simply because nobody cares to investigate its fundaments'. During the second half of the twentieth century work on peat profiles in northwestern Europe refuted the hypothesis (e.g. Ratcliffe and Walker 1958; Barber 1981), and Backéus (1972) investigated one of the mires that von Post and Sernander had mapped 60 years earlier and found that the hummock structures were quite stable. There are circumstances where alternating layers of different peat types have been observed, but such sequences are not predictable, regular and cyclic shifts between hummocks and hollows as the original hypothesis suggested. The above-mentioned studies by Svensson (1988) and Janssens *et al.* (1992) exemplify this.

Sometimes thin (a few millimetres), highly humified streaks or bands are interspersed in slightly or moderately decomposed *Sphagnum* bog peat. These are called *humification bands* and they have been described in northwestern Europe and North America. They may be quite numerous in particular horizons. In one example, Tolonen (1987) found 19 bands in *Sphagnum* section Acutifolia peat, averaging 7.4 cm apart, and between two radiocarbon dated levels (1540 and 2110 BP) the average time interval between bands was some 50 years. Tolonen examined these bands and found remains of *Calluna* with some black streaks and slimy residues of reindeer lichens (*Cladonia*). These were thought to represent a short-cyclic pattern of dry and wet, with the dry characterized by dwarf shrubs and lichens and the wet by a rejuvenation of the surface by hummock-forming *Sphagnum*. It has been suggested that these shifts are non-synchronous, not controlled by climate, and represent uneven growth patterns of the different *Sphagnum* species and *Cladonia* in hummock levels. Rather than a hummock – hollow cycle, they indicate a repeated alterations between *Sphagnum* and *Cladonia* dominance within the hummock.

Watery and mineral horizons

There may be a horizon of watery or mushy peat occurring as a lens between upper and lower denser peat layers. This can occur in the case of floating (quaking) mats (e.g. Van Wirdum 1991) and also beneath floating mats or pools of raised bog. They occur also beneath the raised domes of tropical peat swamps (bogs, e.g. Driessen and Dudal 1991), and when excessively deep canals or roadside ditches are dug through the centres of such bogs the peatland dome collapses by movement of water and mushy peat into the ditch and then drainage out of the bog.

One may also find layers of mineral material occurring in peat deposits, deposited by cycles of flooding or wind erosion, alternating with periods of peat deposition. In the deltas of large rivers where there are movements of channels over time, such as the Mississippi in the USA, there can be found layers of silt and clay deposition alternating with layers of peat (e.g. Kearns *et al.* 1982).

8 Hydrology of peatlands

Hydrology is the science that deals with water, its occurrence, circulation, and distribution; its physical and chemical properties; and its interrelationships with the environment including living organisms. Understanding the hydrology of wetlands, mires, and peatlands is fundamental to the subject of peatland habitats, as it is probably the single most important condition influencing peatland ecology, development, functions, and processes. This chapter provides a basic overview of peatland hydrology and considers both the quantitative and qualitative (chemical) aspects of water in peatlands.

Water quantity

Water quantity refers to the physical presence of water, its fluctuating depths and flow patterns. For fundamental treatments of physical hydrology, we refer the reader to Freeze and Cheery (1979) and Todd and Mays (2005). For treatments focusing on mires and peatlands, we refer to the works by Ivanov (1981), Ingram (1983), and Price *et al.* (2005).

Surface water is water which is exposed to the atmosphere. Under ground, water exists in a variety of spaces, in the pores and cracks of bedrock and sedimentary strata, caverns and underground streams, mineral soil and peat. The word *groundwater* has a more restricted meaning, referring to water beneath the water table. The water table is the level to which water will rise in an open tube or bore-hole, exactly balancing the pressure of the atmosphere. Below the water table the matrix is essentially saturated (trapped air or biogenic gases are sometimes present). Above the water table, the water pressure is less than atmospheric, and water is held by capillarity within the pores of the soil. This is called *soil water*. Immediately above the water table there may be a zone of capillary saturation, called the *capillary fringe*. Above the capillary fringe pores are partially drained, and air is present. This zone is unsaturated. Water that enters the soil (i.e. either rain or snowmelt) is called *infiltration,* and this water may then percolate through the soil water zone to the water table.

Depth to water table

Depth to the water table (DWT) is one of the most important measures relating to vegetational physiognomy, plant occurrence, and growth. Verry (1997) showed that with increasing DWT in a wetland the maximum height of plants increases, and the lowest water tables will allow trees to grow. This is a general trend, and there are exceptions. For example, reeds in standing open water are taller than the sedges in the peatlands back from the water's edge. There are also positive height relationships of vegetation forms with nutrient regime for the same water level, and with different temperature regimes, for example, north to south. Therefore vegetation height is jointly influenced by moisture, nutrients, and growing season temperature.

For measures of DWT, water wells (tubes commonly 2–10 cm in diameter, slotted or perforated along their entire length) are placed in selected locations in transects or clusters. Depth to water can be measured with a plumper (a narrow cylinder on a string, hollowed out at the end, which makes a 'plump' sound when its hits the water surface), or an electronic probe which buzzes or lights up when the probe tip touches the water (e.g. Bodley *et al.* 1989).

In experiments involving hydrological change (e.g. drainage, damming, or vegetation removal), it is essential to collect water levels from both control and treated sites. To document the effects of drainage, water wells are often placed in transects across drainage ditches (Fig. 8.1; Berry and Jeglum 1991; Rothwell 1991). Practical research has been directed to determining the influence of drainage of forested swamps on tree stem growth (McLaren and Jeglum 1998), root growth (Rothwell *et al.* 1996), and total tree stand biomass and carbon content (Laiho and Laine 1997). Long-term data on water table fluctuations in natural peatlands are scarce, and are required to understand long-term trends in vegetation change, especially in relation to impacts of drainage and climate change.

Several kinds of continuous water level recorders have been developed in which a programmable electronic sensor or pressure transducer collects and stores the relevant information. The data are downloaded periodically in the field from the logger into a portable computer.

Hydroperiods are repeating cycles of change in water level over a period of time. These can show cyclic variation over a short periods such as a day (e. g. tidal variations), over medium periods such as weeks or months, and over long periods such as a year to several years. It is important to distinguish absolute water level relative to a fixed reference level, and water level

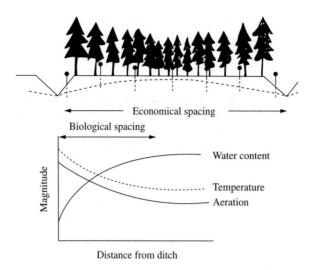

Economical spacing

Biological spacing

Water content

Temperature

Aeration

Magnitude

Distance from ditch

Fig. 8.1   The influence of ditches on water tables, water content, temperature, and aeration. The mean depth to water is least in the centre of the strip, but drops towards the ditches. Pipes for water level measurements are placed at various distances from the ditches. Water content in surface peat is greatest in strip centre, while temperature and aeration are least. Optimum biological spacing for tree growth response is narrower than the optimum for economics (cost/benefit). Modified from Rothwell (1991).

relative to the ground surface. The latter is the depth of the aerated zone, which is a biologically relevant measure. The most extreme variation of the absolute level is found in some marshes and swamps. Fens can have high or low variation of water levels depending on the type of hydrological system in which they are located, but most fens have a small water-table variation because of a continuous supply of groundwater. The smallest variation of DWT is in floating mats, where the surface follows the movements of the water table up and down, although the absolute water level variation can be high. For such a site Roulet (1991) reported surface level changes of several centimetres a day, but the water level was at the peat surface all the time. Bogs often have deeper water tables and are more variable because the only input is precipitation, which can be episodic. In a slightly continental climate Rydin (1986) measured over 30 cm drop in absolute water level in a bog during a dry summer.

Mire breathing

The ground peat surface is not fixed, and it often moves up and down as water tables fluctuate. The deformations have been called *Mooratmung* (Weber 1902), mire oscillation, and mire breathing (Ingram 1983; Schlotzhauer and Price 1999). Glaser *et al.* (2004a), using a global positioning system (GPS) network, recorded a gradual subsidence during a dry period in the glacial lake Agassiz peatlands of northern Minnesota. More dramatically, there were events when the surface abruptly rose and fell as much as 15 cm in a bog and 32 cm in a fen over an 18 hour period. Not only are there vertical oscillations, but also horizontal displacements. These vertical and horizontal deformations can be caused in several ways: floating mats at edges of ponds, lakes, and rivers; expansion and contraction of a peat body owing to changes in groundwater storage; stress from tidal inundation; freeze–thaw cycles under permafrost or seasonal frost; shear stresses from lateral water flow; atmospheric pressure changes; and ebullition fluxes of methane (CH_4) bubbles formed deep in the anoxic peat (e.g. Ingram 1983; Roulet 1991; Quinton and Roulet 1998; Timmermann 2000, 2003; Glaser *et al.* 2004a).

The least amount of vertical fluctuation of the peat surface is found in the features with firmer, consolidated peat, such as fen peats with higher mineral content, the hummock phases in ridges of patterned fen and bog, and wooded phases of fens and bogs, thicket swamps, and swamp forests.

Water flow

Because water enters peatlands intermittently as precipitation, snowmelt, and groundwater inflow there is movement of water in the peat, both vertically and laterally. Lateral flow occurs most rapidly in strongly sloping, soligenous peatlands, decreases as slopes decrease, and is slowest in topogenous peatlands and ombrogenous bogs. Where the water is above the peat surface it flows fastest. Water flow beneath the surface is much slower. Within the peat the greatest water flux is typically at the top of the groundwater, and decreases with depth.

Hydraulic head and direction of water flow

The direction of water flow is determined by the water pressure at different depths and at different places on the peatland. The water pressure at a certain point in the peat, together with the vertical location, gives the hydrostatic energy, which is called the *hydraulic head* (*h*). The hydraulic head is measured with piezometers, which are tubes inserted vertically in the peat. They are open at the top and slotted or perforated at the lower end, so that water will rise to a level in the tube governed by the ambient pressure in the vicinity of the intake at the lower end. The water level in the piezometer (relative to a fixed reference level) will be a measure of the hydraulic head at the point of water intake (Fig. 8.2).

To judge the vertical direction of water flow at a location, piezometers are placed in clusters ('piezometer nests'; see Fig. 8.2), with the perforations at chosen depths. If the highest head is in the deepest piezometer, the flow will be upwards; if the highest head is in the shallowest piezometer the flow will be downwards. To judge the horizontal direction of water flow,

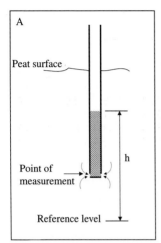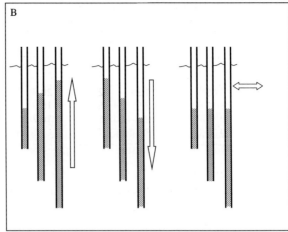

Fig. 8.2   (A) The use of a piezometer to determine hydraulic head. The piezometer is an open-top pipe with holes at the bottom to let in water (but not peat). The water-filled part of the piezometer is hatched. The hydraulic head (*h*) at the level of water intake is the elevation of the water table in the tube above an arbitrarily set level (in metres). (B) Clusters of piezometers for determining vertical flow direction. The arrows show the direction of flow of the groundwater: Left, highest head in the deepest piezometer: flow upward; middle, highest head in the shallowest piezometer: flow downward; right, head equal in all three piezometers: flow (if any) in an indeterminable horizontal direction. Adapted and redrawn, based on Pielou (1998).

piezometers are placed in a grid pattern or in selected spots on the peatland, and flow occurs from positions with higher heads towards those with lower heads.

Hydraulic conductivity

Another important concept in peatland hydrology is the *hydraulic conductivity* (*K*). It describes the permeability and how easily water can flow through the peat. It is strongly related to the degree of humification, which often increases with peat depth. As the peat becomes more humified, the bulk density will increase, the total amount of pore space and the proportion of large pores will decrease, and as a result the hydraulic conductivity will also decrease (Table 8.1). For comparable degrees of humification, the lowest conductivities are generally for *Sphagnum* peat, intermediate for *Carex* (sedge) peat, and highest for lignoid (woody) peat (Päivänen 1982). The reason is that the structure of the peat matters. Dai and Sparling (1973) suggest that sedge peat has a more stratified structure than *Sphagnum* peat, probably owing to the layering of dead leaves each year, and this facilitates the horizontal flow of water. Seasonal water table changes lead to mire breathing which causes peat volume to change, and this can change hydraulic conductivity by several orders of magnitude (Price 2003). The hydraulic conductivity can be quite variable over time: it increases with pore water

Table 8.1 Average pore size distribution of peats with different bulk densities. Tabular values are percentage of volume of water at saturation. From Päivänen (1982)

Bulk density (g cm^{-3})	Pore size μm						
	>300	300–100	100–30	30–3	3–0.3	0.3–0.2	<0.2
0.05	21.3	22.3	20.2	13.7	10.5	3.6	8.4
0.10	3.9	16.7	18.9	27.2	11.1	6.6	15.6
0.15	3.3	5.2	16.7	31.3	13.3	8.7	21.5
0.20	3.0	7.2	11.6	26.2	16.5	9.9	25.6

pressure which leads to pore dilation, and can be reduced by CH_4, generated by anaerobic respiration, forming bubbles that block pore spaces and decrease water flow (e.g. Beckwith and Baird 2001).

Hydraulic conductivity can be measured in the laboratory by means of a device called a permeameter (Fetter 2001). The main problems here are extracting and transporting the sample from the field in a relatively undisturbed condition, and establishing tight connections of the sample with no leakage along the sides of the container.

Hydraulic gradient and hydraulic conductivity determine flow rate

To understand the rate of water flow we need combined knowledge about hydraulic conductivity and hydraulic head. We can express the flow of water as the volume that passes through a cross sectional area per unit time. For this we use Darcy's law:

$$Q = AK(\Delta h/\Delta x)$$

where Q is the volumetric flow of water (m^3 s^{-1}) passing through a cross-section with area A (m^2), and $\Delta h/\Delta x$ describes the change in hydraulic head (Δh) per unit of distance (Δx). $\Delta h/\Delta x$ is the *hydraulic gradient*, which can be seen as a measure of the pressure that drives the water flow. K is the hydraulic conductivity (m s^{-1}). Hence, the flow of water between two points increases linearly with the hydraulic conductivity and with the hydraulic gradient.

Water on the surface may sometimes flow slowly in pools or channels, or as sheet surface flow in water tracks. Below the surface the flow rate will decrease drastically – even weakly decomposed material offers considerable resistance to water movement. The highest saturated conductivities, for example 10^{-4}–10^{-6} m s^{-1}, are for poorly decomposed fibric peats and undecomposed *Sphagnum*. Conductivities of 10^{-6} or 10^{-8} have been reported for moderately decomposed mesic peats and the lowest values, 10^{-7}–10^{-10}, for highly decomposed humic peats with very fine pores (e.g. Paavilainen and Päivänen 1995).

Measures of water flow

The concentrations of cations and anions at different depths can be used to infer whether groundwater is coming from above, below, or laterally. Release of various tracers such as dyes, mobile elements such as sodium (Na) or caesium (Cs), or isotopes have been tried, but a potential problem is that the substance may become adsorbed to the peat. Sometimes unplanned release of elements or pollutants into the environment provides useful markers. An example is the release of tritium (^{3}H) into the atmosphere from nuclear weapons testing in the 1960s, subsequently used to trace water movements in peatlands (e.g. Gorham and Hofstetter 1971). The use of ^{3}H is limited by the fact that it has a half-life of 12.4 years and is rapidly diminishing to undetectable levels, but it was possible to obtain measures as recently as the late 1990s (Sirin *et al.* 1997).

Siegel *et al.* (1995) measured hydraulic head and pore water chemistry on the Lost River Peatland raised bog in Minnesota (see Fig. 10.4), and concluded that groundwater flow systems are altered by climate, with upwards water movement (discharge) of calcium (Ca)-rich groundwater prevailing during droughts, and downward movement (recharge) by higher precipitations occurring during wet periods. After a number of dry years when an upward gradient prevailed, Ca and dissolved organic compounds were advected upwards into the peat profile, and waters with pH > 6.0 and high electrical conductivities around 200 μS cm^{-1} were found as little as 50 cm below the surface. However, after a number of wet years when a recharge regime prevailed (water from precipitation percolating downwards), the pH decreased to < 5 and conductivities to < 200 μS cm^{-1} from the peat surface to almost 2 m depth. Similar decreasing trends occurred for dissolved inorganic carbon (C), Ca, magnesium (Mg), and Na.

Acrotelm and catotelm

One of the important concepts of peatland function and development is that bogs are *diplotelmic*, that is, they have two layers (Ivanov 1981; Ingram 1983). The *acrotelm* is the aerated, 'active layer' which occurs above the lowest level of the water table. Thus it is the upper layer of peat in which the water table oscillates. The depth of this layer reflects the variable depths to water in different microtopographic positions; hence mud-bottoms and carpets can have no or very thin acrotelms, lawns can have acrotelms 5–20 cm thick, and hummocks can have acrotelms of 20–50 cm or more. Below the acrotelm is the *catotelm*, the constantly anoxic, 'inactive layer', usually more humified and darker, and comprising most of the volume of the mire. The wooded part of a bog has a thick acrotelm in which root growth is possible (Fig. 8.3).

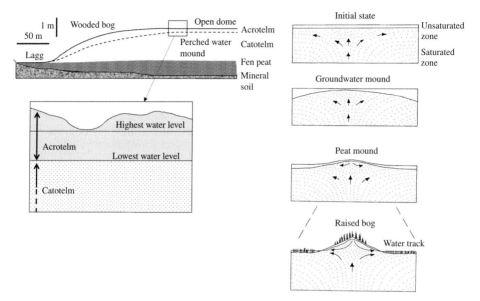

Fig. 8.3   Hydrological development of raised bogs. Left: Cross-section of a raised bog, showing conventional peat–water mound theory (Ingram 1987). The acrotelm extends down to the lowest depth of the water table as indicated by the insert. Right: Peat–water mound/discharge–recharge theory, where during dry periods there is initially upwards groundwater flow creating a groundwater mound, enhancing local peat accumulation and further developing into a rainwater mound. The right pane from Glaser *et al.* (1997). Regional linkages between raised bogs and the climate, groundwater, and landscape of northwestern Minnesota. *Journal of Ecology*, **85**, 3–16, with permission from Blackwell Publishing.

Some of the salient characteristics of the acrotelm and catotelm are presented in Table 8.2. The most intense hydrological and biogeochemical processes occur near mire surfaces in the acrotelm, rather than in the catotelm (Ivanov 1981; Ingram 1983). Since the acrotelm consists of living or recently dead peat material, it tends to be poorly decomposed at the surface, becoming more decomposed with depth. The rate of lateral groundwater flow therefore depends strongly on the position of the water table within the acrotelm, since the larger pore spaces of the less decomposed upper layer of the acrotelm has the highest hydraulic conductivity. Water also flows through the catotelm, albeit much more slowly.

The acrotelm/catotelm is a conceptual model which has been used widely, but perhaps sometimes uncritically. The concept was developed in the context of raised bog, and some workers maintain that the terms should only be used for raised bogs (A.A. Sirin, pers. comm.; Joosten and Clarke 2002), whereas others suggest that it may be a useful concept for other kinds of peatlands also.

There can be considerable variation between dry years and wet years in the lowest position of the water table (Glaser *et al.* 1996), and there may not be

Table 8.2 Some key features of the acrotelm and catotelm peat layers. Modified from Ivanov (1981)

Acrotelm	Catotelm
Intensive exchange of water with atmosphere and surrounding area	Very slow exchange of water with underlying substrate and surrounding area
Frequent fluctuations in level of water table and moisture content	Constantly saturated
High hydraulic conductivity which decreases with depth	Low hydraulic conductivity
Periodic access of air to pore spaces	No access of atmospheric oxygen
Large population of aerobic microorganisms leading to rapid decomposition	Anaerobic microorganisms, and reduced decompositional activity
Dense root systems and a diversity of invertebrates	Much reduced vascular rooting, very few invertebrates

as sharp a distinction between acrotelm and catotelm as commonly assumed (Sirin *et al.* 1997). In contrast with the view of very slow water flows in the catotelm, it can have layers where the flow is relatively fast. High hydraulic conductivity zones have been reported for the catotelm as much as 3 m below the surface, and these may be explained by a layer of less decomposed fibric peat with abundant large pores, or by abundant woody stems and roots which provide channels for rapid flow (Chason and Siegel 1986). For example, Boelter (1965) found high conductivities associated with poorly decomposed woody peat layers over a metre below the surface, and Sirin *et al.* (1998) reported an increase in horizontal conductivities in a 50 cm thick layer at around 2 m depth. If hydraulic conductivity does not show a distinct decline at the acrotelm–catotelm border, it may be possible that there are significant downward and upward exchanges (Baird and Gaffney 1995, 1996; Baird *et al.* 1997; Sirin *et al.* 1997). Charman *et al.* (1994) studied the age of DOC in profiles in a *Picea mariana* swamp forest in northeastern Ontario, and found that younger carbon had been transferred deeply into the profile, and other studies indicate that there is downwards movement of DOC from acrotelm and surface catotelm to deeper layers of the catotelm and then into downstream waters (e.g. Chasar *et al.* 2000).

Water balance

The overall water balance, or budget, of a peatland is an accounting of the inputs, outputs, and storage within the peatland. Building a water balance equation requires information about water movements into and out of the peatland, including groundwater exchanges during a defined time interval. It is necessary to develop water balances for calculation of nutrient budgets, and for prediction of the effects of natural and man-induced changes on hydrological systems (Carter 1986).

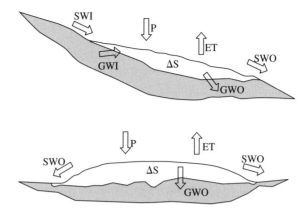

Fig. 8.4   The components in the water balance equation for a soligenous fen (upper) and a raised bog (lower). The mineral soil is indicated by grey patterning. P, precipitation; SWI, surface water inflow; GWI, ground water inflow; ET, evapotranspiration; SWO, surface water outflow; GWO, ground water outflow; ΔS, change in storage.

A general water balance formula, with inputs on the left and outputs on the right, is:

$$P + SWI + GWI = ET + SWO + GWO + \Delta S$$

where P is precipitation, SWI is surface water inflow, GWI is groundwater inflow, ET is evapotranspiration, SWO is surface water outflow, GWO is groundwater outflow, and ΔS is change in storage (Carter 1986). These components are depicted in the simplified hydrological models for a soligenous fen and a raised bog in Fig. 8.4.

Precipitation is the total incoming rain and snow, in millimetres of water, collected with standard gauges in the open at prescribed heights. Not all of the total precipitation reaches the ground surface, because some of it is intercepted by vegetation and evaporated. Some of the precipitation reaches the ground as *throughfall*, the rainfall passing through vegetation, and *stemflow*, running down the stems. In forested peatlands located in areas with frequent low clouds and fog, there can be significant condensation which either evaporates or reaches the ground as fog drip (e.g. Price 1991).

Surface water inflow includes sheet surface flow, channel flow, springs, seeps, and soaks. In the case of lacustrine and riverine peatlands water can come in by surface flooding, and from tides, storm surges, and seiches. For soligenous peatlands it can be upland runoff, marginal springs and seeps giving rise to surface waters which flow over the surface. In the case of topogenous systems, there can be runoff from higher surroundings in towards the centres of the basins, which either pools in the centre or marginal lagg, or drains out if there is an outlet.

Groundwater inflow can come in as subsurface flow, which may be general diffuse flow or focused in more conductive horizons or pipes.

Evapotranspiration is the amount of water transferred from a land cover surface – including ground, water, and vegetation – to the atmosphere by evaporation and transpiration combined. It represents the change of state of water from liquid to vapour and is proportional to the difference between the vapour pressures at the evaporating surfaces and the overlying air.

Evapotranspiration from a wetland is a key component of the water balance, and the most difficult to measure in practice. The evaporation pan is a standard instrument in which the drop in water level in an open, water-filled pan is measured over time. However, the relationship between pan evaporation and field evapotranspiration can vary considerably because of shifting effects of plants, differences in surface temperatures, etc. The *actual evapotranspiration*, ET_a, can be measured with a lysimeter, a box with impermeable walls and bottom, containing an isolated block of peat with its vegetation. The evapotranspiration can then be measured by the weight loss of the material in the box.

While lysimeters give evapotranspiration values representative for small patches, aerodynamic measurements from towers generate values representative for larger areas. Profile measurements of air humidity, temperature and wind speed with electronic equipment at three levels or more can give aerodynamic fluxes of vapour and hence of evapotranspiration.

Because of difficulties in measuring evapotranspiration, it is often estimated from different empirical relationships by *potential evapotranspiration*, ET_p, which is the theoretical evapotranspiration from a surface completely covered by a homogeneous green vegetation (crop) experiencing no lack of water. It is usually estimated from standard meteorological data and some commonly used methods are those of Thornthwaite or Penman (Rosenberg *et al.* 1983). These are mathematical formulae which use measures such as mean air temperature, relative humidity, and wind speed. None of these empirical methods is entirely satisfactory for estimating peatland evapotranspiration. Verry (1997) found that ET_a was about equal to the calculated ET_p when the water level was within 30 cm of the hollow level. From 30 to 40 cm the ET_a decreased rapidly, then less rapidly. At 50 cm the ET_a/ET_p ratio was about 0.4, and at 80 cm it was about 0.2. Another method of estimating evapotranspiration is to use equilibrium evaporation as calculated from the formula of Priestley and Taylor (Rosenberg *et al.* 1983). This expression has shown good correlations with measured evapotranspiration (Stewart and Rouse 1977; Lafleur *et al.* 1997; Kellner 2001b).

Comparative studies of evaporation from open water compared to vegetated wetlands have given mixed results. Some show that the presence of vegetation increases evapotranspiration compared to open water evaporation, while others show that evapotranspiration from vegetated wetlands is less than lake evaporation. Both sides of this controversy are well supported, and differences may be related to differences in vegetation characteristics as well as methods of measurement (e.g. see Lafleur 1990). Nichols and Brown (1980) found that evapotranspiration from *Sphagnum* moss surfaces was significantly greater than evaporation from open water surfaces. Takagi *et al.* (1999) studied the evapotranspiration from a *Sphagnum* bog compared with an adjacent *Sasa palmata* (a short bamboo) peatland in northern Japan. Over a summer period ET was 372 mm from the *Sasa* peatland and 285 mm from the *Sphagnum* bog. The difference is related to the large leaf area in the *Sasa* site.

Evapotranspiration values are dependent on many climate and site conditions – peat moisture, water table, surface (canopy) structure and aerodynamic properties, temperature, and humidity. Since evapotranspiration is tightly bound to the energy balance, both storage and fluxes of water and heat at the mire surface layers should be studied at the same time (Kellner 2001a).

Interception is precipitation that is caught by the vegetation and evaporated. Päivänen (1966), working in Finland, estimated that during the growing season there were interception losses in mature tree stands on drained peatlands of 23–35% in *Picea abies*, 20–25% in *Pinus sylvestris*, and about 20% in *Betula pubescens*. Interception can also be significant for dense layers of high or low shrubs. Päivänen (1966) estimated the retention by the low shrub layer for rain events of different magnitudes: 1 mm precipitation led to 46% retention; 5 mm, 28%; 10 mm, 13%; and 15 mm, 7%. Hence the *relative* amount of precipitation retained is highest for small precipitation amounts.

Surface water outflow is runoff as surface flow or channelized in brooks or streams. *Groundwater outflow* is water moving out of the wetland into an adjacent peatland, mineral matrix, or water body by underground matrix flow, or as pipe flow. It may have horizontal and/or vertical components.

Storage is the total volume of water contained in a peatland. Change in storage, ΔS, is a change in water volume reflected by a rise or fall in water table.

It is often difficult to obtain good estimates of the inflow (SWI and GWI) and the outflow (SWO and GWO) terms in the equation. This is because the pathways of inflow or outflow are too broad to intercept, or they may involve discharge or recharge exchanges with underlying parent materials; hence one must use indirect estimates combined with modelling (e.g. Hooijer 1996; Wilcox *et al.* 2006). Furthermore, water flow may show reversals, that is, move in different directions at different times (e.g. Romanov 1968a, 1968b; Glaser *et al.* 1996; Devito *et al.* 1997).

In water budget studies, the boundaries for which the budget is being determined must be defined. Sometimes it is the peatland body alone, a specific hydromorphologic unit such as a raised bog, a fen water track, or basin fen. However, a particular peatland body is always bordered by other wetland units or uplands, and it is often necessary to include these units in the budget. For examples, a fen unit may receive both surface and groundwater inputs, and hence the unit of interest becomes the catchment containing the fen. If the interest is in the overall impact of drainage, or forest removal, then the appropriate unit becomes a large river basin with its complement of peatlands and wetlands (e.g. Vompersky and Sirin 1997).

Water balance for bogs

The water balance formula for bog is simplified, since by definition there are no water inflows:

$$P = ET + SWO + GWO + \Delta S$$

A comprehensive assessment of water balance for bogs was carried out for the European territory of Russia (Romanov 1968a, 1968b). This area has a continental climate with low moisture during the growth season, and within the region humidity decreases from northwest to southeast, precipitation from 514 to 119 mm, evaporation from 336 to 243, and bog runoff from 250 to 46 mm. Raised bogs cannot develop in the southern regions of European Russia unless mean precipitation during the warm period exceeds evaporation. Therefore, the southern boundary of raised bogs is not determined by the annual precipitation but rather by the precipitation for the warm season. The water table in *Sphagnum* – dwarf shrub communities in wet and average years does not fall below depths of 25–30 cm, but in dry years it continues to fall to 50–55 cm and deeper. Under these conditions, fires may readily occur on raised bogs. This is also a common occurrence in continental Siberia and North America.

A water balance was developed for Clara Bog (Fig. 8.5), a raised bog in central Ireland, by Leene and Tiebosch 1993 (cited in Daly *et al.* 1994). In this area the mean annual precipitation is between 850 and 950 mm, and is rather uniform over the year, but evapotranspiration is much higher in the summer. For the year studied the water balance $P = ET + SWO + GWO + \Delta S$ (in mm) was $922 = 587 + 300 + 14 + 21$. It is noteworthy that there was estimated to be a leakage out of the bottom of the bog of 14 mm, and a net gain in storage of 21 mm.

The peat–water mound theory of bog development proposes that a water mound can develop over a relatively impermeable mineral bed, supplied only by precipitation (Fig. 8.3A). Because of the stratification with a surface layer with high conductivity and a lower, more humified layer with low

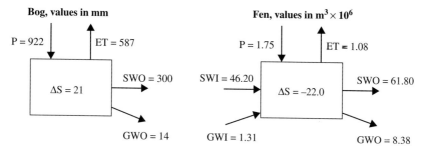

Fig. 8.5 Water balance over a year for a bog and a fen. P, precipitation; ET, evapotranspiration; SWI, surface water inflow; SWO, surface water outflow; GWI, groundwater inflow; GWO, groundwater outflow; ΔS, change in storage. The bog example is for Clara Bog, Ireland (from Leene and Tiebosch 1993, cited in Daly *et al.* 1994). About two-thirds of the precipitation is evaporated, and most of the outflow is as surface water. In this year there was a small increase in storage. The fen water balance is for a 269-ha large water pond contained in a large fen complex near Seney, Michigan (Wilcox *et al.* 2006). The water balance is dominated by surface flows, and precipitation accounts for only 3.6% of the input. The storage estimate contains quite a large error component because it was extrapolated from single surface-water discharges to annual discharges.

conductivity, the vertical and horizontal hydraulic conductivities are not equal, and the general assumption is that seepage (flux) is predominantly at the water table and moves horizontally outwards towards the bog margins, with only a small downward component. The theory has been put forward in detail by Ivanov (1981) and Ingram (1983), and extended to blanket bog by Baird *et al.* (1997) and Lapen *et al.* (2005).

There is now evidence that there can be both upward and downward vertical flow in raised mires (Fig. 8.3B). The evidence comes from the Lost River Peatland in northern Minnesota (Siegel *et al.* 1995; Glaser *et al.* 1997). This study suggests that upwelling minerogenous water from below is strong enough to reach the surface in the spring – fen mound, but that a similar upward pressure is not strong enough to reach the surface in the raised bog mound (see Fig. 10.4). During droughts, when water from precipitation is not recharging the mounds, there is upwards pore pressure beneath the raised bog, but it is not strong enough to reach the surface. During wet periods, the incoming precipitation and ombrogenous waters press the minerogenous upward flow downwards.

Water balance for fens

In contrast with bogs, fens have inputs of surface and groundwater (SWI and GWI) which are often diffuse and difficult to measure. Also, the hydrological settings for fens are much more variable, ranging from flat to sloping, floodplains, and groundwater discharge/recharge. Early studies of water balance of fens were conducted by Romanov (1968a, 1968b). The evaporation from fens for the European area of Russia was found to be 20–25% greater than from *Sphagnum* bogs with dwarf shrubs, and 3–10% greater than that from whole raised bogs. In a large fen studied in detail, during a normal year the peat received moisture from the underlying mineral bed and shores during spring and the first half of summer, whereas there was an outflow of water from the peat deposit in the second half of summer.

Several important water balance studies have been carried out relatively recently for fens (e.g. Price and FitzGibbon 1987; Price and Maloney 1994; Rouse 1998). An example of a water balance is the one done for the Seney fen in the Upper Peninsula of Michigan by Wilcox *et al.* (2006; Fig. 8.5). The approach was to use a large central pond as the water balance reference. The components of water balance here include surface and groundwater inputs in addition to outputs, $P + SWI + GWI = ET + SWO + GWO + \Delta S$, and the respective values for one particular year (millions of m^3) were $1.75 + 46.20 + 1.31 = 1.08 + 61.80 + 8.38 - 22.0$. The large decrease in storage in this case was an effect of a drainage system. In contrast to the bog, the precipitation is only a very small fraction of the total input (3.6%).

Peatlands as regulators of water flow

Peatlands are of particular interest to water resource managers because they occur extensively in the headwater areas of many streams and rivers. Peatlands can have large impacts on the quantity and quality of the receiving waters (e.g. Brooks 1992; Verry 1997). The response of peatlands to large rainstorms is different from that of mineral soil uplands. The lack of topographic relief, the absence of well-defined channels, and the shallow water tables all combine to make peatlands behave hydrologically like unregulated, shallow reservoirs. Some peatlands act to regulate the flow of water in the landscape. Flow regulation would attenuate flow in wet conditions and release it in dry conditions. Some wetlands are good flood attenuators (mid-basin stream or riparian wetlands). Others, for example many bogs, are not. Some of the other influences of peatlands on the hydrology of streams are:

- The main contribution from peatlands is as horizontal flow either on the surface or through the upper layers of peat at the top of the water table. Flow in the catotelm is much less and can normally be considered negligible in terms of contributions to streamflow.
- In climates with snowy winters, the greatest percentage of annual streamflow from natural peatlands occurs in the spring months as a result of snowmelt.

- Streamflow from peatlands is generally reduced during summer months with lower precipitation and higher rates of evapotranspiration.
- Depth to water table governs the magnitude of both streamflow response to moisture input and evapotranspiration losses. When the water table is deeper, more precipitation can be held in the upper layers of peat before there is runoff.

Environmental impacts on water balance

Peatlands around the world are experiencing direct and indirect impacts on their hydrology. Because of the delicate balances between the precipitation, inflow, outflow, and evapotranspiration that controls them, peatlands are sensitive to human interference. The direct effect of drainage, a lowering of the water table, leads to increase in depth of aerated layer, increase of peat oxidation, losses of nutrients and organic matter into receiving waters, and subsidence (Chapter 5). We refer to reviews of environmental impacts on peatland hydrology (Bragg *et al.* 1992; Järvet and Lode 2003; Price *et al.* 2003).

Drainage creates new hydrological conditions that in turn promote vegetational changes. A striking example is in the Sarobetsu Peatland, northern Hokkaido, Japan. The mire has been extensively impacted by a canal for flood control of the Sarobetsu River, drainage, pastureland development, and peat extraction. There has been invasion of *Sasa* bamboo from the minerotrophic margins towards the bog centre. Iqbal *et al.* (2005) have studied the effects of damming of small outflow channels on water quality and on *Sasa*. The damming caused water level rises, return of anoxic conditions, and changes of the quality of the surface waters towards bog water characteristics, that is, low pH, low mineral and nutrient contents, and higher concentrations of dissolved and particulate organic matter. These changes were accompanied by dramatic decreases in *Sasa*.

After ditching for forestry, changes of the water balance of peatland are complex and change with time (Rothwell *et al.* 1996). Initially there is an increased capacity for storage owing to the increased depth of aerated surface layer, causing reduced runoff with small rains, delays in occurrence of peak streamflow after large rains, and increased summer baseflow. With continuing decomposition of organic matter and compaction, the proportion of fine particles and bulk density increases, and the proportion of large and medium pores decreases, resulting in increased water retention (Silins and Rothwell 1998), decreased saturated and unsaturated hydraulic conductivity, and decreased water storage capacity. The higher water retention capacity causes a rise in the thickness of the zone of capillary saturation (Silins and Rothwell 1999).

Drainage for peat harvesting has higher impacts on water balance than drainage for forestry, owing to the more intensive drainage and continuing removal of peat over several decades. Price and Schlotzhauer (1999) focused on the influence of shrinkage and compression in influencing water storage changes in cutaway peatlands. Of course, peat harvesting applies to relatively small areas compared to forest drainage, which is spread over quite large areas.

A future climate warming will likely cause greater seasonal water deficits, and thereby lower water tables, outflows, and evapotranspiration (Clair 1998). There could also be increased frequency of fires (Hebda *et al.* 2000). Even though winter snowmelt may recharge a wetland, it will remain at its capacity for a shorter time in a warmer climate (Rouse 1998). The effects are of course crucially dependent on how precipitation changes in the future climate.

Water quality

In nature water always contains various dissolved substances or suspended particles which come from the soil, the air, and organisms. Water chemistry has a large influence on the kinds of plants and animals that can occur in a peatland, and therefore on the character of the organic matter that accumulates. For general treatments on water chemistry, we refer the reader to the works by Appelo and Postma (1994) and Stumm and Morgan (1996). General references focusing on chemistry of peatland are Gorham *et al.* (1985) and Hughes and Heathwaite (1995).

Saline, ombrotrophic, and minerotrophic waters

The total salt content of seawater ranges between 30 and 40‰. This is not a typical situation for peat accumulation, but it is found in, for instance, mangrove swamps and some marshes. Seawater is diluted in river estuaries and in large bays with freshwater river inflow.

The waters where mixing of salt and freshwater occurs are called *brackish* (1–10‰ salt; Freeze and Cheery 1979), and here peats of considerable depth can develop. Three large brackish water seas are the Hudson and James bays in Canada, the Baltic, and the White sea in northwest Russia. These are areas where land uplift is still occurring since the last glaciation, and their surrounding lands all have high proportions of flat terrain covered by peatland. The salinity of these bays ranges from of seawater where the bay water meets ocean salt water, to fresh water where rivers and streams meet the bays. Peatlands that have formed over previously saline seas often have saline mineral or bottom aquatic sediments, and these can contain algae that indicate saline conditions (Chapter 6; Korhola 1992).

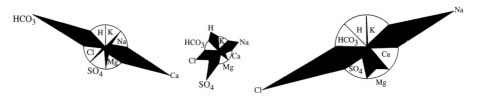

Fig. 8.6   Maucha diagrams can be used to synthesize and visualize the chemical composition of peat water. The circle area is proportional to the logarithm of the total ionic concentration of the sample and the black areas are proportional to the concentrations of the different ions. Left, minerotrophic water with high Ca content; middle, ombrotrophic water; right, water with seaspray origin as indicated by the high content of Na and Cl. Reproduced from van Wirdum (1991).

For a freshwater peatland it is possible to characterize the relative role of the sources of incoming water (Fig. 8.6). Water originating from sea spray has high concentrations of Mg, Na, and chlorine (Cl). Ombrotrophic water has low concentrations of all elements. In minerotrophic waters the amount and proportions of nutrients depends on the chemistry of the underlying bedrock, overlying sediments, degree of washing and sorting of the sediments, and occurrence of floodwater washing over the site. Levels of basic cations, specifically Ca, Mg, Na, and potassium (K), in sediments may be low, intermediate, or high. Extremely high pH and Ca concentration occur where the bedrocks are limestone or the parent materials are base rich; here marl-bottomed lakes and pools are frequent (see Fig. 1.4).

Salt spray can be deposited on peatlands several kilometres from the coast. It influences the chemical composition of the peat water as well as the composition of the vegetation. The influence of sea salt on soil chemistry of coastal peatlands in Hokkaido, Japan, was investigated by Haraguchi *et al.* (2003). The concentrations of Mg, Na, and Cl in precipitation and peat water decreased with increasing distances up to 4 km from the coast. The presence of a *Picea glehnii* canopy was strongly related to the amount of sea spray, as indicated by higher salt concentrations in peats beneath *Picea* than beneath *Sphagnum*. Also, the acidification was stronger under *Picea* than in open *Sphagnum* communities. This was explained by sea salt accumulation and consequent cation exchange (H^+ release) in the peat soil, and organic acid release from *P. glehnii* by stem flow and needle litter accumulation.

Because of the variation in dust, salt spray, and airborne pollutants the inorganic content of precipitation is not fixed; rather, it varies among bogs in different localities or regions (Bragazza *et al.* 2005). Oceanity is reflected by higher total ionic concentration, and Na, Mg, and Cl especially originate from sea spray, whereas the content of Ca is more dependent on addition by wind from local terrestrial sources (Proctor 1992). For instance, Bragazza *et al.* (2003) reported significant differences in Ca^{2+}, Mg^{2+}, K^+, nitrate

(NO_3^-), ammonium (NH_4^+), and sulfate (SO_4^{2-}) in precipitation between a boreo-nemoral mire site in Sweden and an alpine mire in Italy. Areas downwind from industries or located close to farms with manure are influenced by airborne pollutants such as NH_4^+. Continental-scale nitrogen (N) pollution can lead to reduced C/N ratios in peat, better growth of trees, increased grassiness, and reduced *Sphagnum* growth (see Chapter 4). Dustfall with its mineral content, especially Ca, Mg, and Na, can be significant where bogs are adjacent to gravel roads or agricultural areas (Gorham et al. 1985). For ombrotrophic peatlands the different chemical compositions of rainwater mean that several species that are restricted to fens in continental areas can grow on bogs in oceanic areas (Chapter 2).

Sampling of peat water

There is generally a close relationship between the chemical composition of the peat (see Chapter 5) and the water filling the pore spaces in the peat, and both have been used to assess chemistry and nutrient gradients (Bragazza et al. 2003). Water chemistry is more subject to the variations of precipitation and drought than is peat chemistry – the ion concentrations become higher as the amount of water decreases in a dry period. On the other hand there are differences between hummock, lawn, and carpet peat, and groundwater is a good integrator to characterize a site as a whole. Often workers must choose between analysing water or peat chemistry, simply because of limitations in time and resources. Peat waters can be sampled from surface waters or pools within the peatland, which are subject to variation with time since last rain, or from a specific depth (in piezometers). It is also possible to sample water from the peat above the groundwater using porous ceramic cups (see Domenico and Schwarz 1997). Water is extracted from soil by creating a vacuum in the porous cup. Depending on the length of the tubes going from the top of the soil to the samplers, the water is then removed by suction or by gas displacement.

It is normal practice to filter the water before analysing it, normally with pore size 0.45 μm (but 0.2–0.7 μm filters have been used). By filtering, the particulate organic matter is largely removed, and hence the determinations will reflect only the inorganic nutrients plus the dissolved organic substances that pass through the filter. Whether or not the water has been filtered can have important influences on the conclusions of a study, especially for elements such as N and P which are contained in suspended particles (Joensuu et al. 2002).

Measurements for water

Water pH

The pH of water is measured with electrodes and meters, or with indicator paper as for peat pH. In most early work the water was left to equilibrate with the air and the concentration of carbon dioxide (CO_2) in the air. In addition, water has been collected from open pools, from pools formed by pressing down the moss, from squeezed moss, and from wells or piezometers, which has lead to questions of comparability of methods. Tahvanainen and Tuomaala (2003) conducted a detailed study of the pH differences for different methods of collection, and differences owing to diurnal variation and weather (cloudy–sunny), aeration, and open pools compared to pipe wells. They proposed a standard sampling protocol, a combination of unaerated and aerated samples obtained from pipe wells, to give comparable and representative sampling of mire water for pH measurements. They also proposed that the wells be perforated with 3–4 mm holes in an area maximally 10 cm high, located below the water table level. Wells should be emptied on several occasions after establishment and shortly before sampling.

Electrical conductivity

Electrical conductance has been used for a long time to measure the total ionic concentration in natural waters (e.g. Sjörs 1950; Golterman and Clymo 1969; Day *et al.* 1979), and is still used in wetland research (e.g. Richardson and Vepraskas 2001, Sjörs and Gunnarsson 2002; Bragazza *et al.* 2003). As the total ionic concentration of the water increases, the electrical conductance increases. Even though laboratory methods for measuring various ions are now easily accessible, conductivity is a convenient proxy for the chemical conditions, especially for field surveys.

Electrical conductivity, or *specific electrical conductance*, is the conductance of an electrical current through a cube of water 1 cm on a side (Golterman and Clymo 1969; Day *et al.* 1979). The units for reporting specific electrical conductance are reciprocal ohms (mhos), or the preferred SI unit, siemens (S). Normally the values for natural fresh waters are very low, and values are reported in $\mu S\ cm^{-1}$ ($\mu mhos\ cm^{-1}$). There is a positive relationship between temperature and conductance, so the temperature is taken at the time of measurement and corrected to a standard of 25 °C. Modern field and laboratory instruments have built-in functions for temperature corrections, otherwise the following formula can be used:

$$EC_{25} = EC_t/[1+0.02(t-25)],$$

where EC_{25} is conductivity at 25 °C and EC_t is conductivity at the measured temperature (t). Earlier work often reported conductivity adjusted to 18 or 20 °C, and the formula can be used to re-calculate such data for 25 °C.

For solutions that have pH around 5.0 or less, the conductivity owing to the hydrogen ions becomes a significant component of the measured conductivity (Sjörs 1950), and it has become common practice to subtract the hydrogen ion conductivity from the measured conductivity to obtain a corrected conductivity:

$$EC_{corr} = EC_{measured} - EC_{H^+}.$$

High measured conductivity can be caused by high Ca content or by low pH, and therefore corrected conductivity gives a better correlation with the poor–rich vegetational gradient than does the uncorrected conductivity.

The conductivity of the hydrogen ions at 25 °C is calculated as $EC_{H^+} = 3.49 \times 10^5 \times 10^{-pH}$ μS cm^{-1} (Strong 1980). This means that the hydrogen ion conductivity is 349, 35 and 3.5 μS cm^{-1} at pH 3, 4 and 5, respectively. The corresponding equation for 20 °C is $EC_{H^+} = 3.25 \times 10^5 \times 10^{-pH}$.

When the peatland is very dry, it may be difficult to obtain water for pH or conductivity. Peat extracts may be used, bearing in mind that the measurements may not be fully comparable with those from the water samples. For obtaining conductivity of peat, samples may be prepared by adding distilled water using a dilution that provides just enough water for analysis. Different manuals propose dilutions that range from a ratio of 1:2 to 1:8 of peat to distilled water, stirring or shaking and taking measurements of the supernatant (Riley 1989; Richardson and Vepraskas 2001). It is possible to combine the determination of conductivity and pH, taking care to measure conductivity before pH.

Elements and solids

As with peat (Chapter 5), to obtain a reliable picture of the water quality the analysis should, in addition to pH and conductivity, comprise all major ions (Ca^{2+}, Mg^{2+}, Na^+, K^+, Cl^-, SO_4^{2-}, HCO_3^-). For N and P the total amounts are often analysed, but more interesting are the inorganic forms available for plant take-up (NH_4^+, NO_3^-, PO_4^{2-}). In addition, it may be desirable to analyse for other elements or mineral constituents carried in by water or deposited by wind, such as iron (Fe), aluminium (Al), and silicon (Si). The choice of elements or compounds will be determined by the objectives of the study.

In a data set with 15 measures of water chemical variables, all had some degree of correlation (Tahvanainen 2004). There were strong correlations ($r > 0.5$) between pH, Ca, Mg, and total base cations and between K, Na, and sulfur (S). In this data set it is noteworthy that total N and total P, the main macronutrients, had only low significant correlations with other attributes. It could be that available forms of N and P, if they had been measured, might have been more strongly related to pH.

Humic substances derived from organic matter decomposition give the brown or coffee colour to peatland waters and waters of streams, ponds, and lakes ('brown-water lakes') that receive water from peatlands or coniferous forest soils. Humic substances are measured as *dissolved organic matter* (DOM) and *particulate organic matter* (POM). DOM is defined as the organic material that passes through a filter with a pore size of 0.45 μm (Giller and Malmqvist 1998). It is a heterogeneous mixture of sugars, lipids, amino acids, and proteins bound to large humic molecules and colloids. Dissolved organic carbon (DOC) is the carbon component of DOM. Values of DOC have been given for peatlands in a number of studies. For example, Moore (2003) reported DOC seasonal averages of 20–40 mg L^{-1} for pore water of fens and bogs.

An intensive study of the influence of peatlands on acidity and chemistry of lakes in northeastern Alberta, Canada, revealed that the extent and types of peatland terrain in watersheds plays an important role in determining the pH status of lakes (Halsey *et al.* 1997). Water chemistry was characterized for 29 study lakes, and measures of lake morphometry, wetland slope, and areas of wetland types for the surrounding watersheds were made. Detailed samples of surface water chemistry were made for 32 wetlands in three of the 29 watersheds; these wetlands formed 3 units with distinctly different water chemistry – bogs, fens (including wooded fen, open fen), and shallow organic deposit. All acid-sensitive lakes occurred in watersheds with greater than 30% peatland cover. Fen peatlands were particularly important influences, followed in turn by bogs and shallow organic deposits. It was suggested that the difference in lake chemistry between watersheds with high bog compared to high fen areas is a function of how the acidity is generated. High DOC concentrations (average about 30 mg L^{-1}) in lakes with high bog cover in their watersheds reflect generation of humic acids from decomposition processes; this complements the idea that acidity in bogs with pH < 4 is most likely a combination of inorganic cation exchange along with organic acidity produced by decomposition (Hemond 1980). Lower DOC concentrations (about 18 mg L^{-1}) in lakes with high fen cover in their watersheds reflect the generation of acidity that is more influenced by exchange of cations for hydrogen ions.

The high cation binding capacity of DOM in bog waters facilitates metal mobility, especially in acidic wetlands (pH < 5) where metal solubility is high and DOM concentrations are highest (Helmer *et al.* 1990). Vitt *et al.* (1995) presented figures for metals (Al and Fe), both of which decreased along the bog–rich fen gradient. Both Al and Fe are enriched in bogs, and this is a function of high DOC in the waters and organic acid metal binding that occurs at low pH (Helmer *et al.* 1990). High levels of Al^{3+} and certain hydroxyl Al complexes are toxic to fish, and other organic or inorganic Al complexes may be harmful to trees and peatland plants. Sparling (1967) suggested that toxic Al concentrations restricted the occurrence of *Schoenus nigricans* in peatlands influenced by acid rain in the Pennine uplands of England.

Environmental impacts on water quality

It is well documented that the outflow of water from drainage ditches into streams and lakes causes increases in many elements and nutrients, DOC, POM, pH, alkalinity, conductivity, and temperature (Berry and Jeglum 1991; Prevost *et al.* 1999). Lundin and Bergquist (1990) studied pre- and post-drainage water chemistry for surface and groundwater for a bog drained for forestry. The runoff from drainage of the bog had lower concentrations and losses of organic C and N compared to drainage of minerotrophic peatlands. Joensuu *et al.* (2002) documented the effect of ditch cleaning on runoff water chemistry for 40 treated and 34 control peatland forest catchments. They reported increases in pH, electrical conductivity, and base cations (Na, K, Ca, and Mg), and decrease in DOC. Mean concentrations in Al and Fe increased in the first year after ditching. There were no major changes in total dissolved N and P. The largest relative changes in element transport over 3 years following ditching were the increases in NH_4^+ and POM. The authors believed that the substantial increase in POM was the most adverse effect, because the sedimentation of material destroys bottom vegetation and breeding habitats of fish and other aquatic organisms. The same conclusion was reached in studies of impacts of peat harvesting (Heikkinen 1990a, 1990b; Heikkinen and Visuri 1990). Joensuu *et al.* (2002) concluded that the leaching of nutrients was unlikely to cause eutrophication of receiving water bodies.

Aspects of nutrient cycling in natural and drained peatlands are considered in Chapter 9.

Variation in water chemistry along the bog – rich fen gradient

The strongest link between water chemistry (pH, conductivity, and Ca) and vegetation is along the bog–rich fen gradient (see also the section on ombrotrophication in Chapter 7). Whether the sharpest discontinuity in northern peatlands is between bogs and fens or between *Sphagnum*-dominated mire (bogs + poor fens) and richer fens has been debated for decades, for example recently by Wheeler and Proctor (2000), Økland *et al.* (2001a), and Sjörs and Gunnarsson (2002). There are large overlaps of the pH ranges of bogs, poor fens, and rich fens (Fig. 8.7; Sjörs and Gunnarsson 2002). Nevertheless, it is often possible to define boundaries of pH and Ca within which most of the sites of each type occur. The pH for ombrotrophic bog is usually below 4.2, but because of seasonal variation and variation within bogs it can be higher. Calcium for bog is normally 2 mg L^{-1} or less.

For North America, Glaser (1992c) reported comparable ranges of pH and Ca: bog, pH 3.7–4.1 and Ca 0.6–2.0 mg L^{-1}; extremely poor fen, pH 3.8–5.0; a group of poor fens with bog-like vegetation, pH 4.1–4.6 and Ca

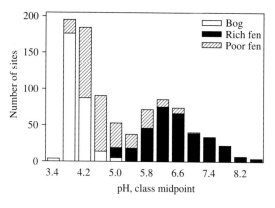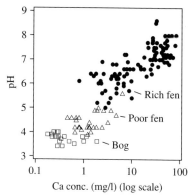

Fig. 8.7   Calcium, pH, and vegetation in mire in central and northern Sweden. Left, frequency of bog, poor fen, and rich fen in different pH classes. For simplification, rich fens include also the group of intermediate fens. Right, relationship between pH and Ca for a subset of all sites. Data provided by Urban Gunnarsson (cf. Sjörs and Gunnarsson 2002).

2.2–5.5 mg L^{-1}; another group of poor fens, weakly minerotrophic, pH 4.3–5.8 and Ca 3–10 mg L^{-1}; transition rich fens, pH 5.8–7.0 and Ca 10–25 mg L^{-1}; extremely rich fens, pH 7–8.5; and spring fens, pH 6.8–7.4 and 20–45 mg L^{-1} Ca.

Another North American example of chemical components along the gradient bog–poor fen–moderately rich fens (forested and open) – extremely rich fen comes from central Alberta, Canada (Table 8.3; Vitt *et al.* 1995). In addition to pH, Ca and conductivity, this example showed that also alkalinity, Mg, and Na increased along the bog – rich fen sequence. In general, the nutrients N (NO_3^- and NH_4^+), P (soluble reactive and total dissolved), and K, were not related to the bog–rich fen gradient (Table 8.3). However, NH_4^+ and P were higher in the forested moderately rich fens compared with other sites, especially in water extracted at depths of 1.0 and 1.5 m. In samples from the British Isles, Wheeler and Proctor (2000) found a considerable overlap between bogs and other *Sphagnum*-dominated peatlands in pH and Ca, so they separated these from the richer fens.

Samples of water taken in transects from bog to adjacent fen can show gradual or abrupt changes in water chemistry. Transitions are gradual where the geological underlay of a region is poor in bases, and where the minerogenous waters tend to be acidic and low in bases. An example is the Precambrian shield in Fennoscandia and Canada where granites, schists, and other base-poor rocks are dominant. In contrast, transitions are more abrupt where the geology of a region is rich in bases, and where the minerogenous waters tend to be less acidic and moderately rich in bases. When there is distinctly groundwater-derived flow in the fen adjacent to the bog, sharp and streamlined borders may occur.

Table 8.3 Summary of chemical data for surface water for five peatland sites in central Alberta, Canada. Values consist of averages of samples taken at 18 sampling times during the ice-free season in 1989. From Vitt *et al.* (1995)

	Bog	Poor fen	Forested moderately rich fen	Open moderately rich fen	Extreme-rich fen
Water temperature (°C)	6.1	8.4	12.2	14.8	11.9
pH	3.96	5.38	6.28	6.00	6.88
Conductivity (μS cm^{-1})	39.0	48.0	91.0	79.0	187.0
Ca (mg L^{-1})	3.0	5.9	12.2	10.1	23.3
Mg (mg L^{-1})	0.7	3.1	5.3	4.8	7.8
Na (mg L^{-1})	2.8	3.2	4.3	3.8	9.3
K (mg L^{-1})	0.3	1.2	0.8	0.7	0.6
Al (mg L^{-1})	0.2	0.1	0.1	0.0	0.1
Alkalinity (μeq L^{-1})	0	198	764	623	1716
Nitrate (μeq L^{-1})	0.41	0.35	0.31	0.33	0.26
Ammonium (μeq L^{-1})	1.26	0.94	1.04	0.70	0.46
SRP (μeq L^{-1})	0.80	5.10	14.52	0.70	0.71

Conductivity is standardized to 20 °C and corrected for conductivity owing to H$^+$ ions. SRP = soluble reactive phosphorus.

An abrupt transition of bog to fen was documented in a 'miniature bog', a hummock of *Sphagnum fuscum* 56 cm high and 137 cm in diameter, in a rich fen in the UK (Bellamy and Rieley 1967). There was a rapid transition in peat types from the base of the hummock upwards:

- 0–4 cm: *Carex* + brown moss (*Campylium stellatum, Palustriella falcata*) + *Sphagnum palustre*
- 4–8 cm: *Carex* + *Sphagnum palustre* + *S. fuscum*
- 8–24 cm: *Carex* + *S. fuscum*
- 24–32 cm: *Carex* + *S. fuscum* + *Phragmites australis*
- 32–56 cm: *S. fuscum* + *Calluna vulgaris, Vaccinium oxycoccos, Pleurozium schreberi.*

Thus, *Carex* and *Phragmites*, the main elements of the fen, were replaced completely by ombrotrophic vegetation over about 30 cm. In this study the most abrupt changes in water chemistry occurred between the water of the fen and the water extracts at the 0–4 cm and 4–8 cm levels (when *S. fuscum* invaded) where pH dropped from 6.5 to 4.1 and Ca from 6 to 2 mg L^{-1}. Thus, the main changes in chemical composition in this switch from rich fen to bog vegetation occurred over a vertical distance of 4 cm at the base of the mound, and were almost complete after 8 cm.

Chemical variation within a bog

Bogs are nourished by rainwater, but there can be some differences in the chemistry of the water relating to spatial variation in vegetation, peat type, depth to the groundwater level, and location relative to the bog crest, slope, and margin. Damman (1988) made a detailed study of a transect across a raised bog, collecting water from open water pools and from just above the water table (using porous ceramic cups). Ionic composition for Ca, Mg, Na, Fe, K, P, and NH_4^+ varied insignificantly among communities and slope positions, but during May–June and September, K and Mg increased downslope in the bog lawns and pools. This trend was missing from the dwarf-shrub-covered slopes. Slight nutrient enrichment was indicated by the appearance of *Nymphaea odorata* and *Utricularia vulgaris* in the pools, and *Calamagrostis pickeringii, Solidago uliginosa*, and *Eriophorum virginicum* on pool banks. All of these species have somewhat higher nutrient requirements than other species in ombrotrophic vegetation. The occurrence of these species could have been due to slightly higher water flow and consequent nutrient supply in these locations. Damman concluded that seasonal changes in nutrients supply are considerable, and that ionic concentrations alone do not adequately describe differences in nutrient supply, which points out the importance of water flow to interpret vegetation changes in peatlands.

9 Nutrients, light, and temperature

Most classical schemes of peatland classification, as detailed in Chapter 1, focus on two complex factor gradients – wetness and acidity – and how they affect the peatland organisms. In this chapter we discuss how nutrients, light, and temperature vary within and among peatland types.

Nutrients

Nutrients are the elements and chemical compounds needed for growth and metabolism. With the exception of potassium (K), metal ions are here not treated as nutrients, although in cases of very low availability they could be regarded as growth factors. As noted previously in Chapter 5, nutrients may be divided into macro- and micronutrients based on the relative contents in organism tissues. As in most other terrestrial ecosystems, nitrogen (N) is often the limiting factor for plant growth, but it is not unusual for phosphorus (P) or K, individually or in combination with N, to be limiting. As noted in Chapter 5, there is variation among the nutrients in the total versus the extractable part, where extractables represent nutrient 'availabilities' for plants. However, extractables are only an approximation of availability, because uptake mechanisms vary from species to species.

When one analyses the living material of vascular plants and bryophytes, the levels of the macronutrients are higher than in the non-living peat material (the opposite is mostly the case for calcium (Ca), however). There is cycling such that macronutrients that are released from litterfall or from root exudates early in the decomposition process are quickly taken up by roots or mosses. As discussed in Chapter 3 (for vascular plants) and Chapter 4 (for *Sphagnum*), macronutrients are also translocated within plants from basal parts towards the living tissue. Thus, macronutrients are preserved in the living skin of mire ecosystems. In the following sections we deal with N, P, and K individually.

Nitrogen

In organisms, N is an essential constituent of amino acids which are combined into proteins. Many other organic molecules also contain N, especially the cell walls of fungi and bacteria. It is one of the most abundant of the elements in peat, along with carbon (C), hydrogen (H), oxygen (O), and sulfur (S), but it is largely tied up in unavailable organic forms. N is a very mobile element, circulating between the atmosphere, the soil, and living organisms, and it is the nutrient that has the most complex interactions with biotic processes.

Figure 9.1 presents a simplified scheme of N cycling in peatlands, showing inputs, outputs, internal fluxes, and storage. Virtually all living or dead organic matter in nature contains N, so that when organic matter is transported or transferred, so is N. All movement of water in peatlands also means, to a large or small degree, a transport of N included in dissolved or particulate organic matter. Nitrogen occurs in organic forms (in living organisms and in litter) and in inorganic molecules in oxidized form (nitrate, NO_3^- or nitrite, NO_2^-) and in reduced form (ammonium, NH_4^+). Nitrogen is largely bound in organic forms, with only a small proportion occurring as inorganic

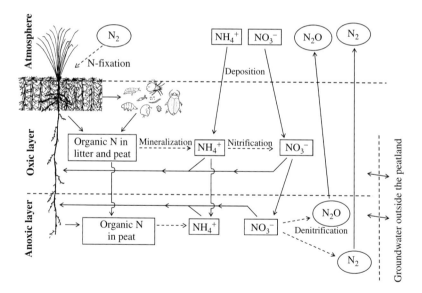

Fig. 9.1 Simplified scheme of the nitrogen cycle in peatlands. Encircled symbols represent gases, dashed arrows represent microbial processes. The largest pool in the peatland is the organic nitrogen in peat which is unavailable to the plants. In the oxic zone, plant uptake includes the ecologically important assistance of mycorrhizal mutualism. Some microbial nitrogen fixation occurs in the anoxic zone, based on N_2 gas transported down through plant aerenchyma. The bidirectional arrows to the right indicate the exchange with groundwater outside the peatland, that is, leaching and inflow of inorganic components and nitrogen in dissolved organic matter.

forms (Vepraskas and Faulkner 2001). The transfer between the inorganic types is dependent on redox potential and pH and involves several microbial processes, some leading to gaseous nitric oxide (N_2O) or elementary N_2.

Air is 78% N_2 gas. A few groups of organisms have the ability to take up gaseous N_2 by *N fixation*, which is the reduction of N_2 to NH_3 and its incorporation into organic molecules. Cyanobacteria can do this, and play an important role in peatlands as free-living species, in association with *Sphagnum*, or when they form a lichen together with a fungus. Important in wetlands is the mutualistic association between some vascular plants and N-fixing organisms, for instance Actinomycetes in partnership with woody plants such as *Myrica gale* and *Alnus* spp. Some free-living bacteria can also fix N, such as the aerobic *Azotobacter* and the anaerobic *Clostridium*. For the latter the N_2 substrate can be transported down to the anoxic zone through plant aerenchyma.

Mineralization is the general term for the breakdown of organic matter leading to the release of inorganic molecules. The specific term for N mineralization is *ammonification*, since the preliminary result is NH_4^+ ions. Ammonification is performed by most heterotrophic bacteria (both aerobic and anaerobic) and fungi as they decompose the organic material. At low pH, fungi are more important than bacteria for the aerobic ammonification. Nitrogen mineralization has also been found to be highly correlated with the amount of microbial biomass (Williams and Sparling 1988). In general, mineralization in peat is slow compared to that in other ecosystems, and normally limited by the availability of easily degraded carbon sources. For fresh litter (with high C quality) decomposition may be N limited. Chapin *et al.* (2003) found that mineralization was higher in fen than bog, but when N was added to a bog it was immobilized, which suggests that the microbial community in this case was N limited. Very high inputs, for instance in the form of anthropogenic deposition, affect many parts of the N cycle and give higher rates of N mineralization, denitrification, and N_2O emission (Verhoeven *et al.* 1996a). Also, when peatlands are drained (or drained and fertilized) the mineralization increases and inorganic forms of N increase. Water samples from a gradient from bog to extreme rich fen (Vitt *et al.* 1995) showed decreasing NH_4^+ and weakly decreasing NO_3^-. This may mean that the vegetation in the rich fen is consuming the supply of mineralized N faster than in the bogs. The C/N ratio of the organic matter affects the decay rate, and this is discussed in Chapter 12.

The forms of inorganic N are influenced by pH and redox potential (see Chapter 5), and governed by microbial processes. Under oxic conditions, *nitrification* is performed by aerobic bacteria, and is the oxidation of NH_4^+ to the intermediary nitrite (NO_2^-, by e.g. *Nitrosomonas* spp.) and then to nitrate (NO_3^-, by e.g. *Nitrobacter* spp.). This is prevented by anoxia and severely hampered at pH <5. Where oxygen is released into the peat from aerenchymatous plants, some nitrification can occur around the roots in an otherwise anoxic zone.

Denitrification is the process in which nitrate is reduced by other anaerobic bacteria to N_2O or to N_2. A large number of bacterial species are involved. The gases can move up through the peat by diffusion and enter the atmosphere.

Plants take up N as NO_3^- or NH_4^+. This represents the *immobilization*. In addition, some plants may take up simple organic compounds (Näsholm *et al.* 1998). Normally, in acid, anoxic peats NH_4^+ is the main form of plant-available N, and NO_3^- is absent or in low amounts. In neutral to calcareous, aerated conditions NO_3^- can be present. NO_3^- is easily leached and lost from the system, whereas NH_4^+, as a cation, binds to the ion-exchange sites of the peat and tends to stay in place. Since peat soils are largely anoxic and acid, the main form of plant-available N in peatlands is NH_4^+.

Phosphorus

Phosphorus is an essential element in living tissue, occurring in several hundred compounds, including DNA, RNA, and ATP (adenosine triphosphate, a molecule that drives energy-requiring synthetic reactions in cells). Next to N, P is the most abundant nutrient in soil organic matter, and particularly high amounts are in microbial tissue, as much as 2% of the dry weight (Stevenson 1986). Under natural conditions, P is only scantily represented in rainwater or dry deposition, and so it requires continual weathering from bedrock and soil to replenish supplies. Thus, because the peat surface is gradually cut off from the mineral soil, the total amount of P tends to be low in peatlands, and especially low in bogs.

The cycles of P and K are simpler than the N cycle, because there are fewer biological transformations involved. Inorganic P does not occur in different oxidation states and therefore cannot be used as an electron donor or acceptor. Nonetheless, the dynamics of P remain poorly understood. Major processes in the P cycle include uptake by plants; recycling through return of plant litter and animal residues; biological turnover through mineralization–immobilization; fixation reactions at mineral, organic, and oxide surfaces; and mineralization to phosphates through the activities of microorganisms. Some P is lost through leaching and runoff.

From the plant nutrition viewpoint three main soil phosphate fractions are important (Mengel and Kirkby 1982). The first fraction is available for root uptake in the peat water as $H_2PO_4^-$ or HPO_4^{2-} depending on pH. The second fraction is the inorganic phosphate held on surfaces so that it is in rapid equilibrium with the soil solution phosphate. The third fraction is the substantial amount which is only slowly released. Most of this is bound in organic matter, but some is bound in clay minerals or as precipitates or occluded forms of iron (Fe), aluminium (Al), and Ca phosphate. Only about 10–30%

of the total P occurs in soluble forms (Vepraskas and Faulkner 2001). The mobilization and availability is determined by several environmental factors (Boström *et al.* 1982):

- *Redox potential*: Fe-bound P is released at potentials below 200 mv when Fe^{3+} is reduced to Fe^{2+}.
- *pH*: In acid soils, phosphate is readily precipitated as highly insoluble Fe or Al phosphates, or occluded in Fe or Al complexes. Both forms are poor sources of P for higher plants. In calcareous soils poorly soluble Ca phosphates are formed. Lucas and Davis (1961) showed, for organic soils, decreasing availability of P for plants below pH 4.5, optimum availability between about pH 5 and 6.5, and decreasing availability above about pH 7 (rising again above 8.5).
- *Temperature:* Increase in temperature leads to increased bacterial activity, which increases oxygen consumption and decreases the redox potential and pH. The production of phosphate-mobilizing enzymes and chelating agents may increase accordingly.

P is often reported to be a limiting factor for tree growth on peatlands. The total content of P is often greater in mires with tree cover and woody peats, and higher in fens than in bogs. However, as indicated above, the available amounts are governed more by pH and redox potential than by total amount. Phosphorus tends to recycle and be maintained in the living plants, and a rapid mineralization to phosphates by microbial biomass means that there is not always a difference in P availability between bog and fen (Kellogg and Bridgham 2003).

Potassium

Potassium is important in physiological and biochemical functions in all living organisms, and is particularly important to the water status of plants. It is preferentially transported to young meristematic tissues (Mengel and Kirkby 1982).

The chemistry and cycling of K is simpler than that of N or P. The main source of K for plants in natural conditions is weathering of minerals. Mineral soils rich in clay and young volcanic materials are generally rich in K, whereas materials which have been highly weathered or washed by fluvial action often contain little. Potassium is also supplied from mineral sources and also as small amounts in precipitation and dry deposition. Therefore it is the mineral soils surrounding a peatland that supply most of it. Potassium occurs as readily available ions for uptake in the peat water, adsorbed in exchangeable form to peat particles, in organic matter – especially in living tissues – and in mineral particles. In peat, almost all K is in the adsorbed exchangeable form, with much smaller amounts free in solution as K^+. Consequently, K is rather mobile and it is readily leached or flushed out of sites.

In general K content is greatest in mires with trees and with mire margin conditions, for instance with inflow from K-rich clay soils or in spring fens. Values are low in treeless rich fens and mud-bottom flark fens, possibly because water continuously flushes through the site and carries away the easily leached K. Under anoxic conditions K is easily leached from the peat, but in the oxic zone it is strongly bound in living rootlets and microorganisms and effectively recycled.

Peat pH has less influence on K than on P availability. The peat particles can also absorb large quantities of Ca at their cation exchange sites. If Ca levels are too high (naturally or as a result of liming), then the mass-action effect and high Ca accumulation by some plants may depress the uptake of K, causing K limitation of growth.

Nutrient controls and limitations in peatlands

Nutrient deficiencies are typical of peatlands, and a large amount of research has been done in connection with peatland forest fertilization (Kaunisto 1987). The productivity of trees may be limited by N, P, or K, individually or in some combination. Which combination depends on the mire type and its nutrient status. For blanket bogs it may be N or P or a combination of the two (Beltman *et al.* 1996). Boeye *et al.* (1997) found a strong P limitation in low-productive calcareous fen (dry mass increased after P addition), whereas in more productive fens N was limiting. This fits with several other reports of low above-ground productivity ($< 100 \, \mathrm{g \, m^{-2} \, yr^{-1}}$) in rich fens because of strong P limitation (El-Kahloun *et al.* 2003). In drained peatlands, K may be lost by leaching and there are also removals owing to increased tree production and sequestering of K in biomass, as well as when the plant biomass is harvested (van Duren *et al.* 1997).

Deficiencies of other major elements, e.g. magnesium (Mg) and trace elements such as boron (B), copper (Cu), Fe, manganese (Mn), molybdenum (Mo), and zinc (Zn) (Kolari 1983), can develop on peat soils. In northern Scandinavia, B is in poor supply, and it can become unavailable by drought, liming, N fertilization, or drainage. Increasing pH followed by co-precipitation with Al and Fe compounds are the main factors behind the fixation of B in soil (Wikner 1983).

Imbalances are manifested when high levels of one nutrient are ineffective owing to low levels of another. As indicated above, high Ca in peats, and applications of lime, may induce deficiencies in some of the elements such as P, K, and Mn. Fertilization trials adding various combinations of N, P, K, and micronutrients have been the traditional method to determine limiting factors. A more diagnostic approach is to fertilize with all elements minus one. This identifies specifically the single or few elements that are limiting.

The negative effects of fertilization on swamp mosses have been studied by Jäppinen and Hotanen (1990). They found that rapidly soluble PK and NPK + micronutrient mixtures have a more destructive effect on *Sphagnum girgensohnii* and *S. angustifolium* than on dry forest mosses (*Pleurozium schreberi, Hylocomium, Dicranum* spp.). 'Burning' occurs when bryophytes come in direct contact with the fertilizer.

The conceptual model presented in Chapter 4 (see Fig. 4.8) can be used to describe the effects on the interaction between *Sphagnum* mosses and vascular plants on a bog under different levels of N deposition. Up to a certain level of N deposition the *Sphagnum* mosses will increase their growth and very little N will become available to the roots of the vascular plants. Without anthropogenic N deposition, *Sphagnum* mosses and other plants in ombrotrophic peatlands are N limited. Nutrients in precipitation are effectively trapped by *Sphagnum*, with very little reaching the vascular plant roots below. With anthropogenic deposition, the initial response is that the *Sphagnum* mosses will take up more N and grow faster. The larger amount taken up will be diluted in a larger plant biomass, and the tissue N concentration will not necessarily increase (Spink and Parsson 1995). With higher input, the growth is no longer N limited, but *Sphagnum* will continue to increase its uptake, and now the tissue concentration will increase. If this continues for some time, or if the input increases even more, the *Sphagnum* filter will be saturated and incoming N will pass down to the roots. Now the vascular plants will expand, and the *Sphagnum* mosses will face severe light competition. Pastor *et al.* (2002) have formalized this in a model in which the balance between mosses and vascular plants depends not only on the input of nutrients, but also on the transfer rates between the compartments involved: live moss, moss litter, peat, inorganic N pool, etc.

An example showing how living *Sphagnum* mosses and surface peat function as a filter for added N was given by Nordbakken *et al.* (2003). In an area of Norway with relatively low N deposition (5 kg ha^{-1} yr^{-1}), experimental addition of N led to rapid increase in N concentration in bryophytes and shallow-rooted vascular plants. The N concentration in deeply rooted plants did not increase, even after 3 years and with a dose as high as 40 kg ha^{-1} yr^{-1}. The ability of *Sphagnum* to capture N also depends on other growth conditions. If the water table is lowered, for instance after drainage or as a result of a drier climate, *Sphagnum* growth will be reduced and hence also its ability to capture N (Williams *et al.* 1999).

Vitt *et al.* (2003) compared *Sphagnum fuscum* bogs receiving different levels of N deposition from oil sand mining over 34 years in Alberta, Canada. Close to the emitting source (deposition 4 kg ha^{-1} yr^{-1}) *Sphagnum* productivity had increased to 600 g m^{-2} yr^{-1}, compared with 100–250 g m^{-2} yr^{-1} at sites with much lower or negligible deposition. Even at the high deposition site the N content increased only in the top centimetre of the mosses. There are now quite a number of studies showing that N deposition at low rates initially leads to increased growth in *Sphagnum*, but higher doses, or low doses over longer time, will reduce *Sphagnum* growth (Gunnarsson and Rydin 2000). The literature indicates that the critical load for *Sphagnum* in bog vegetation is around 10 kg ha^{-1} yr^{-1} (Gunnarsson 2005) – above that level *Sphagnum* growth will not increase (or even decreases) and N will pass through the *Sphagnum* filter down to the roots of the vascular plants. The mosses then become P limited (Aerts *et al.* 1992), or co-limited by P and K (Bragazza *et al.* 2004).

Vascular plants can have a strong impact on nutrient cycling. Silvan *et al.* (2004) showed that the wilting of leaves in the autumn can lead to a marked liberation of P, which is lost in outflow water. They also showed that a plant such as *Eriophorum vaginatum* could be responsible for a large part of the immobilization of both N and P. One reason for its importance in the cycling of nutrients is that a considerable proportion of the nutrients is in the roots, and as they die they are quickly incorporated into anoxic peat and the nutrients are then withdrawn from circulation. In many areas, peatlands with sedge or grass vegetation have been mown (often by scything) for hay in traditional agriculture. Such practices can lead to a reduction in K (Øien and Moen 2001).

It has been suggested that the N/P ratio in the vegetation is a good indicator; values below 14 indicate N limitation, and values above 16 indicate P limitation. Values between 14 and 16 mean that either N or P is limiting or they co-limit growth (Verhoeven *et al.* 1996b). Boeye *et al.* (1997) noted that N/P was 23–31 in P-limited sites and 8–15 in N-limited sites. Note that these values show limitations for the total plant productivity of a site. The diversity and the survival of individual species will be affected differently when a limiting nutrient is added. Most commonly, large, productive species will increase their dominance and bryophytes and other low-growing species will decline when a limiting nutrient is added. This happens in peatlands with, for example, increasing N deposition (Berendse *et al.* 2001; Heijmans *et al.* 2001). The use of the N/P ratio as an indicator suggests that one element is limiting for a specific site. In reality, different species have different demands, and experience the environment differently. Øien (2004) carried out fertilization experiments in boreal rich fens in Norway and found that *Succisa pratensis* was largely limited by P, *Carex panicea* by N, and *Eriophorum angustifolium* by K.

An example of the complex interactions between different nutrients and pH is the changes in moss flora of fens in the Netherlands. Acidifying S deposition peaked around 1965. The deposition of P peaked later, around 1980, and that of N around 1990. Paulissen (2004) compared the frequency of bryophyte species for three periods: 1940–59, 1960–79, and 1980–99.

Campylium stellatum and *Scorpidium scorpioides* declined after 1980. Other brown mosses, notably *Calliergon giganteum* and *Calliergonella cuspidata*, increased in frequency from the first to the second period, and then declined. *Sphagnum* species associated with rich fens, *S. subnitens* and *S. squarrosum*, have declined, whereas the poor fen species *S. fallax* and *S. palustre* have increased, together with *Polytrichum commune*. Acidity and toxic effects of SO_2, together with P eutrophication, probably caused the decline of some species before 1980. Later, the high levels of NH_4^+ seem to have been toxic to brown mosses previously favoured by P pollution, and this favoured invasion of *Sphagnum fallax* and *Polytrichum* (Paulissen 2004). At low pH, high levels of NH_4^+ may accumulate. Some of the increasing species seem to have a strong ability to detoxify NH_4^+ by converting it to amino acids that are stored in plant tissue. In addition to the direct chemical effects, the altered levels of nutrients and their ratios may also alter the competitive relations among species (Kooijman and Bakker 1995). The high inputs of both P and N over a long time are probably why *Molinia caerulea* and *Betula pubescens* have invaded bog vegetation in the Netherlands (Tomassen *et al.* 2004).

Nutrient balance and retention

For a particular peatland, it is of interest to discuss the nutrient budget of the ecosystem. Such a budget would contain inputs and outputs, and the retention is the difference between inputs and outputs expressed as a percentage of the inputs. Net retentions of nutrients and metals will generally decrease across the gradient bog to rich fen, swamp, and marshes, and the amount released into streams and lakes will increase. In other words, the more minerogenous and open the ecosystem, the less efficient is its ability to trap nutrients and metals.

Peatlands show considerable retention of N and P, whereas K is more mobile (Table 9.1). Today, it is important to realize that the high levels of N deposition, caused by human activities, drastically increase the inputs in some areas. For example, in a study including 15 peatlands across Europe, N in precipitation ranged from around 20 kg ha^{-1} yr^{-1} in the Netherlands and the Czech Republic to around 1 kg ha^{-1} yr^{-1} in remote areas of Sweden, Norway, and Finland (Bragazza *et al.* 2004).

It has often been hypothesized that freshwater peatlands are nutrient sinks which efficiently extract N and P, reducing the eutrophication of downstream water bodies. Richardson (1985), however, found that bog, fen, and swamp ecosystems used for wastewater treatment did not conserve P as effectively as mineral soil ecosystems. His data indicated that high initial removal rates of P will be followed by large exports of P within a few years. The peatland with the highest rate of P removal was a fen. The P retention capacity was strongly correlated with extractable Al.

Table 9.1 Nutrient budget of a *Picea mariana* bog in Minnesota (Verry and Timmons 1982). Because of the method used, a portion of mineral soil near the bog edge was included, hence the flow inputs. Retention is the difference between inputs and outputs expressed as percent of the inputs. Since outputs are measured in streamflow, gaseous losses of N count as retention

| | Input (kg ha^{-1} yr^{-1}) | | | Output (kg ha^{-1} yr^{-1}) | Retention (%) |
	Precipitation	Inflow	Total	Streamflow	
Nitrate N	1.74	0.30	2.04	0.28	86
Ammonium N	1.70	0.55	2.25	0.71	69
Organic N	3.85	4.56	8.41	5.38	36
Total N	7.29	5.40	12.70	6.37	50
Phosphate P	0.11	0.27	0.38	0.15	60
Organic P	0.49	0.30	0.78	0.31	61
Total P	0.60	0.57	1.17	0.46	61
Potassium	4.05	7.01	11.06	6.12	45

The annual storage rate of P in temperate freshwater wetlands is very low (Richardson 1985).

Despite the retention characteristics, there are still outputs in the surface and groundwater of all nutrients and metals. These, plus the dissolved and particulate organic matter, influence the receiving streams and lakes. The key factors influencing the chemical characteristics of outflow water are the relative influence of minerogenous groundwater and the composition of botanical components, in particular *Sphagnum*, in the peat. It has been noted previously that *Sphagnum* peat has a particularly strong influence through its high content of uronic acids, and the ability to assimilate nutrients and metals. At the other end of the spectrum marshes, as flow-through systems, fluctuate widely in their input–release patterns of nutrients: net retention during summer and winter alternates with net release of P, N, cations, and organic matter in spring and autumn (Lee *et al.* 1975).

We must mention fire as a natural phenomenon that can influence nutrient inputs and outputs in peatlands, as well as storage and fluxes of water and heat at the mire surface. In a palaeoecological study Korhola *et al.* (1996) found that wildfires acted to induce pH rise in a recipient lake. The pH rise may have been caused by increased turbulent mixing of the water column, as well as increased inputs of ash rich in basic cations.

Nutrients after draining for forestry

Much of our knowledge of nutrient limitations and cycling comes from the vast literature on peatland ditching for forestry. Drainage increases the depth of the oxic layer and increases rates of mineralization and amounts of available nutrients. For instance, the pore water of the peat shows increases of many constituents – Al, Fe, Mn, NO_2^-, NO_3^-, total N, total P, and conductivity, as well as pH (Berry and Jeglum 1991). However, unless accompanied by fertilization the amounts released are often not enough to make significant improvements of growth (McLaren and Jeglum 1998).

Where the peat is shallow (<40 cm) tree growth is influenced to a great extent by the underlying mineral soil and the degree of water movement. The richest shallow organic sites are those with distinct subsurface water seepage or groundwater discharge (upwelling) on silty to clayey mineral soils, and rich alluvial sites along streams and rivers. Shallow-peated sites, especially those that already have a tree cover, are usually more responsive and more profitable to drain than deep-peated open ones. The peat will also subside after drainage, and if the peat is thin to start with this may be enough to give the roots access to nutrients from underlying mineral soil (cf. Westman and Laiho 2003).

Foresters have developed recommendations for fertilization based on the peatland site types that take the natural fertility of the site into account (Westman and Laiho 2003). High pools of Ca and Mg generally imply a potential for high productivity, and P and K are considered the most likely growth-limiting elements in Nordic peatland forestry (Kaunisto 1997; Sundström et al. 2000). Operational fertilization of drained peatlands is either with NPK plus micronutrients on the poorer sites, or PK plus micronutrients on sites that are more nutrient rich to start with, notably the herb-rich swamp forests. Analysis of nutrient content in the tree foliage or in the peat can be used to decide on fertilization schemes (Paavilainen and Päivänen 1995; Paarlahti et al. 1971). As a final aid, it is possible to visually recognize nutrient deficiency symptoms in needles of trees on peatlands, although they are not always specific and can indicate more than one deficiency. Deficiency of N leads to chlorosis and stunting of needles, and in severe cases needles are short, stiff, and yellowish; K deficiency leads to short needles, yellowing starting at the tip, in severe cases purpling and top dieback; lack of P is recognized by slender and bent stems and small needle biomass, older needles distinctly purple tinged, the purple deepening with severity of deficiency (Morrison 1974).

One concern with forestry is whether it will be sustainable, or lead to depletion of the nutrient pools. When trees are growing after ditching, nutrients will be bound in biomass and removed at harvest. In Finland, Laiho and Laine (1994) found that trees had accumulated 300–400 kg N ha^{-1} and 30–40 kg P ha^{-1} in sites drained half a century earlier. These amounts are small in comparison with the stores of N and P in the top 50 cm of peat, and the rates of mineralization of peat and efficient recycling are probably high enough to supply such levels of uptake. For K, however, the stores are smaller, and there is a risk of future depletion of the available pool after tree harvest (Westman and Laiho 2003). Westman and Laiho (2003) conducted a detailed analysis of the effects of drainage on nutrient dynamics of peatland forests. With the exception of Mg, ditching did not decrease soil nutrient pools over the 75 year observation period.

In some cases nutrients from drained peatlands have led to lake eutrophication (Granberg 1986) with increased P concentration and algal blooms. These potential problems can be partly rectified by placing sedimentation pools at the ends of ditches to reduce the levels of nutrients and solid particles that reach the receiving waters (e.g. Paavilainen and Päivänen 1995).

Species composition, nutrients, and other factors

In Chapter 1 we described peatland vegetation types based on wetness, tree cover and pH–base saturation (as indicated by the presence of certain indicator species). In ecological investigations the aim is often to describe and analyse the vegetation in more detail – how it varies within the broad peatland types, how it changes over time, and how the species composition (i.e. the presence and abundance of different species) can be explained by nutrients and other chemical and physical factors. For this purpose, vegetation data are normally collected in sample plots. For each species, a response variable commonly measured is its cover (the percentage of the plot it covers) or more simply presence/absence. Since many species are analysed, and each species yields one response variable, the techniques are referred to as *multivariate analyses*. Several computer packages are available for such analyses, which are called *ordination*, since they order the data in a contiunuum (in contrast to classification). The program calculates the similarity in species composition among the sample plots. The result is visualized in an *x–y* graph (normally referred to as axes 1 and 2) in which plots which are similar in their composition are placed near one another, and plots that are less similar are placed far apart. At the same time different abiotic variables measured in the same sample plots are used to explain the differences in vegetation (see Jongman *et al.* 1995 for these methods). An example for wooded peatlands in Ontario, Canada is shown in Fig. 9.2. The ordination separates the broad classes, bog, fen, conifer swamp, and thicket swamp. In addition, the graph shows a considerable variation in the richness gradient within each of these, and that this variation is most strongly associated to variation in pH and exchangable bases. The pH–base saturation gradient was also linked with N, but this is not always the case in peatlands. The example also shows that much of the variation in species composition within each class can be explained by water level.

It is essential to understand that in multivariate analysis the main factors controlling the vegetation depend on the variation included in the data set. For example, if the data set is for bogs only, the variation will be strongly controlled by the depth to water–moisture–aeration complex gradient, and the pH–Ca–alkalinity gradient will not be so strong.

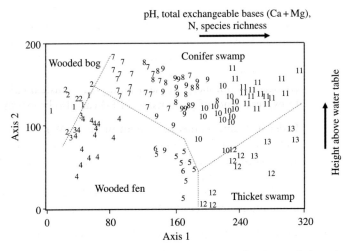

Fig. 9.2   Ordination based on presence/absence of species in 127 plots in wooded peatlands in Ontario. Based on physiognomy and indicator species the following vegetation classes were identified: 1,2, wooded bog; 3,4, extremely poor wooded fen; 5, intermediate wooded fen; 6, rich wooded fen; 7, poor conifer swamp; 8,9, intermediate conifer swamp; 10, transitional rich conifer swamp; 11, rich conifer swamp; 12, transitional rich thicket swamp; 13, rich thicket swamp. The variables most strongly correlated with the axis 1 and 2 coordinates for the sample plots are indicated by the arrows. For axis 1 the correlations were $r = 0.81$ for pH, $r = 0.80$ for total exchangable bases, and $r = 0.56$ for N. This axis was also strongly correlated with species richness ($r = 0.80$). Axis 2 was correlated with microtopographic position of the sample plot ($r = 0.33$ for height above the water table). Extracted from Jeglum (1991).

Light

Light is the fundamental energy source for photosynthesis – the trapping of CO_2 in organic molecules and incorporating these into plant tissue. Photosynthesis is fundamental to the issue of carbon balance, to which we will return in Chapter 12.

Measures of light

Our interest in light stems from the fact that plants capture light for their photosynthesis. A basic principle for all light measurements is therefore to use a method that mimics how plants experience light availability. The methods presented here try to take this principle into account.

Direct measures

Chlorophyll takes up light at wavelengths between 400 and 700 nm, and radiation in this band is called *photosynthetically active radiation* (PAR). Each captured photon (light particle or quantum) can start a particular reaction in the cell chloroplast. To measure PAR, quantum sensors are used which filter out light outside this wavelength band and count the number of photons hitting it per unit area and time. This measure is called *photosynthetic photon flux density* (PPFD), and the unit is $\mu mol\ m^{-2}\ s^{-1}$. (The unit 'mol' is the same as is used to express numbers of molecules: $1\ mol = 6.02 \times 10^{23}$ [Avogadro's number]. In plant biology the einstein (E) is sometimes used to express 1 mol of photons, so PPFD can also be expressed as $\mu E\ m^{-2}\ s^{-1}$.

Older instruments measured light in lux or $W\ m^{-2}$, but only approximate factors are available to convert these measures to PPFD. The incoming radiation reaching the earth on a clear day is about 100 000 lux, or $1000\ W\ m^{-2}$, and the corresponding PPFD is about $2000\ \mu mol\ m^{-2}\ s^{-1}$ (Hall and Rao 1999). A common value on an open field is about $1000\ \mu mol\ m^{-2}\ s^{-1}$, and a tree canopy can reduce this to values in the range of 1–100.

There are several problems with direct measures using quantum sensors:

• The sensor is placed horizontally, and PPFD thus refers to horizontal area. Since not all leaves are horizontal, other angles could be relevant, but then data sets quickly become larger, complex, and difficult to compare.
• The sensor is about 0.5 cm in diameter, and a large number of measurements are needed to represent small-scale variation, for instance in a grass sward. To solve this several sensors are mounted on a narrow bar that can be pushed into the sward from the side without changing leaf orientation (seven sensors on a 1 m bar were used in the example in Fig. 9.3).
• During a series of measurements at a site, overall light intensity will change with time of day and cloud cover. To overcome this ecologists

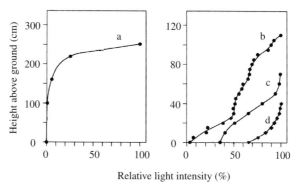

Fig. 9.3   Vertical profiles showing light penetration in fens with vegetation types differing in characteristic species and total biomass. Light is expressed in percent of the intensity above the canopy. a, *Phragmites australis* (above ground biomass 2980 g m^{-2}); b, *Carex acuta* (1160); c, *Typha–Phragmites* (407); d, *Calamagrostis stricta* (217). Redrawn from Fig. 2 in Kotowski, W., van Andel, J., van Diggelen, R. and Hogendorf, J. (2001). Responses of fen plant species to groundwater level and light intensity. *Plant Ecology*, 155, 147–56. With kind permission of Springer Science and Business Media.

often express light as relative PPFD, that is, PPFD as a percentage of what is simultaneously measured in the open, or above the vegetation.

• Measurements at one occasion are not a good indicator of how the plants experience the site over the whole season. A particular spot could be shaded by a tree, but the next hour exposed to direct sunshine. For this reason light measurements should preferably be made on an overcast day with thin but continuous cloud cover which causes a 'soft' distribution of light. Bright, sunny days should be avoided. So should days with thick clouds, as the relevant variation in space then tends to be equalized. Continuous measurements would require many sensors in a comparative study and are rarely used.

Indirect measures

Indirect measures have been developed to simplify data collection and to overcome some of the problems mentioned with direct methods. A drawback is that they do not give a value for PPFD.

The simplest measure is an estimate of the percentage tree crown cover. For large-scale comparison this may suffice, but the choice of the size of area for which crown cover should be assessed is rather subjective, and comparisons among studies are difficult. A useful method is to take a photograph with a fisheye lens that gives a 360° view of the sky. There are computer programs that can calculate the proportion of the sky that is not blocked out by stems, branches, or leaves.

Sometimes the plant composition can be used to assess the light regime. If good indicator lists are available for the region, a light index can be calculated from the abundance of species with different light values. An example is the Ellenberg indicator values in Central Europe (Ellenberg 1979). This method should be used with caution. Only a few of the indicator values have been strictly tested, and values may be different in other parts of Europe. Circular reasoning must also be avoided; one cannot discuss the plants' response to light if the light measure is based on the presence of the same species!

Incoming sun light will warm up the site. Therefore, it is in practice very difficult to completely disentangle the roles of light, temperature, and desiccation in peatlands.

Vegetation and light

Plants respond to variation in PPFD. The light compensation point is the level required for the photosynthesis at that instant to match the plant's own respiration. Species with low compensation point can endure severely shaded conditions. With increasing PPFD the plants increase their photosynthesis up to a point of light saturation. Species adapted to open habitats may require a PPFD of around 500 μmol m^{-2} s^{-1} to reach their maximum photosynthetic rate, but many plants, especially bryophytes confined to wooded peatlands, require much less.

Light availability is obviously controlled not only by time of the day and the year, but also by latitude and weather, and by the plants themselves. Competition for light is strongly asymmetric, and there is a hierarchy among wetland plants related to their size (Gaudet and Keddy 1988). Trees will shade all other plants; shrubs, herbs, and graminoids will shade and be shaded in proportion to their height and leaf area; and mosses will be shaded by all other plants. In general terms, shading is dominated by trees in swamps, by dwarf shrubs and some graminoids in bogs, and by grasses, sedges, and some ferns in fens and marshes.

Since shading is caused by the plants, it will come as no surprise that light availability at different levels above the ground surface is related to plant biomass (Fig. 9.3). Sites with high productivity tend to be dominated by one or a few species of tall-growing grasses or sedges, and their dense cover could be fatal to low-growing species and bryophytes. In fens, light availability for the bryophytes is rather independent of the rich–poor gradient. Poor fens with overall lack of nutrients and strongly P-limited rich fens could both have a sparse cover of vascular plants.

As discussed earlier, the negative effect of N deposition on biodiversity is largely caused by the asymmetric light competition. The strongest effect of N deposition may be from the accumulation of litter on the moss surface rather than by the shade cast by the plants themselves – although the living grasses and sedges allow some light to penetrate, the dead leaves can form a blanket cutting off light totally from what is beneath. This does not happen only after increased N deposition; similar effects will occur when tall species expand after drainage, and when mowing or grazing ceases in naturally fertile fens.

Vascular plants have means to adjust to reduced light levels. They can reduce their root:shoot ratio, which means that they allocate more carbohydrates to expand their aboveground parts at the expense of the root system (Kotowski *et al.* 2001). This response enables them to catch more light. This is a way both to tolerate shade and to survive under a canopy of taller species and a way to exert an influence on the plants beneath. Vascular plants have a tendency to elongate and grow taller in shade, which is also a mechanism to capture more light.

Light availability is related to the microtopographic gradient in bogs and fens. In the low-lying carpets of pure moss and scattered sedges, there is almost unrestricted access to incoming light. As the carpets firm up into lawns, various field-layer herbs and graminoids develop above the bottom layer, and take some of the light before it reaches the ground, especially in tall sedge fens. As the microtopography goes to hummocks, surface peat aeration permits the growth of dwarf shrubs making more shade for the bottom layer mosses and lichens. In both fens and bogs shading tends to be strongest on hummocks and caused by dwarf shrubs, especially the evergreen ones. On the highest hummocks trees can develop and these cause more shading, even to the exclusion of *Sphagnum* and the invasion of *Pleurozium schreberi* feathermoss around the tree bases.

The ground flora of open and wooded peatland differs quite a lot even when comparing sites that are similar in pH and wetness. In Table 9.2 we give

Table 9.2 The following is a list of some of the most obvious of open growing and forest-shaded plants in different nutrient regime series. However, since trees preferentially grow in sites with low water table, it is cautioned that the changes from open to closed vegetation may be owing to one or the other, or both, of moisture-aeration and light. The examples are taken from northern Europe (E) and North America (A)

Open growing plants	Forest-shaded plants
Marsh–thicket–hardwood swamp sequence	
Calamagrostis canadensis A	*Climacium dendroides* AE
Carex rostrata AE	*Cornus canadensis* A
Carex stricta A	*Cornus suecica* E
Eleocharis palustris = *E. smallii* AE	*Gymnocarpium dryopteris* AE
Equisetum fluviatile AE	*Maianthemum bifolium* E, *M. canadense* A
Schoenoplectus lacustris E, *S. acutus* A	*Mitella nuda* A
Typha latifolia AE	*Paris quadrifolia* E
	Rhytidiadelphus triquetrus AE
Open fen–treed fen–conifer swamp sequence	
Carex chordorrhiza AE	*Coptis trifolia* A
Carex lasiocarpa AE	*Equisetum sylvaticum* AE
Carex livida AE	*Hylocomium splendens* AE
Eriophorum angustifolium AE	*Pleurozium schreberi* AE
Many brown mosses AE	*Ptilium crista-castrensis* AE
Menyanthes trifoliata AE	*Ribes triste* A
Rhynchospora alba AE	*Sphagnum girgensohnii* AE
Salix pedicellaris A, *S. repens* and	*Sphagnum russowii* AE
S. myrtilloides E	
Sphagnum fallax AE	*Sphagnum wulfianum* AE
Sphagnum papillosum AE	*Trientalis borealis* A
Trichophorum alpinum AE	*Trientalis europaea* E
	Vaccinium myrtilloides A
	Vaccinium myrtillus E
Open bog–treed bog sequence	
Andromeda polifolia AE	*Carex trisperma* A
Calluna vulgaris E	*Gaultheria hispidula* A
Carex oligosperma A	*Rhododendron groenlandicum* A
Carex pauciflora AE	*Rhododendron tomentosum* E
Drosera rotundifolia AE	*Melampyrum* spp. AE
Eriophorum spissum A	*Pleurozium schreberi* AE
Eriophorum vaginatum E	*Ptilidium ciliare* AE
Sphagnum cuspidatum AE	*Sphagnum angustifolium* AE
Sphagnum rubellum AE	*Vaccinium uliginosum* E
Vaccinium oxycoccos AE	

examples of species that are relatively strictly bound to open or shaded conditions.

Temperature and other climatic factors

Temperature and moisture regimes are key climatic factors explaining the rates of peat accumulation and decomposition, and distribution of peatlands in a global context. In addition, temperature is a key controlling factor for photosynthesis and respiration processes. The effect of temperature on carbon sequestering has become a central topic in ecology today – the issues of greenhouse gases and global warming (Chapter 12).

Climatologists recognize climates at three spatial scales.

- *Macroclimate* is the large-scale, well above-ground climate of a large area or a region. We could for instance describe the climatic differences between the geographic concentric bog zone and the eccentric bog zone in Finland (Chapter 10) at this scale.
- *Mesoclimate* is the intermediate scale, describing the climate of a particular peatland, for instance a raised bog or a swamp forest. At this scale we could also compare the climate of a peatland with the neighbouring upland habitat, and note the differential influence of these habitats on the climate.
- *Microclimate* is the fine-scale variation within a peatland, for instance the temperature variation between hummocks and hollows.

The distinction between the scales is difficult to define strictly, but if we try to link with peatland scales (as presented in Chapter 10) we can propose that macroclimate matches the peatland region (supertope), the mesoclimate matches the synsite and complex (i.e. the landforms or mesotope and macrotope), and the microclimate relates to the feature and site (microform and microtope). In addition, we note strong vertical differences related to vegetation and ground surface.

Macroclimate

Some information on the macroclimate is normally included in peatland studies, because it helps the reader to interpret the results and compare with other studies. Several standard climatic parameters may be reported. A detailed review of climatic indices in ecological research was given by Tuhkanen (1980).

Mean daily temperatures for January and July are much more informative than the mean temperature for the whole year, because they indicate how seasonal the climate is and if it includes winter frost. Other temperature measures could be used that are better related to the response of the organisms than simple mean values. One is the *length of the vegetative season*, which could be expressed as number of days with a daily mean temperature >5 °C. Another is the *temperature sum*, for instance measured as the sum of the mean daily temperature minus a chosen threshold, such as 5 °C, over the whole vegetation season. This has been used in forestry as a predictor of timber production on drained peatlands at different latitudes and elevations above sea level (Hånell 1988).

Mean annual precipitation is of course relevant. For peatlands, especially bogs, it is even more interesting to have data on *humidity*, and this depends both on precipitation (P) and the water losses through evaporation from the ground and transpiration from the plants – often combined as evapotranspiration (ET; Chapter 8). A simple measure is P–ET, and this can be measured as the total runoff in streams and rivers from a region. Another commonly used measure is the ratio P/ET. In North America, the distribution of *Sphagnum*-dominated peatlands could be explained quite well with a combination of annual temperature, annual precipitation, and P–ET (Gignac *et al.* 2000). The *thermal seasonal aridity index* (TSAI) proposed by Tuhkanen (1980) has been used to explain distribution limits of peatlands during the Holocene (Zoltai and Vitt 1990).

Since these measures describe the macroclimate, data should be taken from the nearest official weather station, collected in standardized ways. These are typically presented as mean values for a 30 year period.

One climatic aspect which is important for peatlands is the complex gradient from maritime to continental climate. The maritime (or oceanic) climate is characterized by high humidity, low variation in temperature over the year, and low incidence of frost in relation to latitude. Because of the resistance to temperature change in the seawater, maritime climates have mild autumns and cool springs compared to inland sites. One important feature of strongly maritime climates is that plant growth on peatlands is possible all the year round, for instance in British Columbia (Asada *et al.* 2003a,b), and even at high latitudes as in coastal Norway.

It is not easy to reduce the maritime – continental scale to a measured value, but classificatory schemes have been used. For instance, Moen (1999) described the variation in Norway from the highly oceanic areas with blanket bogs to slightly continental ones with string mires, and in the north palsas. In North America, where there are also more marked continental climates, there is a distinct difference between continental bogs which often have trees and oceanic ones which are almost treeless and have more pools. Vitt and Kuhry (1992) discussed the differences in the development of peatlands in oceanic and continental climates.

Mesoclimate

The temperature regime in the peat differs in some respects from an upland soil. A comparative study of peatland and uplands in Minnesota by Nichols

(1998) showed that peat soils warmed more slowly in the spring, had lower summer temperatures, and cooled more quickly in the autumn. As long as the mire surface is wet, high evapotranspiration will keep down summer temperature in the peat, but even a dry peat surface will keep down temperature in the root zone because dry peat is a good insulator. Nichols also found a strong effect of tree cover; the annual temperature in the peat was about 1 °C higher in open peatlands than in wooded ones. In the summer the difference was 2–3 °C. The energy fluxes and the temperature regime are strongly affected by vegetation structures, and can even be said to be under 'ecosystem control' (Bridgham *et al.* 1999).

Some of the differences in temperature regimes between bogs, fens, and swamps may be due to differences in water movements. In general with moving water the substrate thaws earlier; hence, open fens often warm up earlier than open bogs in the spring and early summer. Swamps may remain cooler than fens in the summer owing to the shading of the tree canopies.

Microclimate

Peatland vegetation influences the microclimate. Several examples can be mentioned: plant species composition affects the *albedo*, that is, the tendency in light-coloured patches to reflect more of the income radiation and thus stay cooler than other areas. The microtopography, so typical for many peatlands, will affect frost insulation directly and also indirectly by controlling the distribution, thickness, and duration of snow cover. The surface temperature during sunny days is reduced by evaporation, and the rate of evaporation is to a considerable degree governed by the plants.

Temperature has a direct effect on plant growth, and the current interest in climate change issues has triggered a great deal of research in which temperature is manipulated on peatland surfaces. The experiments often consist of 'open top chambers'. These are miniature greenhouses and consist of transparent plastic frames, 50 cm high, pushed into the peat to surround plots of 1–4 m^2. The frames often lean inwards, so that the opening in the top is somewhat smaller than the ground area of the plot. These chambers increase summer temperature by a couple of degrees. In theory, raised temperature should increase the rate of both microbial soil processes and plant photosynthesis. In practice the results are seldom straightforward. In *Sphagnum*, for instance, the photosynthesis shows rather little variation with temperature compared with the drastic decrease when water content is reduced (Rydin 1993b). Thus, it appears that a warmer macroclimate can lead to desiccation, which has a stronger effect than the warming itself (Weltzin *et al.* 2001; Dorrepaal *et al.* 2003; Gunnarsson *et al.* 2004).

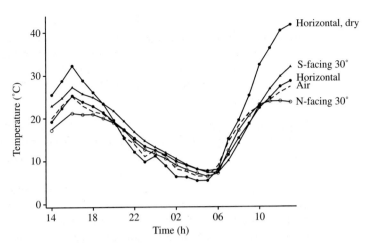

Fig. 9.4   Temperature variation just below the surface in *Sphagnum fuscum* during a clear night and a sunny summer day in eastern Sweden. Compared to the horizontal moss surface, the south-facing sample reached a higher daytime temperature, but the effect of wetness was much larger. In the sample which was completely dry down to a depth of several centimetres, temperatures >40 °C were noted. Redrawn from Rydin (1984).

High surface temperature in the bryophyte canopy is common in open peatlands on sunny summer days. As long as the bryophytes are moist, their transpiration helps to keep temperature down. When they dry out down to a depth of several centimetres, very high temperatures can be reached (Fig. 9.4), but then the mosses are physiologically inactive and are not harmed by the warmth.

Frost and seasonal temperature variation

In boreal peatlands, the peat is mostly frozen during a part of the year. Especially peatlands with low cover of trees and shrubs can experience low night temperatures and late spring and even summer frosts. The distribution and duration of ground frost at a finer scale is governed by the hummock–hollow microtopography, the thickness of the snow cover, and the water content of the peat. In the autumn, the temperature decreases more slowly in the hollows because the higher water content gives them a higher heat capacity. In winters the hollows will have a thicker insulating snow cover. In northern Finland Eurola (1968) recorded ground frost down to 45 cm under a hummock, and this was 15 cm deeper than under the hollows. In a part of the mire without snow cover, frost could reach down to 55 cm and was equally deep under the two microforms. In the spring the snow melts quicker on the hummocks, first on the top and then on the south slope. Warming starts earlier on the hummocks, but spring temperatures are soon equalized and the ground frost thaws a few weeks earlier under the hollows. Large hummocks may, however, remain very cold a few tens of centimetres below the surface. For instance, Sjörs (pers. comm.) found a large ice lens about 30 cm below the surface of *Sphagnum fuscum* and white lichens at the end of July in central Sweden (mid boreal, about 400 m a.s.l.).

In permafrost regions a strong effect of the vegetation on thaw rates has been observed by Camill and Clark (1998). Thaw rates have accelerated over the last decades, but cover of *Sphagnum fuscum* stabilizes the process. This peat moss forms hummocks which will have less snow cover than the surroundings and hence deeper penetration of frost. Snow cover will have the same spatial distribution every winter. Therefore, in the zone with discontinuous permafrost, there is a strong correlation between occurrence of permafrost in the hummocks and duration and depth of the snow cover (Sonesson 1969). At a larger scale, the distribution of permafrost landforms can be related to macroclimate. In continental Canada, Vitt *et al.* (1994) noted that the southern distribution of peat plateaus (see Chapter 10) coincides well with the $-1\,°C$ isotherm. Sometimes the relationship with current climate is not so precise, and relict occurrences of permafrost exist farther south, but they have gradually declined since the Little Ice Age (see references in Vitt *et al.* 1994).

In maritime areas, the variation over the year, and the spatial variation in temperature between hummocks and hollows, are both quite small (Fig. 9.5). Even here, the lower water content of the hummock means that its surface warms more quickly, and its insulating capacity protects the

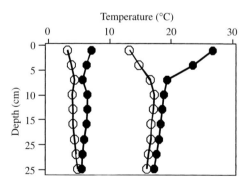

Fig. 9.5   Temperature variation with depth in an Irish bog. Open circles are night temperatures and closed circles are for daytime. The two left profiles are for winter, and the right ones for summer. Measurements were made in hummocks characterized by *Sphagnum fuscum* and *S. austinii* with *Calluna vulgaris*. Drawn from data extracted from van der Molen and Wijmstra (1994).

deeper peat from cooling in the summer nights or heating up in the winter day (van der Molen and Wijmstra 1994). A particularly even peat temperature is seen in fens with upwelling of water from the mineral soil, such as discharge fens or spring fens. Such sites have cold peat in the summer, and can remain frost free during the winter.

10 Peatland patterns and landforms

Surface patterns are one of the most intriguing features of peatlands, appealing not only to the ecologist for scientific explanation but also to the connoisseur of landscape forms and shapes. The geological landform and topography is the template on which the water accumulates and flows. The underlying landform is subsequently modified over time by peat accumulation. The ultimate modification is the raised bog, which is far removed from the underlying landform. Peatland patterning is a response to biotic processes and extremely slow water movements, generally imperceptible on the ground. These patterns are actually composed of plant communities and peat forms that have developed orderly shapes in response to subtle gradients in water movements and water chemistry. The patterns are therefore quite unique among ecosystems, reflecting ecological processes that integrate living plant production, peat accumulation, and hydrological factors.

Hydrologic systems

One of the earliest systems for peatland classification, proposed by Weber (1908), was based on a model of long-term development: *Niedermoore – Übergangsmoore – Hochmoore*. Subsequently, many authors (e.g. Kulczyński 1949) linked this with the origin of groundwater: mires developing in mobile groundwater – transitional (groundwater fed, but with low mineral supply) – rainwater-fed mires. Succow and co-workers (Joosten and Succow 2001) have developed an approach called the *hydrogenetic classification* where mire types are defined by the role of water in peat formation. A recent version (Joosten and Clarke 2002) identifies many hydrogenic types based on peat formation (e.g. infilling or paludification), slope, water storage capacity, effect on landscape water storage, and origin of water (ombrogenous or minerogenous).

In our treatment we refer back to the hydrologic system introduced in Chapter 1, with the main division between mineral soil influenced groundwater and rain-derived groundwater – minerogenerous versus ombrogenous. The minerogenous was then divided into limnogenous, soligenous, and topogenous (von Post and Granlund 1926). In this chapter we add saline and brackish systems.

Saline and brackish systems

Coastal wetlands include salt marshes and mangroves, occurring at the edges of seas with saline and brackish waters, and usually influenced by tides (Chapman 1977). Salt marshes are common in the temperate region, whereas mangroves reach their maximum development in the tropics, extending into the subtropics and slightly into warm temperate regions.

A feature of the wet coastal formations is the widespread occurrence of dominant genera throughout the world. In the salt marshes *Salicornia*, *Spartina*, *Juncus*, and *Plantago* appear in both hemispheres. In the lowest portion of a typical salt marsh, the lower marsh, especially one that is open to the sea, plants generally are flooded twice each day at high tide. These situations do not normally develop true peats with >30% organic matter (e.g. Knott *et al.* 1987; Mitsch 1994). Higher up and in the landward direction there may be an upper marsh zone, where saltwater inundation may not be as frequent. In protected sites and estuaries where fresh water meets salt water, peat with >30% organic matter can develop to considerable depths (Fig. 10.1; Niering and Warren 1980; Warren 1995). On the

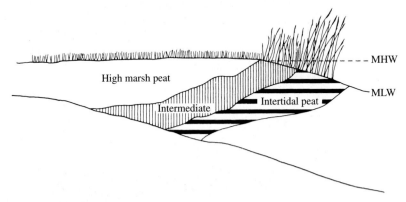

Fig. 10.1 Oceanward salt marsh development in New England, USA, with tall intertidal *Spartina alterniflora* being replaced by intermediate peat and then by high marsh peat. MHW, mean high water; MLW, mean low water. There has been a postglacial rise in sea level, and the tidal marsh has moved outward over the tidal flats and landward over freshwater wetland. With permission from Niering, W.A. and Warren, R.S. (1980). Vegetation patterns and processes in New England salt marshes. *BioScience*, 30, 301–7. Copyright, American Institute of Biological Sciences.

landward, freshwater edge dense reeds (with *Phragmites australis*) can form peat (Niering and Warren 1980). Such conditions appear alongside brackish waters such as the Baltic. Here considerable water level fluctuations are caused not by tides, but by season and weather.

Among the mangroves, *Rhizophora, Avicennia, Acrostichum*, and *Bruguiera* are all widespread (Chapman 1977). Mangrove thickets and woodlands are found over both mineral sediments and peat. The organic matter ranges from 30–70% to as high as 90% (McKee and Faulkner 2000; Cahoon *et al.* 2003). Mangroves have been drastically reduced or disturbed by humans in highly populated countries, and mangrove restoration is a high priority of nature conservation throughout the tropics. There is a large literature on the hydrology and ecology of salt-marsh vegetation (Glooschenko *et al.* 1988; Price and Woo 1990).

Limnogenous systems

Lacustrine

Lacustrine peatlands occur along large or small lakes. They are usually characterized by large water level fluctuations but water levels can also be moderately stable, for instance in quaking mats. Flowage lakes are those connected with rivers or streams, and sometimes both the lakes and their connected rivers can have marginal zones of marshes, fens, and swamps (Fig. 10.2). Beside seasonal fluctuations, large lakes may experience cyclic

Fig. 10.2   Lacustrine and riverine systems. Floating, shore fens along a slow-flowing river entering lake near Geraldton, Ontario, Canada. Southern boreal zone.

high and low water levels over periods of wet and dry years. They also can experience more rapid fluctuations from seiches and storm surges. The fluctuations are governed by the flow regime for the lake, the distribution of precipitation, size of catchment, and inflowing and outflowing streams and rivers. In terms of water chemistry one can envisage a sequence of calcareous, eutrophic, mesotrophic, and oligotrophic lakes.

Riverine

Riverine peatlands appear along rivers, streams, and brooks, including their floodplains. The diagnostic feature is open water in channels, and it includes all sizes of channels, from rills and brooks to rivers. When streams or rivers empty into a lake or sea, or into flatter plains, the water slows, sediment is deposited, and multi-channelled deltas form. Peatlands can develop on the floodplains and oxbow channels in these deltas.

When sites are frequently flooded and receive a lot of dissolved or particulate materials, clay or silt, on the surface, the systems are often marshes or other wetlands with little peat. In other parts of floodplains the floodwaters may not carry as much particulate mineral matter, and these can develop true peats beneath marshes, thicket swamps, and open fens (Fig. 10.2). Further removed from the channels, closer to the uplands, zones of wooded fens or swamp forests can occur. Along streams and drainage ways with continuous or intermittent flow, beaver dams can block the flow and cause paludification.

Soligenous systems

These systems consist of minerogenous sloping peatlands (Fig. 10.3). There are no distinct open water channels, but laminar flow at and below the peat surface, and occasionally (at a time of rapid snow melt or very heavy rains) even sheet flow at the surface. The water carries dissolved minerals derived from mineral soils. Other dissolved and particulate matter, mainly derived from organic matter decomposition, is also be carried in the flowing water. The water in soligenous systems may be supplied in several ways:

- from a headwater lake or wetland emptying into a confined drainage way
- from water off the flat crest or upper slopes of a hill or mountain, converging into a confined drainage way or water track
- from an extensive area of wetlands located upslope and supplying a water track or a broad unconfined sloping peatland
- from meltwater seep from snow packs
- from groundwater discharge (see below).

Groundwater discharge

Groundwater discharge systems consist of springs, seeps and more diffuse groundwater upwellings where water comes to the surface from aquifers below the ground (Chapter 8). Understanding groundwater discharge

Fig. 10.3   Soligenous fen in central Norway with poor lawn level vegetation with, for example, *Eriophorum angustifolium, Scirpus cespitosus*, and *Molina caerulea*. The water enters upslope by a combination of runoff and groundwater discharge.

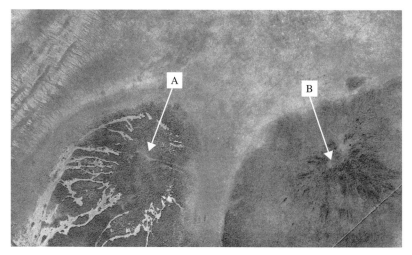

Fig. 10.4   (A) Groundwater discharge: a spring-fen mound with soligenous fen water track. (B) Raised domed bog with radiating lines of *Picea mariana*, resembling spokes of a wheel. Lost River Peatland, Minnesota. Photo provided by Paul Glaser.

requires knowledge of the interplay of groundwater aquifers, geology, landforms, and physiography (Heeley and Motts 1976; Siegel 1992). Common physiographic occurrences for groundwater discharge are on the sides or toe slopes of eskers or glacial till uplands where the aquifer encounters relatively impermeable material, such as bedrock or clay. The aquifer discharges to form a spring or spring-seep line, which then flows over the surface downslope, contributing to a sloping wetland (Fig. 10.3). Such seepage peatlands occur in hilly and mountainous country with humid climatic conditions. Another discharge situation is where the water wells up from below, forming an elevated spring-fen mound (Fig. 10.4). The most common situation, however, is on the lower slopes of hills where the upper level of the groundwater comes close to the surface on a broader front.

Topogenous systems

These are hydrologic systems with relatively flat hydrotopographies, occurring in basins, kettle-holes, and depressions (Fig. 10.5). They may begin as open water, which may be saline (in prairies and steppes), calcareous (limestone bedrock regions), eutrophic, mesotrophic, or oligotrophic. With enough water supply they tend to become filled with vegetation and peat, sometimes with floating mats progressing out over the central open water. In other cases they may start from the beginning with peat-forming plants, which then continue to fill up with peat. Topogeneous mires include

Fig. 10.5   Topogenous fen with floating mat in landscape basin.

floating or quaking mats over water or aquatic peat slurry, as well as completely peat-filled basins with spongy or firmer, stagnating peat. The water table may be high or low, and usually fluctuates seasonally.

These systems can be completely isolated, but more often they have some inflow and outflow. There can be some groundwater discharge into these basins from marginal springs or from the bottom of the basin, but during dry periods the storage water in them can act to recharge the surrounding regional aquifers (Lissey 1971). Topogeneous peatlands of very different kinds have a global distribution (except for deserts), and in many semiarid areas are the only peat-forming systems.

An example of how hydrology can interact with nutrition in a topogenous system comes from the floating fens in the Netherlands (Fig. 10.6; van Wirdum 1991). These develop in ponds, some 30–50 m wide, a few hundreds of metres long, and 1.5–3 m deep, from which peat has been excavated. The ponds connect with water-filled ditches that carry water from varying sources, including weakly saline river water. The relative importance of water and nutrients from the pond and rainwater changes over very small distances within these floating mires. Even at the pond edge, where the peat is relatively thick, deep-rooted vascular plants can reach down into layers in contact with the pond water.

Timmermann (2003) has described three kinds of topogenous conditions differing in the way that the peat surface follows water level fluctuations: *floating* or *quaking mire* where the vegetation surface follows water level fluctuations, *spongy mire* with some surface oscillation owing to expansion

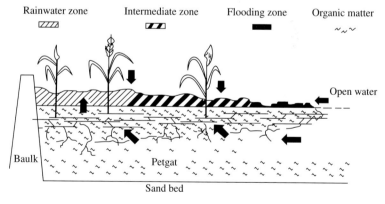

Fig. 10.6 In the Netherlands a *petgat* is a pond originating from excavation of peat from below the water level with dredging tools. Here fens with a floating mat (quagfen) develop. Three zones of water origin are portrayed, representing decreasing influence of flooding and increasing influence of rain water as the distance from the body of open water increases. From van Wirdum (1991).

and contraction of the peat mass (mire breathing; see Chapter 8), and *stagnating mire* in which the surface is more stable even when the water table drops.

Ombrogenous systems

The conventional, but not the most widespread, model for bog develop-ment is a topogenous system with the infilling pathway (Fig. 10.7). As out-lined in Chapter 7, this model has to be expanded to include primary peat formation and especially paludification. Bogs can also develop from lacus-trine, riverine, and soligenous systems in locations removed from the influ-ence of flooding or surface flow (Fig. 10.4 B). From soligenous systems, bogs are established along the margins of interfluves between streams and rivers, positioned up against the levees; they are also established in the centres of interfluves as rounded, ovoid, or linear islands surrounded by a matrix of soligenous fen. In extensively peat-covered areas, such as the Hudson Bay Lowland and western Siberia, some of the interfluves are almost completely covered by ombrotrophic bogs.

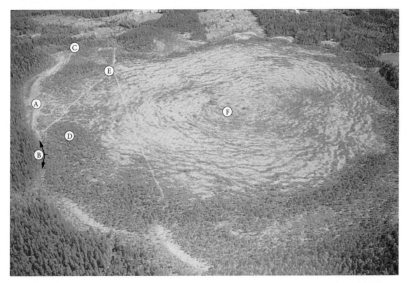

Fig. 10.7   The concentric raised bog Ryggmossen, Sweden, boreo-nemoral zone. A, the lagg fen in which the darker areas are mud-bottoms. B, water divide with arrows showing direc-tion of water flow. C, outlet of water from the lagg fens. D, the marginal forest zone of the bog (pine-bog) with high cover of shrubs (*Rhododendron tomentosum* and *Vaccinium uliginosum*). E, track used several decades ago for timber transport with sleigh. F, the slightly domed treeless expanse with *Sphagnum fuscum* strings with *Calluna vulgaris* and *Empetrum nigrum*. Most hollows are lawns dominated by *S. balticum* and *Eriophorum vaginatum*, but the largest contiguous hollows are carpets with *S. cuspidatum*. Photo H.-G. Wallentinus, provided by Uppsala kommun.

Landform and hydromorphology

The shape of a peatland, and the patterns within it, reflect the interactions between underlying terrain form, climate, and hydrology. *Hydromorphology* is a term used to express these interactions. The concept is most relevant for bogs and fens, less so for marshes and swamps.

An approach to wetland and peatland classification developed in Canada emphasizes landform and physiographic location (National Wetlands Working Group 1997). This approach, although 'landform' based, is in fact strongly related to hydromorphology (Glaser 1992b; Brinson *et al.* 1994). It is based on the fact that there is a close relationship between the surface form of the peatland and the form of the groundwater table. Furthermore, the hydrology of peatland is controlled by the general water supply system in which it is located. Hence, peatlands such as 'raised bog', 'sloping fen', and 'riverside swamp' have been named after their landform or physiographic location. Landform is the fundamental template which determines how the ground interacts with incoming precipitation, and with surface and groundwater movements in the landscape, and how this influences the development of the peatland. The Canadian system first recognizes five classes of wetlands – shallow open water, marsh, swamp, fen, and bog – and then wetland forms (hydromorphological types) within each of the classes. The system is rather subjective and open-ended, and new forms can be added.

Permafrost landforms

In addition to hydrology, permafrost landforms are shaped by freeze – thaw dynamics. Permafrost is a perennially frozen layer in the soil, typical of arctic, continental boreal, and alpine regions. More technically, it is material that has a temperature continuously below 0 °C for at least two consecutive summers. As one progresses towards the polar region, permafrost is at first discontinuous, but towards more continental parts becomes continuous. In-depth treatments of permafrost and peatland landforms in Canada have been given by Zoltai (1972, 1995) and Brown (1977). The structures included in the descriptions below are palsas, peat plateaus, collapse scars, and polygonal peat plateaus.

Mire descriptions and classifications at different scales

Since surface structures are a striking feature of temperate and boreal bogs and fens, it is natural to describe and classify mires accordingly. There are structures of all sizes (from hummocks a few decimetres across to landscape features extending over many kilometres), indicating that hydromorphology can be described and classified at several scales. The hierarchy

Table 10.1 Different levels of description of mire hydromorphology. The text follows Moen's terminology. 'Nanotope' is sometimes used instead of microform. In some Russian and Estonian texts 'massif' is used instead of type and 'system' instead of complex

Description of level; scale	Example	Moen (2002)	Ivanov (1981)
Microtropographic structure, 0.25–100 m², sometimes larger	Hummock, lawn	Feature	Microform
Site type, uniform or patterned, physiognomic group; 100 m² to few km²	Lagg fen; wooded bog, open bog (on a raised bog)	Site	Microtope
Peat landforms, uniform hydromorphology; 0.5–20 km²	'Raised bog with concentric strings and hollows'; 'Patterned fen with strings and flarks'	Type	Mesotope
Systems of peat landforms; few to >20 km²	Raised bog with radiating forest patterns and lagg fen'; 'Patterned sloping fen and swamp forest'	Complex	Macrotope
Combinations of systems in climatic and physiographic areas; 1000s of km²	Finnish concentric raised bog region	Region	Supertope

of complexity has been described in several countries, with different systems developed for peatland inventories and mapping. Comparing and translating between systems is not always easy. Moen (2002) has made attempts to reconcile the Scandinavian and Russian classifications, and the structures he described at different scales of mapping (cf. Sjörs 1948) are feature, site, type, complex, and region (Table 10.1). The 'type' level is the one often used for mapping, and is largely comparable with the Canadian peat landform system.

Mire features

Peatland features are the smallest units of uniform vegetation, peat, and microtopography. 'A mire feature represents the local topographical situation where any particular plant community is growing' (Moen 2002). The features are parts of a mire site and do not usually exist on their own. *Patterned mire* is the general term for sites that have ridges and depressions that alternate in ladder-like or net patterns. They can be bogs, fens, or a combination of the two. Where the features are elongated they are always stretched out perpendicularly to the slope, and the patterning can help us to interpret the direction of water flow.

The basic features in the patterning were described in Chapter 1: hummocks (or strings if elongated), and lower levels divided (Sjörs 1948) into lawn (firm), carpet (soft), mud-bottom, and open water. Bog pools are permanently water-filled depressions (Fig. 10.8) which were initiated and

Fig. 10.8   Bog pool in concentric raised bog, mid boreal zone, Ontario, Canada. A string bank with shrubby *Picea mariana* is behind the pool, and water flow is through the ridge towards the background.

deepened after the peatland was formed, mostly as a result of peat accumulation between the pools.

In gently sloping fens the surface is often split into low damming strings and elongated wet depressions with sparse vegetation (Fig. 10.9). These depressions are called *flarks* (from a Swedish dialect word; cf. *rimpi*, Finnish), and they are often a couple of metres wide, and 5–20 m long, although flarks several hundred metres wide and 1 km long occur in the north. Even such large flarks are entirely horizontal. For most of the time a flark is waterlogged or even flooded. The water level is maintained by the damming ridge on the downslope side, and the flarks and the strings are arranged perpendicularly to the slope. Low strings are minerotrophic, but high ones (perhaps 50 cm) can be almost or completely ombrotrophic, with the same vegetation as on bog hummocks. A secondarily deepened flark is a flark pool.

Fig. 10.9   Patterned fens with flarks and strings. (A) Two flarks divided by a 40 cm wide and 20 cm high string. Northern Quebec, Canada, northern boreal zone. (B) Large flark and string with fen lawn vegetation. The string is raised about 1.5 m above the next flark (on the right). Västerbotten, Sweden, mid boreal zone.

Fig. 10.10   Palsa and collapse scar. When a palsa melts it leaves a pool of water (foreground) which shows that most of the palsa consists of ice. Northern Sweden. Photo Hugo Sjörs.

Spring fens can be small and at the scale of mire features. The vegetation in them is quite variable, with regard to the often scanty vascular component. In the boreal and alpine zones, where springs are conspicuous, the bryophyte component is characteristic but differs according to pH and calcium (Ca) content. The water is often above the freezing point during winter, but stays cold in summer and is richer in oxygen than other mire waters.

Palsas are striking features in regions with discontinuous permafrost (Fig. 10.10). These are peat mounds 2–4 m high (occasionally 7 m) consisting of fen peat with lenses of almost pure ice but with ombrotrophic (or nearly so) vegetation. When a palsa melts it leaves a water-filled depression, showing that the palsa is not created by strong peat formation, but is instead a patch with weak peat formation lifted up by the development of the ice lenses (Lundqvist 1951; Sjörs 1961). Often only the uppermost 20–40 cm melting in summer consists of pure peat. In areas with continuous permafrost in Canada *pingos* occur, which may be up to 100 m high, consisting mainly of pure ice. Pingos require access of liquid water, as in river deltas, but a permafrost climate.

Mire sites

'Mire site . . . is a combination of those mire features which, at any particular location, are under influence of fairly homogeneous hydrological conditions' (Moen 2002). A site is thus an identifiable hydromorphological entity combining the mire features included in the area. This definition is consistent with the concept of ecosite used in Canada, that is, primarily a unit integrating a consistent set of environmental factors and plant community structure and composition. The site level is at a scale that one would sample in an ecological survey, with a general plot size of something like 5 × 5 m or 10 × 10 m (Harris *et al.* 1996). For example, in a raised bog such as in Fig. 10.7, one would sample as different mire sites the wooded bog margin, the open bog expanse or bog plain, and also the *lagg*, which is a narrow fen surrounding a bog, receiving water both from the bog and from the surrounding mineral soil.

A *soak* or *flush* is a sloping strip of fen with seepage of moving water, crossing bogs or separating bog areas (Gore 1983a). In sloping bogs it is common to have such soaks in which water flows from the mineral soil and becomes more and more diluted downstream towards the bog expanse so that the species indicating minerotrophic conditions are gradually lost (Fig. 10.11).

In Ireland soaks have been described as areas of minerotrophic vegetation occurring in ombrotrophic bog and usually associated with an internal drainage system (Connolly *et al.* 2002). They may be linear or round, and may consist of open pools, or be filled with floating mats, and are completely surrounded by bog vegetation. Some of them resemble boreal bog pools, but they contain some minerotrophic indicators, and are probably in connection with underlying minerotrophic peat. Birch woodland soaks also occur in the centres of raised bogs in Ireland, and possess minerotrophic indicators (Cross 1987, 2002). Soaks within raised bogs could represent mineral groundwater upwelling from below, as found by Sjörs (1963) in a very large bog in the Hudson Bay Lowland.

Fig. 10.11   Sloping bog with hummock-strings (left) and a fen soak (centre). Southern boreal zone, Sweden. Photo Hugo Sjörs.

Mire types

The mire type, or *synsite*, is the combination of mire sites that are usually found together. This is often the level that is used when classifying and mapping mires. For the purpose of mapping, different countries have definitions of types recognized at this level. In the Canadian system the term *form* applies at this level (National Wetlands Working Group 1997), and this is also termed landform (Glaser 1992b) and hydro-morphic type. In Russian literature, the term *massif* is used. There is no agreement on an internationally accepted classification of mire types. In the text we describe mire types under the headings of bogs, fens, and mixed mires, and present in tables two of the more complete classifications for the northern hemisphere, Sweden (Table 10.2) and Canada (National Wetlands Working Group 1997). We draw attention also to the excellent synthesis of peat landforms for northern Minnesota (Glaser 1992a).

Bogs

The bog shape is often rounded if the minerogenous system is topogenous or only very weakly soligenous, or elongated especially if the encompassing minerogenous system is distinctly soligenous. The bog body may also be elongated adjacent to an upland edge, or confined by a lake or river.

There are several kinds of patterns of tree cover on bogs, reflecting the depth to water table. They may be completely treeless, wooded on the margin and

Table 10.2 Wetland types recognized in the Swedish wetland inventory (Lonnstad and Löfroth 1994). In addition to the bogs, fens, and mixed mires the inventory covers shores, moist heath, wet meadows and swamp forest. If one wetland type dominates the object, this is a 'single wetland type'. The 'complex wetland types' are named after the dominant type(s). The classification stresses types that can be identified from aerial photographs

Series and classes of wetland	Wetland types
Single wetland types	
Bogs	
Ombrotrophic mires. The bogs can be open or forested; the later usually with pine (pine bogs).	Concentric raised bog
	Eccentric raised bog
	Plateau bog
	Weakly raised bog
	Unilaterally sloping bog
	Bowl-shaped bog
	Bog of north Swedish type (transitional to fen)
Fens	
Minerotrophic mires. The fens can be open, covered with shrubs, sparsely or densely-covered with trees; in the later case they are sometimes regarded as swamps	Level to weakly sloping fen
	Sloping fen (unpatterned)
	Flark fen
	Spring fen
Mixed mires	
At least some of the hummocks are ombrotrophic micro-bogs, and the type is a mix of both ombrotrophic and minerotrophic microtopographic phases	String mixed mire
	Mosaic mixed mire
	Island mixed mire
	Palsa mire
Shores and other wetlands	Wetland shore by sea
	Shore by watercourse
	Lake shore
	Periodically flooded land
	Wet meadow
	Wet heath
	Overgrown lake
	Water area covered with vegetation
	Swamp forest
Complex wetland types	
The wetland complexes are named after the type that constitutes 75% or more of the area. If one wetland does not constitute 75%, more than one type is used to determine complex denomination	Complex of bogs
	Complex of fens
	Complex of mixed mires (with string and mosaic mixed mires)
	Mire complex (can contain bogs, fens and mixed mires)
	Freshwater shore complex
	Marine wetland complex
	Shore complex (freshwater and marine)
	Wetland complex (shores and mires)

Fig. 10.12   Raised plateau bogs, round to ovoid, with high tree density on margins and medium tree density in centres, in unconfined, weakly drained soligenous fen matrix with ridges and flarks. In Hudson Bay Lowland, Ontario, mid boreal zone.

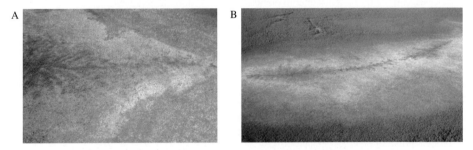

Fig. 10.13   Raised domed bogs with *Picea mariana* and dwarf shrubs in centre, and lighter *Carex oligosperma* sedge bog on margins, with distinct change to darker fen with *Sphagnum majus*. (A) Wheel spoke bog, with spoke-like wooded ridges, and wetter drains in between. (B) Crested feather bog, with elongated wooded crest and feather-vein wooded ridges coming off the crest; open wetter drains in between. Occurring in broad, shallow topogenous basins, or unconfined soligenous drainageways with weak directional flow. Ontario, southern boreal zone.

open in the centre (Fig. 10.7), wooded throughout (Fig. 10.12), or wooded in the centre and open on the margin (Fig. 10.13). In the case of the open centres, they can be *centre wet* (Fig. 10.7) dominated by *Sphagnum* lawns or *centre dry* with dwarf shrubs, *Cladonia*, and *Sphagnum* hummocks. In the case of wooded centres, the wooded part can be quite uniform, or patterned with radiating lines of trees like spokes in a wheel, hence *wheel spoke bog*

(Fig. 10.13A), or patterned with a linear crest of trees with lines coming off like veins of a feather, hence *feather bog* (Fig. 10.13B) (see also figures in Heinselman 1963; Glaser 1992b; National Wetlands Working Group 1997; Glaser *et al.* 2004b). These patterns, and the degree of tree cover, are related to climatic variation: more trees in continental and northern temperate climates, fewer trees in oceanic and northern boreal-subarctic climates. However, the tree cover is also influenced by other factors such as wildfire and wetting/drying cycles.

Raised bogs

Raised bogs are ombrotrophic mires that are noticeably raised above the level of the surrounding fens or swamps. They normally have an adjacent lagg fen. On the basis of their hydromorphology we can separate between domed bogs and plateau bog.

Domed bogs are usually at least 500 m in diameter, with a convex cupola that can be several metres higher than the edges of the bog and the surrounding peatlands. Drainage radiates outwards from the centre, the highest part of the bog. In concentric bog (more precisely, 'concentrically patterned domed bog'), strings and hollows form concentric patterns (Fig. 10.7), sometimes with pools. An eccentric bog is a sloping domed bog with the highest part of the bog off centre, and the larger part sloping off in one direction (Fig. 10.14), sometimes fan-like. It often has strings, hollows and pools across the direction of the water flow (National Wetlands Working Group 1997; Laine *et al.* 2004). The slope on these bogs is very subtle, of the order of 1–2°.

Plateau bogs are almost flat in the central bog expanse, and the bog slopes rather steeply down at the margins. In humid climates they can have concentric patterns of strings, hollows, and pools, but the hummocks in the central part are low and have no clear orientation, and the ridges are best developed toward the periphery. Very large bogs often tend to be plateau-like. Some are broadly ridge-shaped.

Another landform in regions with discontinuous permafrost is the *peat plateau*, a large elevated bog about 1 m above surrounding unfrozen fens (Fig. 10.12). These are thought to have developed under non-permafrost conditions, and subsequently became elevated and permanently frozen. *Polygonal peat plateaus* (Fig. 10.15) also rise about 1 m above a surrounding fen or adjoining upland. They have a polygonal pattern of trenches over ice wedges. The permafrost and ice wedges developed in peat originally deposited in a non-permafrost conditions. During summers there is superficial thawing, and plants can photosynthesize and produce organic matter. The raised permafrost features are usually covered with insulating vegetation, such as dwarf shrubs, mosses, and lichens, and are underlain by fibric peat that is unsaturated with water and hence also insulating. Where the

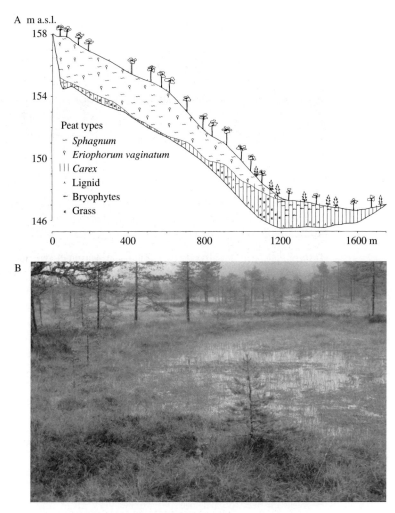

Fig. 10.14 The eccentric Lakkasuo bog, Finland, middle boreal zone. (A) Longitudinal section. From Laine *et al.* (2004). (B) Hollows and strings; the view is downslope.

mounds meet the lower areas, there can be thawing faces and collapsing vegetation and the fen tends to eat into the higher mounds, making *collapse scars* (Fig. 10.10) which are features within peat plateaus and alongside palsas. At first these scars are filled with water, but then they can become invaded by wet fen vegetation.

Non-raised bogs

Non-raised bogs are flat or sloping and hardly raised above adjacent fens, swamps, or mineral ground, but isolated at the surface from inflow of water. The distinction between raised and the non-raised bogs is not sharp, and Moen (2002) discusses the Atlantic raised bog (often without the typical

Fig. 10.15   Permafrost features: polygonal peat plateaus. Near Hudson Bay, Ontario, southern arctic zone.

lagg fen) as a transition between the raised bogs and the oceanic blanket bog. In some regions it is common to find level, only slightly raised bogs. They are often tree-covered and sometimes surrounded by a lagg fen, but in contrast to the truly raised bogs they do not have well-developed surface patterns of hummock and hollows.

Three types of flat, non-raised bogs that occur in boreal or temperate climates are:

- *flat bog* (plane bog) in shallow basins on rather flat terrain
- *basin bog* in well-defined basins such as ice block kettle-holes
- *riparian bog* which are flat mats floating or firmly anchored at the edges of ponds, lakes, streams, or rivers.

They generally have mosaic or diffuse surface patterns, and they may be treeless or tree-covered depending on region and latitude. Often the surface is close to the groundwater and hence the ombrotrophic peat is thin. In places there are a few fen species which can be rooted deeper down in the minerotrophic peat, making the area transitional between fen and bog.

There are also non-raised unilaterally sloping bogs, which are sometimes fan-shaped. Water from the surrounding terrain can enter the bog area in sloping fen soaks which become gradually diluted with bog water. The slope can be as weak as $<1°$ and sometimes up to about $10°$. The gradient is probably related to degree of effective water supply, generally the precipitation/evaporation ratio. They are closely related to the eccentric bogs mentioned earlier, and to the blanket bogs dealt with below.

Blanket bogs

In strongly humid oceanic climates the blanket bogs (Fig. 10.16) are typical and even landscape forming. Here the water surplus is so high that ombrotrophic vegetation covers much of the terrain, even low hills in the landscape. The terrain often has minerotrophic flushes or drainage ways running between bogs or around streams or lakes, and with thinner peat the blanket bog gradually changes into wet heath. Blanket bogs are found in northwesten Europe (British Isles, Norway, Iceland), along northern Pacific coasts (northern USA, Canada, Russia, Japan), the Canadian Atlantic coast, Tierra del Fuego, and New Zealand.

Fens

Classifications of fens used for mapping and inventories are based on the poor–rich gradient and the degree of woody cover, in combination with the distinctions we make here based on hydromorphology. The most important distinctions are between horizontal (topogenous) fens with stagnant or very slowly flowing water, and visibly sloping (soligenous) fens with seepage water, and between patterned and non-patterned fens.

Fig. 10.16   Blanket peatland. In lower left is fen vegetation with *Juncus* and *Carex* species, in mid and background is blanket peat, mostly bog, extending onto the top of the hills. The dark lines are peat cutting banks and piles of hand cut peat blocks. Co. Connemara, Ireland, oceanic temperate zone.

Topogenous fens

Horizontal fens occur in broad, poorly defined basins such as glacial lake beds, loess plains, and river floodplains with subtle gradation into adjacent upland, often through thicket or swamp forest. They are generally feature-less, with level carpet or lawn vegetation. Where surfaces are periodically flooded they may develop a mosaic structure of closely packed tussocks or narrow hummocks above the base level. The tussock-hummocks are built by certain species of *Carex* or by shrub stems which provide support for various bryophytes. The hollows are litter-covered or mud-bottoms, or occupied by some of the wetter occurring brown mosses or *Sphagnum* (depending on pH).

Several types of topogenous fens have been described, but the distinctions are often more based on their genesis than on the current hydromorphol-ogy. Basin fens have formed in distinct basins (see Fig. 10.19) such as ice block depressions called kettle-holes, in the swales between beach ridges on rising coastlines, or in embayments of seas or lakes which have become cut off by beach ridges. Channel fens occur in cut-off channels or oxbows of a river, or in old erosion channels of a glacial river, and are sometimes soligen-ous. They may be floating fens (*Schwingmoor*; see Fig. 10.5); shore fens or riparian fens which are on the banks of ponds, lakes, or slow-flowing rivers (see Fig. 10.2); and stream fens which are along streams. Flood mires or inundation fens can be distinguished as topogenous fen types, being peri-odically flooded by rivers or lakes. Finally, there are permafrost fen types comparable to the bog types: lowland polygon fens with ice wedges in poly-gon patterns, surrounding low, wetter centres; palsa fens; and collapse scar fens.

Soligenous fens

Where fens expand over a slope they may or may not have patterns. Smooth slope fens have no peat ridges or pools, but wet seepage tracks may be present. Patterned fens are also called flark fens, string fens, ribbed fens, ladder fens, and ripplemark fens by different workers. The degree of patterning can be striking, with combinations of different mire features as strings, lawns, flarks, and flark pools. Several kinds of patterning can be recognized, such as flark fens in which the strings are at lawn level (Fig. 10.9A, 10.17A), hummock level (Figs 10.9B), and hummock level with low-growing trees (Fig. 10.17B). Mostly only the central parts with the most distinct flow of water are patterned.

The *net fen* (*reticulate fen*) has a web-like pattern of low peat ridges separ-ating shallow ponds (mostly mud-bottoms), but no real flarks. The fen surface is almost flat.

A  B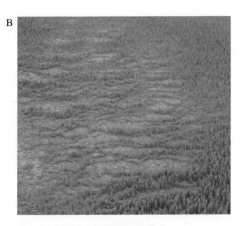

Fig. 10.17   Variations in patterned mires. (A) Lawn level strings separating wide flarks. Sweden, southern boreal zone. Photo H. Sjörs. (B) Hummock level strings (probably minerotrophic) with *Picea mariana* and *Larix laricina*, with hollows occupied by lawn level fen. Northeast Ontario, Canada, mid boreal zone.

Fig. 10.18   Steeply sloping rich fen with the brown moss *Scorpidium scorpioides*, *Carex flava*, *C. panicea*, *Bartsia alpina*, the orchid *Gymnadenia conopsea*, and the spikemoss *Selaginella selaginoides*. Central Norway, mid boreal zone.

In humid areas, also as one progresses from lower elevations to more humid ones in hills or mountains, one can find strongly sloping fens (Fig. 10.18). With very strong slopes the surface is usually smoothly unpatterned, but ridges with a few deep pools behind them may occur; in these situations high water can break through strings, and irregular channels may be more typical than flarks. The relative steepness of slopes of soligenous peatlands depends on the constancy of supply of water from upslope. However, many slopes even in humid climates are rather slight, normally less than 2°. As the climate becomes effectively wetter with increasing effectiveness of precipitation and P/E ratios, or as water supply from higher elevations increases, the degree of slope for soligenous peatlands can become much higher, for example, 15° or more in Scandinavian mountains and in coastal Alaska.

Spring fens are often used as a mapping category (mire type) when they occur in isolation. Otherwise they could be seen as a feature within, for instance, a sloping fen.

Mixed mires

A mixed mire is a mire type with bog and fen features or sites in close connection. Starting from a patterned fen, the strings may reach such a height that they develop into purely ombrotrophic vegetation with for instance *Sphagnum fuscum*. This is the *string mixed mire* (Fig. 10.19A), with elevated strings of bog vegetation separating minerotrophic lawns or flarks. The strings often develop stunted trees; *Picea mariana* in North America, and *Pinus sylvestris* in Europe. Another type is the *mosaic mixed mire* which has a more irregular close mixture (no linear patterns) of bog and fen sites (up to several hectares in size). The mosaic mixed mires are more typical for completely topogenous situations where there is no unidirectional movement of water. Areas with level fens in which 'miniature bogs' (bog mounds) are irregularly dispersed can be referred to as *island mixed mires* (Fig. 10.19B). The mounds, 1–3 m in diameter and up to 1 m in height, are isolated in the upper parts from mineral soil water influence. Palsa mires are technically mixed mires since the palsas with their ombrotrophic vegetation (but mostly originally fen peat) are surrounded by fen.

A

B

Fig. 10.19   Mixed mires in the mid boreal zone, Västerbotten, Sweden. (A) String mixed mire; strings with bog hummock vegetation separating fen lawns. (B) Island mixed mire; bog mounds (with a slight degree of string elongation) with fen lawns and carpets.

Mire complexes

A mire complex (Fig. 10.20) is an area consisting of several hydrologically connected but often very different mire types (i.e. several landforms; Glaser 1992b; Moen 2002). In the Russian literature, the term *system* is often used for this level. Sometimes mire complexes are clearly separated by mineral soil uplands, but in cases of regions dominated by peatlands, as in the Hudson Bay Lowland (Canada), the Glacial Lake Aggasiz peatlands (Minnesota), and the Siberian lowlands (Russia), vast expanses of continuous mire complexes may be recognized.

The Finnish term *aapa mire* originates from Cajander (1913), and was originally used to denote large patterned fens or ribbed fens. Following Ruuhijärvi (1960; 1983) it has taken on a wider meaning (Laitinen *et al.* 2005) to denote a mire complex dominated by wet flark fen in its centre (Pakarinen 1995). The complex may often contain *Carex* fens, wet forests, flat bogs, and string mixed mires. They are vast peatlands, typical of slightly sloping or concave areas of the boreal zone, where they may even dominate the landscape.

Glaser (1992b) interpreted 11 mire-complex types from the peatlands in northern Minnesota. Each complex type is a combination of several

A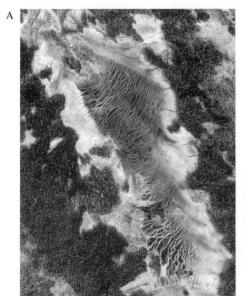

B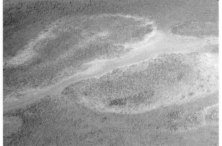

Fig. 10.20 Mire-complexes. (A) Weakly soligenous basin with patterned fen – parallel fen strings and flarks – and marginal featureless fen. Västerbotten, northern Sweden, mid boreal zone. By permission from Lantmäteriverket MEDGIV-2005-8934. (B) Mire complex. Large raised plateau bogs in wooded fens, with internal water track. Hudson Bay Lowland, Ontario, Canada, northern boreal zone.

kinds of peat landforms: raised bog with radiating forest patterns, ovoid bogs with diffuse or sharp margins, smooth (non-patterned) fen water tracks, patterned water tracks with strings and flarks, patterned water tracks with fields of tree islands, internal water tracks, and swamp forests.

When rivers meander, cutoff channels and oxbow lakes may become isolated from all but extreme floods. This may create a complex with a mosaic of peatlands. Along river, levees (banks) can build up, which when overflooded trap water behind on the floodplain, and promote peat accumulation by infilling, primary peat formation, or paludification (Fig. 10.21A). Along the active channels levees may build up, partially isolating the floodplain from the effects flooding, but also blocking the drainage of flood waters off the floodplain. On the poorly drained floodplains and inundation terraces marshes, fens, and swamps develop, such as for the callows along the river Shannon in Ireland (Heery 1993). On the parts of the

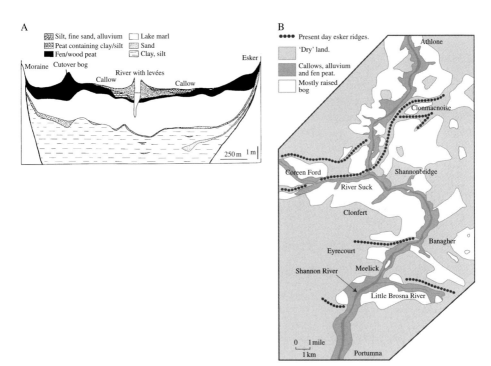

Fig. 10.21   Peatland complexes along the Shannon River floodplain and its tributaries in central Ireland. Oceanic temperate zone. (A) Cross section of the Little Brosna River showing the floodplain with fen peat, and the levee of the river composed of mineral deposition by river flooding. The rise of the surface on the floodplain to the left is the remnant of a raised bog that has been harvested. (B) Peatland complex with raised bogs between the Shannon River floodplains, the callows, and the uplands. Redrawn from Heery (1993, after Hooijer 1996).

floodplain back from the river and close to the upland margin where there is less influence of flooding, ombrotrophic raised bogs can develop (Fig. 10.21B).

Peatland regions

Peatland regions are a geographical concept, defined by similar climate, parent material, and physiography. Within a peatland region one finds a particular combination of peat landforms. Regional maps have been drawn up for several countries or areas. They have different scopes – for example in Finland (Fig. 10.22) and Norway (Moen 1999) the focus is on bogs and fens, and the result could be referred to as mire regions. The Canadian Wetland Regions include a broader scope of marshes, swamps (including non-peaty ones), fens, bogs, and lacustrine vegetation (National Wetlands Working Group 1997). Attempts have been made to depict peatland regions

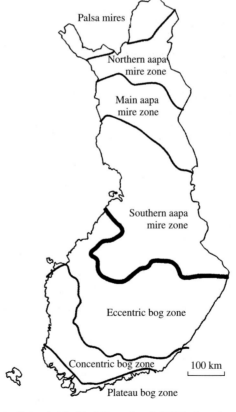

Fig. 10.22   Peatland regions in Finland. Modified from Seppä (1998, based on earlier work by Ruuhijärvi and Eurola).

in Europe (Jeschke *et al.* 2001), but this is fraught with difficulties (Gore 1983a). A regional division for the former Soviet Union was published by Botch and Masing (1983).

The formation of peatland patterns

Why do peatlands have such characteristic patterns? This question has baffled scientists for a long time, and there is probably no single mechanism that explains the many kinds of patterning. Thus, for example, bogs are patterned in the Baltic area, whereas patterned fens occur only in the northern parts and at higher elevation in southern Scandinavia. Some of the primary causes of the patterns are:

- lines of debris that catch on vegetation or small rises and expand across the direction of flow, similar to how litter collects on sloping paths (probably only of very local significance)
- in a weakly drained basin with high water in spring floods, there is a tearing and wrinkling of a mat caused by the shearing forces of lateral water flow, creating lawn level ridges and swales
- chance development of a hummock-developing moss species that is able to create a slight rise and develop into a hummock, which then expands crosswise to water flow.

There are then secondary mechanisms that serve to enhance the patterns, for example

- biotic processes, such as different rates of plant growth and decay in ridges and depressions (Sjörs 1990)
- higher temperature and oxygen access in depressions promote decay and accentuate their depth (Sjörs 1961)
- frost upheaval in hummocks.

The role of biotic processes is illustrated by the hummock – hollow differentiation observed on many bogs, with *Sphagnum* of the Acutifolia section often dominating the hummocks and species of the Cuspidata section growing further down. Several biological mechanisms help to explain this pattern (Rydin 1993b, see also Chapter 4):

- Hollow species decay more rapidly than hummock species.
- Hummock species have higher capillarity, and can therefore grow higher above the water table than the hollow species.
- Woody dwarf shrubs reinforce the hummocks and prevent compaction, for instance from snow (Malmer *et al.* 1994).

It may even be that the initiation of a hummock is promoted by *Sphagnum* mosses 'climbing' to overgrow shrub branches on the mire surface (Kenkel 1988). Thus, the vascular plants may act both as armouring and as scaffolding.

For the development of flarks and bog pools, we can envisage as a starting point a hummock – hollow microtopography dominated by different *Sphagnum* species. Patches of dead *Sphagnum* are often seen, and may be caused, for instance, by fungal attacks, trampling, animal faeces, or bird droppings, perhaps in combination with an expansion of liverworts and lichens. On hummocks, the necrotic patches are slowly revegetated, whereas in the hollows they may be flooded and colonized by an algal mat (Karofeld and Pajula 2003; Karofeld 2004). The algae decay easily and do not form peat, and in addition produce oxygen by photosynthesis. Lacking a moss cover, the surface can become well aerated, and the dark surface leads to higher temperature. All of these conditions promote peat oxidation. Exposed to desiccation, *Sphagnum* mosses are slow to colonize and regain such patches.

In regions with snow cover, spring flooding may hamper *Sphagnum* growth, and since cold water can dissolve much oxygen it can further promote surface decomposition (Sjörs 1990). So, once the peat growth is reduced in a patch the process could be self-perpetuating (Charman 1998), and the peat accumulation may be retarded so much that the patch deepens into a pool. Boatman *et al.* (1981) suggested that the wet hollows were unstable on the Silver Flowe (south-west Scotland) and that they could develop into pools if not invaded by *S. papillosum*. This is also supported by fine-scale peat growth models (Belyea and Clymo 2001).

In addition to biotic processes, hydrology, frost and ice activity, and gravity acting along the slope have been invoked to explain the patterning (see reviews by Seppälä and Koutaniemi 1985; Sjörs 1990; Glaser 1998). A basic, and important, observation is that the water table slopes in the strings, whereas the intervening flarks or hollows have a horizontal water surface. Since the hydraulic conductivity is lower under the hummocks, they tend to raise the water table just upstream, and this could be enough to reduce the probability for the initiation of a new hummock here (Swanson and Grigal 1988). Instead, new hummocks could form beside the existing ones, and these could also expand sideways. The intervening depressions could deepen by the above-mentioned processes hampering *Sphagnum* growth and peat accumulation. With time both the hummocks and hollows grow and coalesce sideways. Belyea and Lancaster (2002) showed that bog pools expanded more rapidly in length than in width, and many authors have shown that large flarks and pools with the same water level can coalesce along the topographic contours. Pools can also merge if the string between them erodes (Foster and Fritz 1987). The most well-developed string fens appear with only a slight slope. For obvious reasons it would be impossible for wide, horizontal flarks to be maintained on steep slopes, and here we find instead narrow flarks separated by strings in a staircase pattern (Sjörs 1990).

Pools are initiated over a long time during the development of a bog. On a raised bog in Sweden, Foster and Wright (1990) recorded that pools were younger towards the periphery, and that the marginal, younger, and more strongly sloping part of the bog had no pools. Once formed, the patterning persists for long time (see Chapter 7).

Early literature focused on the role of frost upheaval and lateral pressure from expanding ice in the flark, but these factors were probably overemphasized (Seppälä and Koutaniemi 1985). In a remarkable study over 21 years (Koutaniemi 1999) measured string movements of 2–5 cm yr^{-1} (up to several tens of centimetres for some years), both uphill and downhill. For downhill movement ice expansion and solifluction was more important than the water pressure, and an uphill trend was caused by freezing of the mire. The ice pressure may, however, increase the height of the strings, rendering them more ombrotrophic in flattish, somewhat continental areas.

Sjörs (1998) argued that bog pools occur in a wide variety of climatic situations, both oceanic and continental, in Scotland, central Scandinavia, southern and eastern Finland, Estonia, parts of Russia, some Pacific areas, northern Ontario, Labrador, Newfoundland, Tierra del Fuego, and the South Island of New Zealand. It is difficult to relate their distribution to some climatic variables, and since they were formed a long time ago, today's climate may not be the best predictor. We think of patterned mires mostly for the northern temperate and boreal zones, but Mark *et al.* (1995) give a general discussion on the occurrence of patterned mires in the southern hemisphere. It appears that the tendency to form patterns is stronger in bogs than in fens, and that greater water logging is needed in the latter for flark formation (Sjörs 1990). Flarks are more bound to cool climates, especially snow-rich ones, (Sjörs 1998), but are not entirely restricted to such conditions.

11 Peatlands around the world

The focus of this book is on the northern peatlands, reflecting the large areas they cover in Europe, Asia, and North America, and the great number of studies performed there. However, there are subtropical, tropical, and southern peatlands where the patterns and processes differ to some degree from those of the northern peatlands. To offer some insights into this, we here briefly describe the distribution of peatlands around the world, and enclose sections contributed by colleagues describing peatland types and processes in the tropics and the southern hemisphere.

Areas of peatland

It is not possible to provide accurate estimates of the worldwide areas of peatlands. The accuracy of the land or soil survey information from the many countries of the world is highly variable. Each country defines its own land and soil terms, and the information must be interpreted in terms of standard categories of peatlands. Each country is confronted with problems of limits, for example between peaty and non-peaty tropical wetlands, between swamp forests and moist upland forests, and between bog and wet heath or tundra. Subantarctic tussocky grassland forms organic soils which could be regarded as peat, but lack a uniform depth. Hence, the estimates will depend on definitions of peatlands, as well as on practical problems of delimiting these on aerial photographs and in ground inventories. The differences between peat underlain and non-peat underlain systems are difficult to judge, especially in the case of the continuum of wooded peatlands bordering wooded uplands where the dominant tree species is the same. Another problem is that shallow peatlands may subside to less than 30 cm depth after drainage, and thus be classified in a mineral soil category. For example, in Finland where extensive drainage has occurred, and where continuing forest inventories monitor the amounts of peatland and upland sites, there has been a decrease in the area of peatland and an increase in mineral soil sites that were previously peatlands.

Global peatland estimates have been steadily increasing over the last century as more and better inventory results have become available, especially in tropical countries. A very ambitious attempt to assemble data is the Global Peatland Database of the International Mire Conservation Group (IMCG). Because of the difficulties discussed above even these data are uncertain, but they are undoubtedly the best available. In the database 'peatland' refers to areas with at least 30 cm peat thickness. This means that many arctic and alpine mire areas with shallower peat layers are excluded, as are large areas with peat-forming vegetation along the edges of many mires. Given this approach, the total peatland area is now estimated to be 4.16×10^6 km^2, or about 3% of the globe's total land area (Fig. 11.1).

Peatland areas for the 44 countries with at least 2000 km^2 of peatland are listed in Table 11.1. Russia and Canada each have roughly a third of the global peatlands, and together with the USA and Indonesia they contain 85% of all peatlands. The countries in the list comprise > 99% of all peatlands. The table indicates that at least 80% of the peatlands are in areas with northern temperate or cold climates, 15–20% are tropical or subtropical, and only a few are in a southern temperate or cold climates. Large tropical peatland areas are found in the Amazon basin, in south-east Asia (Indonesia, Papua New Guinea, and Malaysia), and in the Congo River basin.

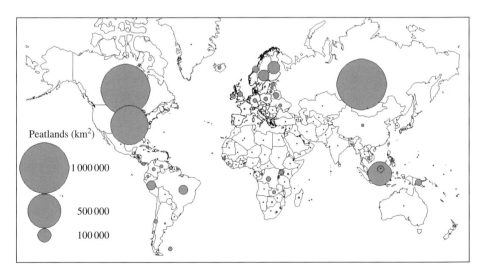

Fig. 11.1   Peatland area per country. The estimates are from the Global Peatland Database of the International Mire Conservation Group (IMCG) and refer to areas with peat depth >30 cm, as described in Table 11.1. Note that the circles in some large countries are not located at the actual largest peat areas, e.g. Alaska in the USA and Tierra del Fuego for Argentina.

Table 11.1 Countries with more than 2000 km^2 of peatlands. The estimates are from the Global Peatland Database of the International Mire Conservation Group (IMCG), and refers to areas with a minimum peat depth of 30 cm. Since limits other than 30 cm are used in national databases in some countries the IMCG recalculates the data to the 30 cm criterion. For detailed methods, see the IMCG website (*www.imcg.net*), where new and better estimates are continuously posted. Percentage peatland cover is given with decimals for countries with <0.5% peatland

Country	Peatland area (10^3 km^2)	Total area (10^3 km^2)	Peatland cover (%)
Russia	1390	17075	8
Canada	1235	9971	12
USA[a]	625	9629	6
Indonesia	270	1904	14
Finland	85	338	25
Sweden	66	450	15
Brazil	55	8547	1
Peru	50	1280	4
Papua New Guinea	29	463	6
Norway	28	386	7
Malaysia	25	330	8
Belarus	24	208	11
United Kingdom	18	244	7
Democratic Republic of the Congo	14	2345	1
Uganda	14	241	6
Germany	13	357	4
Poland	13	313	4
Falkland Islands[b]	12	12	95
Ireland	12	70	16
Chile	10	757	1
Colombia	10	1142	1
Estonia	10	45	22
Venezuela	10	912	1
Zambia	10	753	1
Guyana	8	215	4
Iceland	8	103	8
Ukraine	8	604	1
Panama	8	76	10
Cambodia	7	181	4
China[c]	7	9571	0.1
Latvia	7	64	10
Cuba	6	115	5
Ecuador	5	272	2
Honduras	5	112	4
Congo	4	342	1
Nicaragua	4	129	3
New Zealand	4	272	1
Lithuania	4	65	5
Antarctica	3	14200	0.02
Botswana	3	582	1
Argentina	2	2780	0.1
Netherlands	2	42	6
Japan	2	378	1
Mozambique	2	799	0.3

[a] Including Alaska, where estimates are uncertain.
[b] Tussocky grasslands included.
[c] Poorly known, and probably underestimated.

In terms of percentage peatland coverage, we note that the Falkland Islands are largely covered by peat, and very high values are noted for Finland (25%), followed by Estonia, Ireland, Sweden, Indonesia, and Canada (Table 11.1). The percentage is regionally very much higher, for example in Russia (northwestern Siberia), Canada (Hudson Bay Lowland), and the USA (Minnesota, Florida, and Alaska). If mires with thinner peat are included the percentage is higher, and 31% has been suggested for Finland in the leading position (Paavilainen and Päivänen 1995).

Peatland areas used for agriculture, forestry, and peat harvesting

In many European countries the original peatland coverage was considerably larger than that shown in Table 11.1. The IMCG database estimates the total losses of peatlands in Europe to be about 100 000 km^2, with the largest losses (in thousand km^2) in Russia (30), the Netherlands (13), Finland (11), Denmark (9), Poland (8), Belarus (6), Sweden (4), Germany (3), Ukraine (3), and Norway (2). Several European countries have lost more than 80% of their original peatland areas. Even among the areas that still are peatlands (with at least 30 cm peat), many are partly or completely damaged and have lost their peat-forming vegetation – that is, they are no longer mires. Joosten and Clarke (2002) estimate that Europe has lost over 300 000 km^2 (about 50%) of its original mire area.

Large peatland areas are drained or altered for agriculture and forestry, but the statistics are highly uncertain because some of the drainage was done in the distant past. Available inventory information (Okruszko 1996; Ilnicki 2003) shows that 14% of the total peatland area in Europe is used for agriculture, with the largest areas of peatland agriculture being in Russia, Germany, Belarus, Poland, and Ukraine. In the rest of the world, about 5% of the total original peatland area is used for agriculture, the largest areas being in Indonesia, the USA, West Malaysia, China, Canada, and Sarawak (Table 11.2). The most intensive conversions for agriculture by countries are indicated by areas of peatland used for agriculture as a percentage of total peatland area. The countries with the highest use are Hungary (98%), Greece (90%), Germany and Netherlands (85% each), Denmark, Poland, and Switzerland (70% each), and Ukraine (50%). For the rest of the world the countries with the highest conversion rates are Israel (100% of 60 km^2), Thailand (83% of 680 km^2), and West Malaysia (71%).

The greatest areas of drainage for forestry have been in Finland, Russia, and Sweden, and the most intensive use is in Latvia, Finland, Estonia, the UK, and Sweden (Table 11.3). In comparison, rather small areas have been drained for forestry in North America. Most recently, extensive drainage has been carried out in the tropical forested peatlands of south-east Asia.

Table 11.2 Peatland drained or altered for agriculture in Europe (Ilnicki 2003) and the rest of the world (Okruszko 1996). Data given for the countries with the leading areas of peatland for agriculture. These data are not fully compatible with Table 11.1 since they are not based on the same precise peatland definition

Europe	Peatlands used for agriculture (km²)	% of country's peatland	Rest of the world	Peatlands used for agriculture (km²)	% of country's peatland
Russia	70 400	12	Indonesia	37 200	18
Germany	12 000	85	USA	17 100	16
Belarus	9631	40	West Malaysia	5760	71
Poland	7620	70	China	1930	13
Ukraine	5000	50	Canada	1410	1
Sweden	3000	5	Sarawak	1370	12
Finland	2000	2			
Netherlands	2000	85			
Norway	1905	8			
Lithuania	1900	39			
Iceland	1300	13			
Estonia	1300	12			
Total[a]	125 000		Total[b]	65 000	

[a] Other European countries listed in order of decreasing agricultural peatland areas are: Denmark, Latvia, Hungary, Greece, Ireland, Great Britain, France, Switzerland, and Czech Republic plus Slovakia.

[b] Russia is included in Europe. Other countries listed in order of decreasing agricultural peatland areas are: New Zealand, Thailand, and Israel.

Table 11.3 Areas of peatland drained for peatland forestry purposes (Paavilainen and Päivänen 1995). These data are not fully compatible with Table 11.1 since they are not based on the same definition of peatland. For instance, for Lithuania the figure includes 4000 km² of forested wetlands, not all of which had 30 cm of peat to start with. Therefore it is risky to calculate the percentage of peatland area used from these numbers

Country	Area drained for forestry (km²)
Finland	59 000
Russia	38 000
Sweden	14 100
UK	6000
Lithuania	5900
Latvia	5000
Estonia	4600
Norway	4200
USA	4000
Belarus	2800
Ireland	2100
Poland	1200
Germany	1100
China	700
Canada	250

Overall, it has been estimated that the total losses of non-tropical peatlands have been 250 000 km^2 to agriculture, 150 000 km^2 to forestry, and 50 000 km^2 to peat extraction (Joosten and Clarke 2002).

A brief global survey

The peatlands of the world have been mapped in a general way by Gore (1983b) and Lappalainen (1996), but we still lack a modern world survey of peatland topography, hydrology, and vegetation with regard to landforms, climate, and other ecological factors. The development of peatlands can be explained in relation to major state factors – climate, relief, parent material, biota, and time (Gorham 1957). Of these, relief is fundamental because waterlogging is strongly correlated with flat to gently sloping ground. The other fundamental factor is climate – specifically, precipitation and temperature. Parent material exerts its influence via the richness of the mineral soil groundwater, and thereby governs the type of minerotrophic vegetation developed. Here we give a brief global overview, and we acknowledge the assistance from Hugo Sjörs in providing background for large parts of this section.

The northern peatlands range from north to south, from arctic and alpine tundra mires, to boreal forest (taiga) and temperate deciduous forest (nemoral) fens, bogs, swamps, and marshes, to marshes of semi-arid prairie and steppe zones.

Tundra mires are usually shallow-peated and, in continental areas, underlain by permafrost. These mires are widespread in arctic Canada and northernmost Alaska, as well as northernmost Siberia, and exhibit large polygonal patterns, with fens in lower, wetter places. Those of arctic European Russia mostly lack continuous permafrost, but discontinuous permafrost occurs, often as high palsas and peat plateaus, again alternating with fens (Botch and Masing 1983). These kinds of patterns also occur within the northernmost fringes of boreal forests in Fennoscandia, Russia, and Canada.

The typical boreal peatlands include fens, often with flark patterns, locally bogs, and frequently mixed mires and swamp forests. Boreal peatlands extend from Fennoscandia eastwards throughout northern European Russia, with extremely large expanses in western Siberia (Botch and Masing 1983). Areas of fens and poor swamp forests occur in the Sangjiang plains and Heilongjiang province, respectively, in north-east China, and there are high elevation subalpine peatlands in the high plateau of north-west China (Lu and Wang 1994). Boreal peatlands also occur in a belt from central Alaska and across most of Canada (National Wetlands Working Group 1997), with practically continuous cover in the lowlands adjacent to the Hudson and James Bays, where some are minerotrophic fens, others very large bogs with bog-pools, and yet others in the southern swamp forest, and all very wet. Patterned mires extend as far south as northern Minnesota (Wright *et al.* 1992), another peat-rich area, and also into northern Michigan (Heinselman 1965). In much of the boreal, except in very flat country, nearly all mires are sloping, most only slightly, but indicating origin by paludification. More locally, in hilly terrain, there are strongly sloping fens, and these are associated with run-off from upper slopes, groundwater discharge, and springs.

Maritime bogs occur especially in western Ireland and Scotland (Taylor 1983) – here there are also some patterned fens in the northeast – and in coastal British Columbia (Asada *et al.* 2003a,b) and southern Alaska, as well as in Japan (Iqbal *et al.* 2005). Large areas of bogs and patterned fens occur on the shores of the Gulf of St Lawrence and into Nova Scotia (Damman and Dowhan 1981), New Brunswick, and Labrador (Wells 1996). Central Newfoundland is also rich in large (but not continuous) peatlands with pools, both minerotrophic and ombrotrophic (Wells 1996).

Further south in Europe mires become small and scattered, but quite variable (Jeschke *et al.* 2001). In the Mediterranean countries mires are very rare. Further south in eastern North America shallow peatlands, mostly swamps and marshes, are found in the Appalachians and on the coastal piedmont. A peculiar type is the *poccasin*, a local term for a largely wooded or shrubby oligotrophic peatland. There are coastal cypress swamps along the south-east coast, around to Louisiana (Dennis 1988), and inland swamps and marshes such as Georgia's Okefenokee Swamp, Florida's Everglades, and Louisiana's Achafalaya swamp (Dennis 1988; Myers and Ewel 1990), which form organic-rich mineral soils and sometimes peat. In the semi-arid steppe and prairie zones, peatlands are confined as marshes to depressions with standing water all year, as the summer drought prevents peat formation elsewhere.

The subtropics are in large parts too dry for peat, but there may be peat-forming wetlands – *Papyrus* marshes in Africa, for example – mostly along rivers. Tropical peatlands occur mainly in rainforest areas and may be extensive, as in south-east Asia (see below), the Congo, and the Amazonas (Junk 1983). There may be evergreen tall forests on shallow or even deep peat, but also areas where enormous fluctuations in water level prevent peat formation. On the high elevation *tepui* plateaus in Venezuela (the summits inhabited by dinosaurs in Arthur Conan Doyle's famous novel *The Lost World*, 1912) one finds shrubby vegetation, completely different in physiognomy from northern mires, but forming true peatlands with peats 40 cm to over 1 m deep (Rull 2004). Tropical mangrove swamps are sometimes peat-forming (Chapter 10).

The wetlands of more southern latitudes are usually shallow and rarely qualify as peatlands. However, there are peatlands in Africa (Thompson and Hamilton 1983), for examples fens in the Okavango Delta in Botswana. There are sloping highland fens in Lesotho, even with flarks (Backéus 1989), but most mountain slopes are too steep to produce more than a surface of shallow peaty soil.

Mires, often even somewhat patterned, with remarkable vegetation features, are found in the very maritime situations of southern Chile, Tasmania, and New Zealand (see below). As described in a section below, there is also a larger peatland area in the eastern (Argentinian) flatlands of Tierra del Fuego with distinct patterning. A subantarctic peatland covers most of the Falkland Islands under 'tussac' (*Poa flabellata*), an extremely tussock-forming grass (Smith and Clymo 1984), and similar vegetation is reported from a few other subantarctic islands.

Peatlands in Tierra del Fuego

Contributed by Dmitri Mauquoy and Keith D. Bennett

In this short section we review the diverse range of peatlands occurring in Tierra del Fuego, describe their principal vegetation composition, and indicate their similarities to peatlands in Europe and North America. Deep peat deposits occur in the region (up to about 10.5 m depth) and the micro-, macro- and megafossils preserved in them offer valuable archives of environmental change on millennial timescales. This is reflected by the rapidly expanding palaeoecological research undertaken in the region, which forms the subject of the second section of this review.

Peatland types

A thorough classification and description of Fuegian bogs was made by Auer (1958), who travelled extensively in the region. Many of these bogs are found in valley bottoms and low-gradient slopes, and have developed over lacustrine sediments (Roig *et al.* 1996). A general review of Tierra del Fuego peatlands, which cover about 500 km^2 of Isla Grande, Tierra del Fuego and comprise about 95% of Argentina's peatlands, has been made by Rabassa *et al.* (1996). The vegetation of the entire region has been detailed by Moore (1983), and the peatlands mentioned in this, and those described by Auer (1958) and Pisano (1983) are summarized in Table 11.4, in addition to their principal vegetation and annual precipitation regimes. Broadly, rainfall is high on the western and southern sides of the mountains, reducing sharply towards the north-east. Raised bogs with *Sphagnum* occur within the predominantly deciduous woodland areas in the central part of this range. The wetter western and southern regions are characterized by local patchy woodland within the intervening landscape covered by blanket peat formed from cushion plants (see below), sedges, and rushes.

The microrelief elements (hummocks, lawns, and hollows) of the (raised) *Sphagnum* bogs resemble European and North American ombrotrophic raised peat bogs, although no *Eriophorum* species are present in Tierra del Fuego. The largest area of these ombrotrophic peat bogs coincides with the

Table 11.4 Tierra del Fuego peatlands (after Auer 1958; Moore 1983; Pisano 1983)

Peatland type	Occurrence	Precipitation (mm yr^{-1})
1 Steppe – region peatlands (geogenous), *Carex*, *Juncus*, and *Marsippospermum grandiflorum* in elevated depressions protected from wind. Fossil examples can be found under aeolian sand deposits	Northern Isla Grande, Tierra del Fuego	300–400
2 Raised *Sphagnum* bogs, dominated by *Sphagnum magellanicum* with *Empetrum rubrum*, *Carex banksii*, *C. curta*, *C. magellanica*, *Gunnera magellanica*, *Marsippospermum grandiflorum*, *Perezia lactucoides*, *Pernettya pumila*, and *Tetroncium magellanicum*	Within the Deciduous Forest zone (*Nothofagus pumilio*)	400–800
3 Raised *Marsippospermum* bogs, dominated by *Marsippospermum grandiflorum* with moss cushions (*Sphagnum*), *Carpha alpine*, *Cortaderia pilosa*, *Festuca contracta*, *Rostkovia magellanica*, and *Schoenus antarcticus*	Towards the northern limits of the Deciduous Forest – drier conditions	400–800
4 Raised *Sphagnum* bogs, dominated by *Sphagnum magellanicum* and similar associated vegetation as in bog type 2, but also *Donatia fascicularis*, *Oreobolus obtusangulus*, *Schoenus andinus*, and *Senecio trifurcatus*. High hummocks may be colonized by *Pilgerodendron uvifera*	Evergreen Forest zone (*Nothofagus betuloides*) to the south and west of Tierra del Fuego	800–850 (up to 4000)
5 *Bolax–Azorella* cushion peatlands (geogenous), represented by *Bolax gummifera* and *Azorella selago*. Hummocks and low mounds, more eutrophic than raised bogs	Submontane and montane areas	900–2000
6 *Donatia–Astelia* cushion peatlands (geogenous), with a dense, low covering of cushion plants *Astelia pumila*, *Bolax caespitosa*, *Caltha dioneifolia*, *Donatia fascicularis*, *Drapetes muscosus*, *Gaimardia australis* and *Phyllachne uliginosa*.	Magellanic Moorland zone–hyperhumid, seaward archipelago	2000–5000
7 Graminoid peatlands (geogenous), dominance of *Poa flabellata*, *Schoenus antarcticus*, *Tetroncium magellanicum*, and *Uncinia kingii*.	Magellanic Moorland zone	2000–5000

distribution of *Nothofagus pumilio* (lenga) deciduous forests (Roig *et al.* 1996). Where they have formed in moraine basins, the vegetation succession recorded in their stratigraphy shows a change from limnic mud and clay strata, then a *Carex* vegetation stage, before *Sphagnum* spread over their surfaces (Auer 1958). Two of these ombrotrophic raised bogs about 30 km from Ushuaia, Isla Grande, Tierra del Fuego, have been described by Mark *et al.* (1995). Their surfaces can be raised 4–6 m above the surrounding land surface, and the niche of *Empetrum rubrum* resembles its northern hemisphere equivalent, *Empetrum nigrum*, in that it is restricted to drier (hummock) microforms. Shrubs of *Nothofagus antarctica* (nirre) and the rush *Marsippospermum grandiflorum* characterize the crests of hummocks. There are pools 3–60 m long which are generally arranged in parallel crescentic lines and are similar to mires in southern New Zealand. Fringing these pools, *Sphagnum fimbriatum* was recorded. The range of pH values recorded from small pools in these ecosystems (pH 4.3–4.6) is consistent with those recorded from northern hemisphere *Sphagnum* raised bogs (Mataloni 1999). There is a lower diversity of *Sphagnum* species in Fuegian bogs compared to European and North American peat bogs, and their surfaces can be entirely covered by *Sphagnum magellanicum*, extending from high hummocks (which can exceed 1 m in height) to pool margins (Fig. 11.2). The wide range of water table depths *S. magellanicum* occupies in Tierra del Fuego bogs is similar to the broad habitat niche it displays in northern hemisphere peatlands.

Cushion bogs formed by *Astelia pumila* (Asteliaceae) and *Donatia fascicularis* (Donatiaceae) occur in the hyperhumid seaward archipelago (annual precipitation of 2000–5000 mm, and high wind velocities with a mean value of 12 m s^{-1}; Tuhkanen 1992). Both have elongated stiff leaves, and

Fig. 11.2   Raised *Sphagnum* bog, Valle de Andorra, Tierra del Fuego.

form tightly packed, dense hummocks that can occur contiguously over wide areas, together with rushes (e.g. *Marsippospermum grandiflorum*) and sedges (e.g. *Schoenus antarcticus*). They resemble some New Zealand cushion bogs (Godley 1960; Moore 1979) and are notable in that *Sphagnum* species are largely absent. Communities of *Donatia* occur under ombrotrophic conditions, whereas *Astelia* cushions benefit from a higher nutrient supply in groundwater-fed depressions, although they are not solely restricted to these sites (Ruthsatz and Villagran 1991).

Palaeoecological studies

Numerous palaeoecological studies have been made using peat bog archive records in Tierra del Fuego, and reconstructions of local (Mauquoy *et al.* 2004) and regional vegetation changes (Heusser 1989; Markgraf 1993), fire regimes (Huber *et al.* 2004), pollution loading (Biester *et al.* 2002), explosive volcanism (Kilian *et al.* 2003), and climate change spanning the entire Holocene Epoch (the last 11 500 years or so) have been made (Pendall *et al.* 2001). Due

Fig. 11.3 Raised *Sphagnum* bog stratigraphy exposed during commercial peat extraction, Valle de Andorra, Tierra del Fuego.

to excellent preservation of a diverse range of micro-, macro-, and megafossils and relatively high vertical accumulation rates of organic matter in Fuegian/Patagonian peat bogs (between about 0.5 and 0.7 mm yr^{-1}, Rabassa *et al.* 1989), detailed and high-resolution palaeoecological records have been produced. The peat bog stratigraphy visible in peat cuttings (Fig. 11.3) is strikingly similar to those visible in peat sections from north-west Europe. The layered, horizontal banding visible in the peat stratigraphy represents changes in preservation and the local peat-forming vegetation, which can be traced for perhaps tens of metres. Much of the palaeoecological literature available for the study region has been reviewed by Rabassa *et al.* (2000). A short selection of some of this research is presented in Table 11.5 as an illustration of some

Table 11.5 Selection of palaeoecological research undertaken using Tierra del Fuego peat deposits

Author	Analysis techniques	Results
Ashworth *et al.* (1991)	Beetle and pollen analyses	Neither the beetle nor the pollen data shows evidence for colder conditions during the Younger Dryas chronozone (c. 11 000–10 000 ^{14}C yr BP). From 5500–3000 ^{14}C yr BP the climate was drier (expansion of *Empetrum* heath and a reduction in mesic habitats)
Biester *et al.* (2002)	Geochemical analyses to determine pollution loading	High accumulation rates of mercury (Hg) in the Magellanic Moorlands have occurred in the last c. 100 years (^{210}Pb and ^{14}C chronologies)
Roig *et al.* (1996)	Tree-ring chronologies from subfossil *Nothofagus* wood	A fragmented floating tree-ring chronology was constructed, covering the last 1400 years. The oldest piece of wood was dated to 7700 ^{14}C yr BP
Heusser (1995)	Pollen, spore, and charcoal analyses	Core taken from a Magellanic Moorland, *Donatia-Astelia* cushion bog, c. 7000 year record, cushion bog established c. 2600 ^{14}C yr BP due to climate change (possibly colder, wetter and stormier)
Mauquoy *et al.* (2004)	Pollen, macrofossils, non-pollen microfossils, peat humification, and testate amoebae analyses	Evidence for warmer drier conditions during the Medieval Warm Period in a ^{14}C wiggle-match dated sequence
McCulloch and Davies (2001)	Pollen, diatoms, and humification analyses of lacustrine/peat bog deposits	Between c. 14 500 and 10 300 ^{14}C yr BP the central Magellan region was a cold/dry tundra environment. At c. 8550 ^{14}C yr BP an increase in available moisture led to the eastward spread of *Nothofagus* forests.
Pendall *et al.* (2001)	Deuterium/hydrogen ratios in *Sphagnum* moss cellulose and pollen analyses	Long record (basal age at 998 cm, 13 360 ^{14}C yr BP), the isotopic D/H ratios from the peat core are similar to changes in the Antarctic Taylor Dome ice core, and therefore circum-Antarctic temperature changes (possible changes in the extent of sea ice)

of the techniques that have been used in peat bog investigations and the principal conclusions derived from the data regarding environmental changes. The density of palaeoenvironmental sites does not, however, compare to northwest European and North American sites, which makes it difficult to identify the timing and nature of regional climate change in Tierra del Fuego and its possible relationship (if any) to reconstructed northern hemisphere climate changes (see Bennett *et al.* 2000 regarding the late-glacial–Holocene transition in the wider study area, and its relationship with the same event extensively recorded in the northern hemisphere).

Restiad bogs in New Zealand

Contributed by Beverley R. Clarkson and Bruce D. Clarkson

Bogs characterized by the family Restionaceae (restiad bogs) are most extensively developed on the temperate oceanic South Pacific islands of New Zealand. The Restionaceae, a predominantly southern hemisphere family and sister group to Poaceae (grasses), comprise jointed rush-like herbs with rigid or flexuose above-ground stems and reduced scale-like sheathing leaves growing on low fertility, seasonally moist or wet soils. In New Zealand, restiad bogs are most common on flat, poorly drained lowlands, but also occur in montane and subalpine settings. There are three main restiad bog species: *Empodisma minus* (North, South, and Stewart Islands), *Sporadanthus ferrugineus* (North Island), and *Sporadanthus traversii* (Chatham Island).

Empodisma minus is the least robust species, with intertwining wiry stems reaching about 50 cm tall in open sites but scrambling to 2 m or so where supported and shaded by shrubs. It is a vigorous peat former, producing a dense mass of upward growing fine roots and root hairs (cluster roots) at the bog surface, which form the bulk of the peat (Fig. 11.4). These cluster roots have similar water-holding properties to *Sphagnum* moss and are very efficient at absorbing nutrients from rainfall. *Empodisma* also has adaptations to conserve water, allowing bogs to form in areas of high seasonal water deficits where typically bogs do not occur. Strict stomatal control of plant transpiration, and a dense, mulch-like mattress of decay-resistant stems, contribute to exceptionally low daily evaporation rates (Thompson *et al.* 1999). *Empodisma* is tolerant of a wide environmental range (e.g. nutrients, pH, water), establishing early in minerotrophic wetlands to initiate restiad bog development, and persisting in significant amounts through to late ombrotrophic phases. It is killed by fire, but recruitment occurs relatively quickly from seed.

Sporadanthus ferrugineus and *S. traversii* are more robust and more erect species (de Lange *et al.* 1999). They are also important peat-formers; however,

Fig. 11.4   Cluster roots of *Empodisma minus* growing at the bog surface. Photo E.W.E. Butcher.

their cluster roots are not developed to the same extent as *Empodisma*, and remain below the bog surface. *Sporadanthus ferrugineus* is the taller, producing clumps of stout cane-like stems up to 2.8 m high. It is a late successional species, establishing after an initial *Empodisma* phase to overtop existing vegetation and ultimately become the physiognomic dominant. *Sporadanthus ferrugineus* is not fire-resistant, and it can take several years for *S. ferrugineus* cover to re-establish after fire. It is tolerant of very low nutrient conditions, having high nutrient use efficiency, low tissue nutrient concentrations, and other features typical of stress-tolerant species. *Sporadanthus traversii* has dense, somewhat flexuose stems normally reaching 1.5 m, but, if supported by other vegetation, it will attain heights of more than 2 m. Its ecological role in restiad bog development on Chatham Island is similar to *Empodisma* on mainland New Zealand, being the main peat former, having a dense decay-resistant litter layer, and re-establishing relatively rapidly after fire.

Sporadanthus ferrugineus/Empodisma minus bogs

Bogs co-dominated by *Sporadanthus ferrugineus* and *Empodisma minus* characterize the mild climates north of latitude 38° S on North Island (mean annual temperature 13–15 °C, mean annual rainfall 1100–1400 mm) (Clarkson *et al.* 2004b). *Sporadanthus ferrugineus* forms an upper tier, 2–2.8 m tall, above a dense understorey of sprawling *Empodisma minus* (Fig. 11.5). Heath shrubs are usually present, mainly *Leptospermum scoparium* and *Epacris pauciflora*. The fern *Gleichenia dicarpa* is often abundant, along with a limited number of sedges (e.g. *Baumea teretifolia*, *Schoenus*

Fig. 11.5   Clumps of flowering *Sporadanthus ferrugineus* overtopping *Empodisma minus* at Moanatuatua Bog, Waikato, North Island, New Zealand. Photo J. Greenwood.

brevifolius), ground-cover herbs (e.g. *Utricularia delicatula*), mosses (e.g. *Campylopus acuminatus* var. *kirkii*), and liverworts (e.g. *Riccardia crassa*, *Goebelobryum unguiculatum*). *Sphagnum cristatum* also occurs, but does not thrive in the shade of the much taller restiads and in drier places. Initiated in the postglacial period (after about 12 000 yr BP), these bogs typically formed extensive domes covering up to 15 000 ha, with peat 10–12 m deep.

Empodisma minus bogs

Empodisma minus is relatively common on mainland New Zealand between latitudes 35–47° S (it also occurs in eastern and southern Australia). Its stature becomes shorter and stems finer with increasing latitude and altitude; it is subdominant to *Sphagnum* (mainly *S. cristatum*, *S. australe*, and *S. novo-zelandicum*) in wetter areas. Strongholds for large blanket and raised bogs dominated by *Empodisma* are lowlands on the west coast of the South Island (Wardle 1977) and southern, cooler areas, including Stewart Island (Wilson 1987). *Empodisma* is also common in mountain bogs up to 1350 m altitude. Typically, dense springy carpets of *Empodisma* averaging 50 cm high cover large expanses, associated with heath shrubs (*Leptospermum scoparium, Dracophyllum* spp.), *Gleichenia dicarpa*, sedges (e.g. *Baumea teretifolia, B. tenax*), sundews (*Drosera* spp.), bladderworts (*Utricularia* spp.), and mosses (*Sphagnum cristatum, Dicranum robustum*). The peats were initiated in the postglacial period but are shallower (usually 1–5 m deep) than in most northern North Island lowland peatlands.

Sporadanthus traversii bogs

Sporadanthus traversii blanket and raised bogs occur on Chatham Island, a sparsely populated, isolated island 90 000 ha in area, lying 870 km east of South Island at latitude 44° S (Clarkson *et al.* 2004a). The island has comparatively low relief, with a cool and windy climate (mean annual temperature 11 °C), low rainfall (mean annual 700–1000 mm) and frequent cloud cover. Extensive areas are peat-covered, with 60% of soils being peat or derived from peat. *Sporadanthus traversii* dominates the raised bogs and many blanket peatlands (Fig. 11.6), in association with the heath shrub, *Dracophyllum scoparium*, and a shrub daisy, *Olearia semidentata*. The two mainland New Zealand restiad bog species are absent; however, several species in common include *Gleichenia dicarpa*, *Baumea tenax*, *Utricularia delicatula*, *Goebelobryum unguiculatum*, *Riccardia crassa*, *Sphagnum australe*, *S. falcatulum*, and *S. cristatum*. Compared with mainland New Zealand peats, Chatham Island peats are deeper (>10 m thick) and older, often dating back to the last previous interglacial period (40 000–30 000 years ago). They have lower pH and higher nutrient contents than northern New Zealand peats (Table 11.6), which may be partly due to the oceanic influence and a long history of seabird nutrient inputs.

Conservation

Restiad peat bogs throughout the lowland zone have been destroyed by widespread drainage and conversion mainly to pasture during the last

Fig. 11.6   *Sporadanthus traversii*-dominated restiad bog vegetation at Lake Rotokawau, Chatham Island. Photo Bruce Clarkson.

Table 11.6 Comparison of peat properties and species richness (per 4 m² quadrat) of *Sporadanthus traversii*-dominated Chatham Island restiad bogs with *Empodisma minus*- and *Sporadanthus ferrugineus*-dominated Waikato (North Island) restiad bogs. Means with standard deviations in parentheses are given for the vegetation types as defined by cluster analysis. (from B. R. Clarkson *et al.*, 2004, *New Zealand Journal of Botany*, **42**, 305, Table 2; reprinted with permission, and with corrected units for available P and total P).

Vegetation type	*Sporadanthus traversii*	*Sporadanthus ferrugineus*	*Empodisma minus*
Region	Chatham Island	Waikato	Waikato
Number of plots	18	9	22
Total K (mg cm^{-3})	0.083 (0.022)	0.013 (0.009)	0.027 (0.023)
pH	4.0 (0.1)	4.4 (0.2)	4.8 (0.4)
Available P	18.6 (10.7)	3.4 (3.2)	6.2 (6.1)
(μg cm^{-3})			
von Post	4.0 (0.5)	1.8 (0.4)	2.8 (1.4)
Bulk density (g cm^{-3})	0.101 (0.024)	0.059 (0.022)	0.065 (0.026)
Total N (mg cm^{-3})	1.20 (0.46)	0.53 (0.16)	0.78 (0.46)
Total P (mg cm^{-3})	0.057 (0.030)	0.019 (0.014)	0.035 (0.003)
Available N (μg cm^{-3})	17.2 (13.1)	8.9 (4.1)	19.0 (12.6)
Non-vascular species (number)	1.8 (1.7)	1.1 (1.7)	1.9 (2.0)
Vascular species (number)	7.6 (3.2)	4.8 (1.6)	6.2 (1.1)
Total species (number)	9.4 (3.0)	5.9 (3.1)	8.1 (2.8)

100 years. The *Sporadanthus ferrugineus/Empodisma minus* type, in particular, has been severely reduced in extent (from more than 100 000 ha to 3000 ha; de Lange *et al.* 1999) and is now largely confined to three peat domes in the Waikato region of North Island between 37° and 38° S latitude. Other threats to natural condition include introduced weeds and pests, domestic livestock, nutrient inputs, fire, peat mining, and blueberry farming. The restiad bogs provide habitat for several threatened species of flora (e.g. the swamp helmet orchid *Corybas carsei*) and fauna (e.g. species of the mudfish genus *Neochanna*), and efforts are now concentrated on protection, maintenance, and restoration. For example, three large Ramsar sites, Kopuatai peat dome (9000 ha raised bog) and Whangamarino wetland (7000 ha wetland complex) in Waikato, and Waituna wetland (soon to be expanded to 20 000 ha of estuary, lagoon, and blanket bog) in southern New Zealand, have statutory protection and contain representative examples of *Sporadanthus ferrugineus* (at Kopuatai) and *Empodisma* (at all sites) ecosystems. Smaller areas of *S. ferrugineus* and *Empodisma* on mainland New Zealand, and *S. traversii* on Chatham Island, are also protected. Maintaining small fragments such as Moanatuatua bog (Waikato) will require continuing active management (Clarkson *et al.* 1999). Some privately owned bogs are being mined for horticultural peat, and at one of the largest mines restoration trials for rehabilitation of *S. ferrugineus* bog vegetation

and invertebrates are currently proceeding with considerable success (Schipper *et al.* 2002).

Tropical peatlands in south-east Asia

Contributed by Aljosja Hooijer

Peatlands can develop where the rate of organic accumulation is higher than the rate of decomposition. Two factors limiting biological decomposition rate are permanently waterlogged soils and low temperatures – that is why peatlands are mainly found in cool climates. In the tropics, however, peatlands can also develop where high rainfall rates combine with poor drainage conditions – this is the case in large lowland areas in south-east Asia, where over 250 000 km^2 of forested peatland was found originally. Most of this, over 200 000 km^2, is located in Indonesia (Rieley *et al.* 1997). Tropical Asian peatlands (often called peat swamp forest) are located in coastal plains, often overlying marine clay deposits formed in mangrove environments, and in alluvial plains where drainage is impeded. This brief overview pertains to south-east Asian lowland peatlands.

South-east Asian peatlands compared to temperate peatlands

Besides their often coastal location, these peatlands differ from temperate bogs in many ways:

- They are covered by tropical rainforest rather than *Sphagnum* and herbaceous vegetation.
- The peat material is woody and only partly decomposed, and the peat surface is covered by a thick leaf litter layer. A clear distinction between a 'fresh' acrotelm layer overlying a decomposed catotelm layer is not as apparent as in temperate and boreal peatlands.
- Although these peatlands are dome-shaped, as in temperate regions, their gradient is often barely noticeable, with gradients often under 0.5 m per km and only above 1.5 m per km near river channels and coasts. This is because the woody peat material has a higher hydraulic conductivity than temperate catotelmic peat; higher rates of groundwater flow result in lower groundwater table gradients in periods of low rainfall, and the peat surface reflects this as it follows the water table.
- Although peat depths over 10 m are rare in temperate regions, they are common in south-east Asia – depths over 25 m have been reported. Peat formation conditions have been regarded as similar to those producing coal deposits millions of years ago (Staub and Esterle 1994).

• In tropical regions, peat formation continued during the last Ice Age – some peatlands in Borneo are reported to be over 40 000 years old (Anshari *et al.* 2004), but most of the lowland peatlands probably formed during the last 5000 years.

Vegetation and structure of south-east Asian peat swamps

Anderson (1964, 1983), working in forested and wooded peatlands in Sarawak, Brunei, Kalimantan, and Sumatra, distinguished a number of peatland types; the precise species composition is based mainly on findings in Sarawak and Brunei, but the structural characterization in zones holds true in other areas as well. The following is a simplification of the types, presented in sequence from coastal to inland, from highest to lowest trophic status, and shallowest to deepest peat (see also Bruenig 1990).

Mangrove woodlands

These occur in littoral zones of coastal salt water, developing over mineral alluvium in bays, on deltas, or on sheltered shores. With continued deposition of sediments, mangroves develop seawards, and the more inland mangroves are progressively replaced by peat swamp communities.

Swamp forests on margins of peatlands

Immediately inland from mangrove the forest is dominated by mixed swamp forest of *Campnosperma coriacea*, and the palms *Cyrtostachys lakka* and *Salacca conferta* are very abundant. Further inland, mixed swamp forests occur at the outer margins of larger peatlands, in zones near rivers and bordering uplands. The tallest trees may be 50–70 m or more in height. Some principal dominants in the Sarawak and Brunei mixed swamp forests are *Copaifera palustris*, *Dactylocladus stenostachys*, *Gonystylus bacanus*, and four species of *Shorea* (Anderson 1983). One type is characterized by *Shorea albida*, occurring as scattered very large trees that may achieve girths of 8 m and heights of 60 m. Shrubs are few, consisting mainly of *Pandanus* spp. and a few palms, of which the spiny *Salacca conferta* may form dense, almost impenetrable thickets. There may be abundant herbs (especially aroids), ferns, or sedges in some open places. Epiphytes and climbers (both large lianas and small root climbers) tend to be abundant.

Padang, ombrogenous bog with good tree development

Padang is a Malaysian/Indonesian term for forests composed of trees with a pole-like aspect, with small dense, even-sized crowns. The even upper canopy varies in height between about 45 and 60 m, and trees rarely exceed 180 cm in girth. One type is completely dominated by *Shorea albida*, whereas another has other dominants or mixtures of dominants such as *Calophyllum obliquinervum*, *Cratoxylum glaucum*, *Combretocarpus rotundatus*, and *Litsea crassifolia* from Sarawak and Brunei (Anderson 1983). The middle storey is sparse and the lower storey moderately dense. Herbs, terrestrial ferns, climbers, and epiphytes are rare or absent and are replaced by small, often prostrate shrubs, species of *Euthemis* and *Ficus*. Pitcher plants (*Nepenthes* spp.) are frequent.

Padang, ombrogenous bog with poor tree development

These bogs occur in the most interior locations in large peatlands. They are ombrogenous bogs consisting of extremely dense bog forests of small trees, or open bog woodlands, occurring in the centres of large raised peat domes. One type has closed canopy, 15–20 m high, and a few taller emergents (Fig. 11.7). Principal species are *Palaquium cochleariifolium*, *Parastemon spicatum*, and *Tristania obovata*. Another type is wooded to open bog, where the plants exhibit a high degree of stunting and xeromorphism. Only one species, *Combretocarpus rotundatus*, attains 12 m. A patchy layer up to 3 m in height is composed of small stunted trees. In permanently inundated places a green algal scum is found, and where the surface is periodically inundated *Sphagnum junghuhnianum* is abundant. The sedge *Thorachostachyum bancanum* and pitcher plants (*Nempenthes* spp.) are widespread.

Several changes accompany the sequence from the mixed swamp forests on margins of peatlands to the interior bog (Anderson 1983): (1) decreasing luxuriance of growth in the understorey; (2) decreasing species richness and an almost complete change in the flora; (3) reduction in the number of tree species from 29–62 species per 0.2 ha plot in the margin, to 22–47 in padang with good tree growth, to less than 5 in the central padang; (4) a change in the density of tree-size stems from 600–700 ha^{-1} in the margin, to 650–800 in padang with good growth, to 1200–1350 in poor growth padang with dense trees, to very few stems in the central padang; and (5) a general decrease in the average size of species.

The size structure of these forests shows the highest numbers in the lowest girth class. Hence, there is a naturally regenerating understorey providing for continuous stand regeneration below the canopy. This indicates that continuous cover forestry could be practiced by selective cutting of the largest individuals, with continuous regeneration from below. However, the predominant type of forestry is clear-cut and plant, using several non-indigenous species such as *Acacia mangium* and *A. crassicarpa*, used because of their nitrogen-fixing capabilities, fast growth, and conventional use in tropical regions. Conversion to oil palm plantations is also widespread.

The forested peatlands probably show the same conservation characteristic for nutrients as other tropical rainforests. The nutrients are tightly held and recycled in the surface biomass, and hence are maintained in the forest ecosystem for long periods as the marginal swamp develops progressively into padang types. The total mineral content, particularly that of phosphorus and potassium, has been shown to decrease significantly towards the

(a)

(b)

Fig. 11.7   Tropical peat swamps: (A) Aerial photo of an open 'mixed swamp forest' with tall emergent trees over a lower height canopy taken from Sebangau peat swamp forest area, Central Kalimantan. Photo Viktor Boehm. (B) Ground shot of a cut edge of 'padang', or ombrogenous forested bog, with uniform tree heights of up to around 20 m and diameters around 12 cm, Sumatra. Photo Aljosja Hooijer. These types of forests are common in Indonesia and Malaysia on vast peat domes that can be 10s of kilometres across. Such forests are rapidly being drained, clearcut, and converted to *Acacia* plantations.

centres of the peatlands. It is difficult to say where the limit between minerotrophic and ombrotrophic occurs, or even if it is valid or useful to attempt to define sensitive indicators of minerotrophy as in boreal and temperate peatlands.

The future of tropical peatlands

In south-east Asian peatlands, abiotic and biotic factors are completely interdependent. The morphology, soil, and hydrology are shaped by the vegetation through accumulation of organic material, and this vegetation in turn can only exist in the environment it has created. Therefore, a change in vegetation cover will have an effect on the morphology and hydrology of the swamps, while at the same time any change in hydrology will change the vegetation (Hooijer 2005). Moreover, these peatlands are mostly located in regions with impoverished and rapidly expanding human populations. Finally, peatlands occur in a region where logging is a major economic activity, and forest resources are rapidly depleted. The combination of these three factors renders these peatlands extremely vulnerable, in the following ways:

- Peatlands are now seen as a last timber resource rather than impenetrable wastelands, and almost all south-east Asian peatlands are a target of widespread logging (legal and illegal); valuable timber species such as ramin (*Gonystylus bancanus*), meranti (*Shorea parnifolia*), and alan (*Shorea albida*) are commercially most interesting, but the remaining trees are increasingly used for paper pulp production.
- Canals and drains for log transport are easily created in peat soils, instantly altering the hydrology of large areas with immediate effects on the soils (through decomposition and subsidence) and vegetation (forest die-off is common). Peat fires can now easily spread as a result of this.
- Clear-cutting of tropical forested peatlands threatens to reduce diversity in the short term, and peatland drainage and conversion to plantations is making this process irreversible in the long term. To make matters worse, conversion to plantations often fails because of inadequate knowledge of peatland hydrology, and does not result in livelihood improvements, further increasing the pressure on remaining peatlands.

At present, the threats to south-east Asian peatlands are such that an almost complete loss of this habitat in the near future – except for a few protected remnants–appears a real possibility. Although there are many different estimates of the area lost to date (depending on definitions and monitoring techniques), it may be assumed that less than half of the original cover survives in a more or less intact state. In the dry El Niño year of 1997 alone, over 24 400 km^2 of peatland was lost to fires in Indonesia alone (Page *et al.* 2002); the smoke from these fires created haze and health problems throughout south-east Asia. Overall, loss of peatland habitat in central Kalimantan is found to be 3.2% per year (Boehm and Siegert 2002); loss rates in Sumatra are even higher. Conservation measures to date have had little effect, owing to rampant logging, continuing fires, and conversion to agriculture. The future of many unique peatland species, and other species which have found a last refuge in the peatlands (such as the orang-utan) is at stake. Another argument for conservation of these peatlands is the very significant contribution their loss makes to increased atmospheric levels of carbon dioxide, and hence to global warming; the output in the 1997 fires in Indonesia alone was estimated between 0.81 and 2.57 Gt, or 13–40% of global output from fossil fuel burning (Page *et al.* 2002).

12 Productivity and carbon balance

It has been estimated that around a third of the carbon (C) in the world's soils is stored in peatlands, an amount that is more than half the current atmospheric stock of carbon dioxide (CO_2). This contrasts with the approximate 3% of the world's land area that is covered by peatlands (Chapter 11). Because peatlands are of major importance for global C circulation, their C balance – the allocation and movements of C between gases in the atmosphere, living biomass, soil, water, and peat – is a critical current issue in peatland and global ecology.

Large-scale environmental changes – drainage, peatland forest harvesting, peat harvesting, global warming, and airborne nitrogen (N) pollution – have triggered questions about the role of peatlands: How do draining, forest removal, peat removal, and afforestation affect the carbon balance? Will global warming lead to increased decomposition and convert the peatlands from peat-forming ecosystems to enormous carbon sources? What happens with peat accumulation when the peatlands are fertilized with N through air pollution? Considerable progress has been made recently in understanding carbon cycling processes (Blodau 2002; Vasander and Kettunen 2006). In this chapter we introduce the basics of C cycling in peatland, and briefly discuss the role of peatlands in current environmental scenarios.

Biomass and productivity

Strictly speaking, biomass is defined as the total mass of living material within a certain area. For plants this is a definition that is sometimes difficult to uphold, since the border between living and dead parts is often blurred. Apparently dead rhizomes can re-sprout from dormant buds when exposed to the right conditions. In *Sphagnum* and other mosses it is practically impossible to draw the line between dead and alive. For practical reasons some authors have used the green part of *Sphagnum* as a measure of living material, but this is an uncertain assumption. When it is possible to distinguish between live and dead, the terms *standing biomass* (*standing crop*) and *necromass* (or *litter*) are used. However, when the distinction is not clear, biomass in many studies includes dead parts still attached to the plants.

Productivity is the rate of formation of new biomass, and most interesting is the biomass produced by plants through photosynthesis, the *primary productivity*. Since the plants themselves use up some of the assimilated carbon, we define net primary productivity as the difference between the rates of photosynthesis and autotrophic respiration. Net primary productivity can be measured for short periods as the difference between uptake and release of carbon dioxide (CO_2) in plants enclosed in a chamber (transparent for photosynthesis and black for respiration), but often it is measured as the change in biomass over a time period, plus the losses due to abscission, mortality, and herbivory (Calow 1998). Adding to the difficulties in defining biomass is the fact that these losses are often extremely difficult to measure – some plant material is always lost without trace between the biomass samplings. To standardize, plant material is always dried before weighing, and depending on the scale of interest biomass is expressed as $g\ m^{-2}$ or $kg\ ha^{-1}$, and productivity as $g\ m^{-2}\ yr^{-1}$ or $kg\ ha^{-1}\ yr^{-1}$. To relate biomass and productivity to C cycling, the rough conversion is 0.5 g carbon in 1 g dry biomass (Calow 1998).

The total plant biomass is strongly correlated with the presence and abundance of trees, but a high biomass does not necessarily indicate high productivity. In woody plants a large part of the biomass is structural (stems, branches) as opposite to functional (green photosynthetic biomass). Productivity is controlled by water level, water flow rate (which affects the oxygen content and thus the growth form and species composition), and nutrient level. In very broad terms productivity in peatland ecosystems ranks as marsh > swamp > fen > bog, but there are many reversals of this order.

Above-ground productivity

The combined productivity of mosses, herbs, graminoids, and shrubs in fens and bogs is often in the range of 200–400 $g\ m^{-2}\ yr^{-1}$ (Table 12.1, Fig. 12.1). The relative contribution of the layers varies tremendously among peatland types. Shrubs often dominate the productivity in bogs, and herbs and graminoids in fens. Bryophyte productivity can be substantial in both bogs and fens (Table 12.1). Tree productivity varies from nil in open peatlands, increasing through the wooded fens and bogs, and peaking in swamp forests.

Vascular plants

Vascular plant productivity is low on open bogs and poor fens (especially where the shrub layer is sparse), and tends to increase with increasing pH

Table 12.1 Above-ground biomass and productivity in bogs and fens. Based on data compiled from the literature by Moore *et al.* (2002). Mean values with range in parenthesis and sample size in italics

Component	Biomass (g m^{-2})		Productivity (g m^{-2} yr^{-1})	
	Bogs	Poor fens	Bogs	Poor fens
Shrubs	478 (80–1020) *16*	359 (21–1615) *19*	180 (43–338) *8*	50 (7–157) *8*
Herbs + graminoids	40 (0.1–130) *14*	193 (52–164) *14*	16 (3–34) *6*	244 (52–965) *13*
Bryophytes	NA	NA	188 (17–380) *8*	122 (27–287) *14*

NA = data not available.

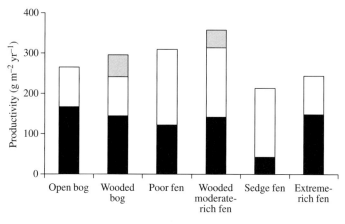

Fig. 12.1 Above-ground productivity of mosses (black bars), trees (grey), and field layer vascular plants (open) in fens and bogs in Alberta, Canada. Herbs and graminoids dominated the productivity in the sedge fen and the extreme-rich fen, whereas shrubs dominated in the bogs. In the wooded fen, herbs + graminoids and shrubs contributed roughly equally. Redrawn from Szumigalski, A. R. and Bayley, S. E. (1996). Net above-ground primary production along a bog-rich fen gradient in central Alberta, Canada. *Wetlands*, 16, 467–76, with permission.

and nutrients, but it can again be quite low in rich fens (see Fig. 12.1) where growth is limited by unavailability of phosphorus (P).

Productivity in the trees is modest in wooded bogs and wooded fens (about 40–50 g m^{-2} yr^{-1} in Fig. 12.1), but can be substantial in swamp forests. As the tree biomass (cover) increases, the understorey contribution to total biomass decreases. According to Reinikainen *et al.* (1984), tree and shrub productivity can be almost 500 g m^{-2} yr^{-1} in mesotrophic spruce swamps and up to 1300 g m^{-2} yr^{-1} in eutrophic ones (data from the southern boreal zone in Finland).

Tree productivity is of obvious commercial interest and is expressed as volume of stemwood rather than biomass. Sites with <1 m^3 ha^{-1} yr^{-1} are of little interest for forestry, and are in some countries considered as 'impediment' rather than forest. This limit corresponds roughly to $400 \, \text{kg ha}^{-1} \, \text{yr}^{-1}$ dry mass of tree stems, but the total tree productivity is several times higher when bark, branches, leaves, and roots are included. Productivity of stemwood is extremely low in sites with stagnant water close to the surface, but in drained eutrophic swamp forests it can be comparable to highly productive mineral soils.

Bryophytes

Because of its importance in northern peatlands, particular attention has been paid to the productivity of *Sphagnum*. To measure it, the most commonly applied method is to insert a vertical, thin stainless wire and then measure the vertical shoot elongation along this wire. This is referred to as the *cranked wire method*, since in the original design the wire was bent to form a turning handle to attach it in the peat some centimetres below the surface (Clymo 1970). In a modification, a small brush is attached to the bottom end of the wire (constructed by B. Wallén, see Gunnarsson and Rydin 2000 for an application). To convert to biomass at the end of the study, a sample of the *Sphagnum* carpet is taken adjacent to the wire so that the mass per unit length can be measured in an appropriate section (often 2–5 cm below surface). Because of the large variation, several wires are needed even in seemingly homogeneous moss carpet.

The highest productivity ($>400 \, \text{g m}^{-2} \, \text{yr}^{-1}$) is reached in wet oligotrophic carpets with full cover of species from section Cuspidata (*S. cuspidatum, S. majus, S. fallax,* or *S. riparium*). In a review of the literature, Moore (1989) found that lawns and carpets had on average about $100 \, \text{g m}^{-2} \, \text{yr}^{-1}$ higher productivity than hummocks. There was, however, a large scatter in the data, some of which could be attributed to climate: productivity decreases with northern latitude, but a stronger effect is that it increases with oceanity (Gunnarsson 2005). Weltzin *et al.* (2001) controlled water table in monoliths collected from a bog in Minnesota and divided the microtopography into high, medium, and low. The productivity of *Sphagnum* (*S. fuscum, S. capillifolium,* and *S. magellanicum* combined) at the high, medium, and low levels was 162, 236, and $311 \, \text{g m}^{-2} \, \text{yr}^{-1}$, respectively.

A main determinant of *Sphagnum* productivity is water availability, which leads to large variation among years with different rainfalls. In concert with their ability to maintain high water content in the capitula (Chapter 4), hummock species show less variability than hollow species (Table 12.2). Temperature also matters. When *Sphagnum* productivity is followed over several years with similar precipitation, there can be a rather strong correlation between mean annual temperature and productivity (Thormann *et al.* 1997), and over the geographic range of northern peatlands there is a positive relationship between *Sphagnum* productivity and the mean annual temperature (Moore 1989).

Table 12.2 Annual growth and production of two *Sphagnum* species in subarctic Alaska (Luken 1985) and subarctic eastern Canada (Moore 1989) in a wet and a dry year. The hummock former *S. fuscum* is less affected by dry weather than the lawn species *S. angustifolium*

| | Annual growth | | Wet/dry quotient |
	Wet year	Dry year Unit	
Bog forest (Luken 1985)			
S. fuscum	4.3	2.4 mm yr^{-1}	1.8
S. angustifolium	11.3	1.3	8.7
Andromeda bog (Luken 1985)			
S. fuscum	8.2	4.7 mm yr^{-1}	1.7
S. angustifolium	22.3	9.1	2.5
Poor fens (Moore 1989)			
S. fuscum	75.3	83.5 g m^{-2} yr^{-1}	0.9
S. angustifolium	127.4	29.2	4.4

Below-ground productivity

Many studies only report above-ground biomass and productivity, but a substantial part of the productivity of vascular plants is below ground in roots and rhizomes where special techniques are required to measure biomass and losses. Harvesting techniques are extremely laborious, because of the difficulties in separating fine roots from the peat matrix, and because fine roots are continually growing and dying, so there is a great deal of turnover during a growing season. Harvesting should therefore be repeated several times over the growing season for a reliable result, but even then spatial variability makes the estimates rather uncertain (Finér and Laine 1998).

One way to measure below-ground productivity is the *ingrowth core method*. Holes are cored in the peat, mesh bags with peat from which all roots have been removed are inserted, and the ingrowth of new roots is measured. With this method Backéus (1990a) found that fine roots constituted up to 60% of the total productivity of vascular plants in a bog vegetation dominated by dwarf shrubs, *Eriophorum vaginatum*, *Rubus chamaemorus* and *Scirpus cespitosus*. The objections to this method are that roots are cut off and the peat medium disturbed, and that the inserted peat probably has different properties from the matrix (Wallén 1992). Another method is to allow the plants to take up ^{14}C-labelled CO_2 (inside a transparent cuvette) and after some days harvest both above and below ground parts. The incorporation of ^{14}C below ground can be used to calculate fine-root biomass (Wallén 1986). With some modifications Saarinen (1996) used this method to estimate below-ground productivity in a fen in Finland and found that in *Carex rostrata* 88% of the productivity was below ground (rhizomes, 8%; coarse roots, 6%; fine roots, 74%).

To further illustrate the importance of fine roots (diameter <0.5 mm), the biomass in a subarctic peatland was divided into the three components: above-ground, rhizome and coarse roots, and fine roots. The values (g m^{-2}) were for *Andromeda polifolia* 17, 150, and 1100; for *Empetrum hermaphroditum* 80, 213, and 600; and for *Rubus chamaemorus* 31, 220, and 350 (Wallén 1986). Biomass data for *Carex lasiocarpa* and *C. limosa* in a fen showed only about 3% above ground, and thus an even greater contrast (Sjörs 1991).

Decomposition

Impaired decomposition following from waterlogging is the primary factor maintaining the net gain of carbon fixed in the peatlands. Decomposition is the breakdown of organic matter into inorganic substances like CO_2, methane (CH_4), and ammonia (NH_3), primarily by bacteria and fungi, but also by soil invertebrates. Decomposition can be measured with several techniques. Johnson and Damman (1993) reviewed the methods that may be used to measure the rate of decomposition. The most common one is the *litter bag method*. Litter bags are porous small containers (like tea bags, but made of material that does not decompose). Material with 0.2–1.0 mm holes have been used to allow soil fauna to enter and yet prevent litter from escaping. Various litter types are placed in the bags, which are then buried in the peat, and the mass loss is measured when they are retrieved after one or several seasons.

Another method is the respirometer technique – here the pre-weighed litter or peat samples are kept in small flasks in the laboratory under strictly controlled temperature, and the rates of CO_2 evolution are measured. This gives a measure of the decomposability of the litter or peat. One can also measure the evolution of CO_2 after addition of an easily degradable substrate such as glucose. When the decomposers are no longer limited by easily decomposed carbon sources, the rate of CO_2 evolution is an indirect measure of the microbial biomass (Anderson and Domsch 1978), which of course is the primary cause of the decomposition.

In general terms the rate of decomposition is determined by aeration status (redox), quality of the substrate, and by several key abiotic factors, especially pH and temperature. These are briefly discussed in the following sections.

Aerobic versus anaerobic decomposition

The key abiotic factor influencing rate of decomposition is the position of the water table (Fig. 12.2) and its influence on the type of microbial respiration – aerobic versus anaerobic. Redox potential is standard measure of

Fig. 12.2   Decomposition of a fence pole in a bog. In the lowest part the wood is fresh, with intact year rings. This part has been in the catotelm, the constantly anoxic layer. The dark zone higher up indicates the depth of the acrotelm. The pole has broken just above the mire surface, where it has been well aerated and kept moist from the mosses below. The fence itself had disappeared several decades before. Photo K. Mälson.

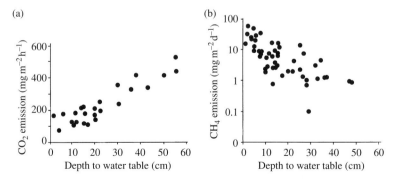

Fig. 12.3   With decreasing water level, the emission of CO_2 increases and that of CH_4 decreases. The thicker aerated zone leads to increased aerobic respiration and higher rate of oxidation of CH_4 to CO_2. Note the logarithmic scale for methane. From Laine and Vasander (1996).

the relative oxidation–reduction condition along the aerobic–anaerobic gradient (Chapter 5). In a pristine mire, there is decreasing redox potential from the surface to below the water table. The redox profile is not static, but changes because of vertical changes in the water table. Two main gases are emitted from peatlands, CO_2 and CH_4. They show an inverse relationship: emission of CO_2 increases as the water level drops, while that of CH_4 decreases (Fig. 12.3). The mechanisms for this are discussed below.

Substrate quality

Only a fraction of the microbial biomass is active, with the main restriction being shortage of easily decomposed organic substrate. The limitation by substrate is shown by the fact that when glucose or starch is added to peat under oxic conditions, there is rapid increase of decomposition as indicated by increased production of CO_2 (e.g. Bergman *et al.* 1998).

Substrate quality refers to the ease with which organic material can be decomposed by microorganisms. Organic compounds can be ranked in order of increasing resistance to decomposition: (1) sugars, starches, and amino acids; (2) carbohydrate polymers, amorphous cellulose, and hemicellulose, which make up the main part of plant biomass; (3) crystalline cellulose, aromatic polymers such as lignin and tannin; and (4) lipids. The specific kinds and proportions of these organic compounds are of course determined by the botanical make-up of the peat, namely, *Sphagnum* peat, sedge peat, woody peat, and their various subtypes (Bohlin *et al.* 1989). In general, sedge peats decompose more rapidly than *Sphagnum* (Verhoeven and Toth 1995). *Sphagnum* peat is quite resistant to decay because of its peculiar biochemistry (see Chapter 4). *Sphagnum* litter is of low quality for soil invertebrates; *Cognettia sphagnetorum* (Enchytraeidae) is one of the few that are able to feed on *Sphagnum* moss remains (Wallwork 1991). This leads to *Sphagnum* litter making up a much greater share in the peat than would be expected from the primary production. Woody peat, with its usually high contents of cellulose, lignin, and tannin, is also quite resistant to decomposition, and larger pieces such as stumps and logs can be preserved in the peat for thousands of years. Most non-woody litter above ground is readily decomposed. Roots are usually more easily decomposed than rhizomes.

Physical structures of the litter, such as hard waxy cuticles on conifer needles and ericaceous shrub leaves, may also inhibit decay. Many invertebrates are detritivores, and their activity facilitates for the bacteria and fungi by breaking down structures in the litter, 'pre-digesting' material, and improving aeration. On peatland sites, most of the soil mesofauna are found near the surface, and the water table is key to regulating their abundance: the higher the water level, the smaller the soil fauna.

When litter, dead roots, or exudates are deposited in the acrotelm, microorganisms quickly begin to work. The simplest organic compounds are decomposed faster than the more resistant ones. Therefore, as a particular layer passes through the acrotelm, by new litter being laid down at the surface and the water table slowly rising upwards, the proportion of easily decomposed compounds decreases while that of the resistant ones increases. Aerobic ammonification is also faster than anaerobic (Vepraskas and Faulkner 2001). Thus, the greatest activity of both production and decomposition of organic matter is in the upper oxic surface layer of the peatland. The degree to which the peat is changed to more resistant materials depends on microtopography. Less conversion occurs beneath the saturated hollows than in the aerated hummocks, so peat beneath hollows may have higher quality substrates than peat beneath hummocks.

The nutrient content, and quality of the litter, is often expressed as the C/N ratio. This is linked to litter structure, as hard evergreen leaves and woody material with large amount of structural compounds have a high C content. The content of N is low in *Sphagnum* but considerably higher in herbs (Aerts *et al.* 1999) which also often have thin, easily consumed leaves. Table 5.5 gives C/N values around 50–60 for *Sphagnum* peat and around 20 for sedge peat. Hence, a *Sphagnum* bog can have lower total productivity than some fens, but the long-term peat accumulation is on average higher in bogs (Turunen 2003) because the fen litter decomposes more quickly (Thormann *et al.* 1999).

As the plant material ages, the C/N ratio changes. The C/N ratio in sedge leaves could be between 10 and 15 (Saarinen 1998), but it is higher in the litter as the plants withdraw N from the leaves before they wilt. The initial decay of the litter could also involve a further loss of N and increase in C/N. Kuhry and Vitt (1996) observed that C/N in *S. fuscum* litter increased drastically (from about 80 to 300) in the upper part of the acrotelm. During further decomposition in the acrotelm, the C/N ratio will decrease with depth, since C is lost and N is recirculated by the organisms (Malmer and Wallén 2004; see Fig. 12.9). Later on, the changes are much smaller. Kuhry and Vitt (1996) noted that the C/N ratio continued to decrease with depth, whereas Malmer and Wallén (2004) noted a small increase when the material was passed on into the catotelm (Fig. 12.9).

Methane production and emission

Emissions of CH_4 in open mires have been measured by many workers (see review by Vasander and Kettunen 2006). The CH_4-producing organisms, the *methanogens*, are Archaea that grow anaerobically beneath the water table, and their main zone of activity shifts up and down with the water table. Their most common substrates are $H_2 + CO_2$ (to form $CH_4 + H_2O$) and acetate (to form $CH_4 + CO_2$). The primary control on CH_4 production is the amount of high-quality organic material (fresh litter and root exudation) that reaches the anoxic zone. This means that plant primary productivity and depth to water table are the two most important factors, with alternative electron acceptors, temperature, and pH as secondary controls.

The peat type affects CH_4 production. A minerotrophic *Carex rostrata* lawn, supplied with nutrients from surrounding mineral soils, had higher rates of CH_4 production than a nutrient-poor, ombrotrophic *Sphagnum*-dominated sites, probably owing to the higher supply of easily degradable litter and root exudates (Bergman *et al.* 1998). Within a bog, *Sphagnum majus* peat produced 1.5 times more CH_4 than *S. fuscum* peat. This was explained by the higher substrate quality in the *S. majus* peat that had formed below the water table compared to the *S. fuscum* peat that had formed above the water table and underwent aerobic decomposition before passing into the anoxic layer. Sundh *et al.* (1994) found that the most active zone of CH_4 production follows the depth to water, and is located just below it. Farther down, the peat becomes depleted of high-quality substrate and the CH_4 producers become resource limited. Where the water table is close to the surface the upper anoxic zone contains fresh, resource-rich material. Production of CH_4 is hence higher under the hollow than under the hummock (Bubier *et al.* 1993). Although CH_4 is produced under anoxic conditions when no other competing electron acceptor – nitrate (NO_3^-), ferric iron, (Fe^{3+}), or sulfate (SO_4^{2-}) – is available, any occurrence of one or more of these will lead to decreased CH_4 production. Deposition of SO_4^{2-} can therefore reduce CH_4 production (Granberg *et al.* 2001).

The CH_4 moves upwards through the peat by diffusion or bubbles. As it passes through the oxic layer it is taken up by CH_4-consuming bacteria, the *methanotrophs*, which are diverse group of aerobic bacteria. A considerable fraction of the CH_4 is transported from the anoxic zone to the atmosphere in the aerenchyma of plants (Thomas *et al.* 1996), bypassing the attacks of the methanotrophs (see further Saarnio *et al.* 1997; Nilsson *et al.* 2001b). Hence, within a peatland, the highest emissions of CH_4 will be from surfaces close to the water table (Fig. 12.3) and from surfaces with a high cover of aerenchymatous plants.

Peat formation and carbon flow

There are many ecosystem components and processes involved with the flow of carbon in peatlands, and the best understanding of how these processes interact comes from studies of bogs, with a permanently anoxic catotelm beneath the acrotelm which is gradually more reducing with depth, and in which the aeration also varies as the water level varies over the season and from year to year.

The most important parts of the C cycle are depicted in Fig. 12.4. Photosynthesis takes place at the surface and the net primary production

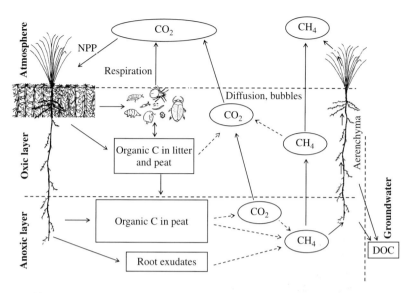

Fig. 12.4 Simplified description of carbon flow and peat formation in a peatland with an oxic upper part and an anoxic layer beneath. Encircled symbols represent gases, and dashed arrows show microbial processes. NPP is the net primary production (i.e. the difference between photosynthesis and the plants' respiration). DOC is dissolved organic carbon leaching out from both the oxic and anoxic layers of the peatland via the groundwater. Peatland fires will also lead to losses of CO_2 to the atmosphere.

Table 12.3 Components of annual carbon budget for a *Sphagnum* pine fen in Finland for the year 1993. Based on compilation from various sources by Alm *et al.* (1997). A minus sign means that carbon is lost from the peatland. Transparent chambers were used to measure the net CO_2 exchange, dark chambers for total CO_2 emission, and gross photosynthesis calculated as the sum of the two. The measurements in darkness summarizes plant respiration, animal respiration, and aerobic decomposition (including methane oxidation)

	Flux of carbon (g m^{-2} yr^{-1})				
	Flark	*Eriophorum* lawn	*Carex* lawn	Hummock	Whole mire
Gross photosynthesis	225.2	300.3	317.3	314.3	302.3
CO_2 emission	−165.1	−195.2	−239.2	−261.2	−204.2
CH_4 emission	−17.1	−24.0	−36.0	−13.1	−22.1
Leaching	−8.0	−8.0	−8.0	−8.0	−8.0
Net accumulation	34.9	73.0	34.0	31.9	68.0

yields above-ground and below-ground litter in the acrotelm, and some in the catotelm. A substantial part of this material, perhaps as much as 90%, is lost as CO_2 through aerobic decay. Some C can also be lost from the peat by leaching of dissolved and particulate organic matter. The role of the different components for the C budget is exemplified in a study by Alm *et al.* (1997) (Table 12.3). The example had a positive C balance (i.e. accumulation of C) as a whole, and for each of the mire features. The study year was somewhat colder than average, but with quite normal precipitation, and the net balance may well shift to net release in a warmer and drier year; it has been suggested that a temperature increase of 2 °C would increase CO_2 emission by 30% and a drop in the water table of 15–20 cm would increase it by 50–100% (Silvola *et al.* 1996). The balance is quite delicate, and Alm *et al.* (1999) observed a net loss of 90 g C m^{-2} yr^{-1} from a bog in Finland following the strong summer drought in 1994, even though the total annual precipitation was well above the long-term average. The drought strongly reduced the *Sphagnum* photosynthesis and the primary production was exceeded by the total ecosystem respiration; it would take at least 4 years with normal summers to compensate for the C losses.

Peat accumulation and its limits

Peatlands can have several metres of peat accumulated over thousands of years. Now we will discuss at what rate the peat has been formed during the development of the peatland, and how different factors set limits to peat growth. The growth of a peatland can be described by its growth in height, its lateral expansion, and by its accumulation of dry mass (sometimes expressed in amount of C). To describe the theories and models we use a *Sphagnum* bog with a well-developed distinction between acrotelm and catotelm as an example, since this is by far the most studied system.

The height growth in a dense *Sphagnum* lawn or hummock is often around 1 cm yr^{-1}. A large part of the volume of the growing *Sphagnum* mat is pores and space filled with air or water. During the passage through the acrotelm, as the water table slowly rises, a proportion of the biomass decays, and the remains are compacted by the weight of new material and by snow. When the material passes into the catotelm the decomposition and compaction have reduced the vertical annual increment to perhaps 1 mm yr^{-1}. If we measure the age at several levels in the catotelm, the height increment further down may be even lower (Fig. 12.5). This does not mean that growth was slower back then, but shows that the longer the material has been in the catotelm, the more it has been compacted, and the more has been lost by slow anaerobic decomposition.

To understand the processes of peat accumulation, a model of mass growth, rather than height growth, is required. Clymo (1984) has developed such a model, based on a number of assumptions: (1) the annual input of new matter into the catotelm (i.e. net primary productivity minus aerobic decay) is constant over time; (2) a constant proportion of the material in the catotelm decays every year; (3) the decay in the catotelm occurs at the same rate at all depths. In the model (Fig. 12.6) the rate of peat accumulation is the difference between the formation of new peat at the acrotelm–catotelm

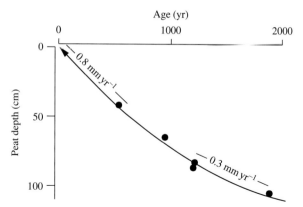

Fig. 12.5   Depth-age relationship of a peat profile at Bolton Fell Moss, England (calculated from data in Barber 1981: Fig. 48). The profile was ^{14}C dated at five depths. The mire surface is added as a point representing depth = 0 cm and time = 0 yr. The regression line is Depth = −0.707 + 0.090863(Time) − 0.00001822(Time^2). The inserted values show that much less material is left per year in the deeper part of the profile.

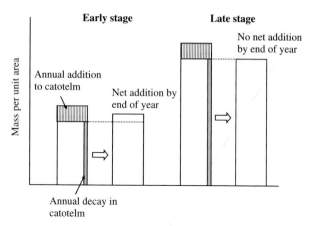

Fig. 12.6   Conceptual model explaining why a peat bog finally stops growing in height. Each year a certain amount of material is added to the catotelm, where a small proportion of all material is lost by anaerobic decay. In the early stage of bog development, addition is larger than the decay and the bog grows in thickness. Later on, the anaerobic decay acts through a thicker peat column and equals the input. The bog has ceased to grow in thickness. Modified from Clymo (1992).

border and the losses in the catotelm. The decay in the catotelm is very slow. The reason is not so much that the decay is anaerobic; more important is that all easily degraded molecules have been used up when the material reaches the catotelm. Realistic values for the fraction of peat in the catotelm decaying per year are in the order of 0.0001, which is a hundredths or even a thousandth of the level of aerobic decay in the acrotelm, but it is not zero.

It acts all the way down to the basal peat, and integrated over the whole depth of the catotelm in an old, deep mire it can be substantial. In fact, according to the model, it is the continuous anaerobic decay, the total of which increases with increasing catotelm depth, that eventually sets a limit to the height that the peat stack can reach. At the early stage of the development of a peatland, the annual addition of matter to the catotelm grossly exceeds the combined losses in the catotelm and the peat stack grows. When the peatland has grown in height for thousands of years, the anaerobic decay occurs over a greater depth and finally it equals the annual addition into the catotelm. When this happens, the peatland no longer has a net accumulation of peat and is no longer growing in height (Fig. 12.6). Formally, the maximum peat depth (in mass per unit area) is set by p_c/α_c, where p_c (g m^{-2} yr^{-1}) is the rate at which material is added from the acrotelm to the catotelm and α_c (yr^{-1}) is the decay parameter for the catotelm. (Fig. 12.7).

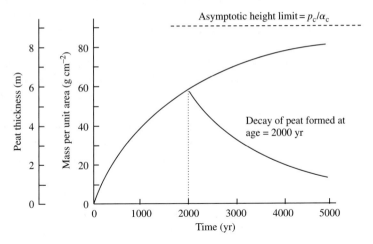

Fig. 12.7 A model showing the accumulation of peat in the catotelm during the development of a bog over 5000 yr. The yearly input from acrotelm to catotelm is p_c = 45 g m^{-2} yr^{-1}. The parameter describing anaerobic decay in the example is α_c = 5 × 10^{-5} yr^{-1}, and the rate of decay in the catotelm is $\alpha_c x_c$, the decay parameter multiplied by the total amount of material per unit area. When the total peat mass in the catotelm has reached 90 g cm^{-2}, the decay is $\alpha_c x_c$ = 5 × 10^{-5} × 90 × 10^4 = 45 g m^{-2} yr^{-1}, which equals the input p_c, and the peat column stops to grow. In other words, the maximum peat depth (as mass per unit area) is set by p_c/α_c, as shown by the asymptotic limit at the top. All calculations are made on a dry mass basis, the depth scale is added on the assumption that the peat bulk density is 0.1 g cm^{-3}. The dotted concave line shows the fate of peat present at the age 2000 yr. At that time the catotelm was 5.5 m thick, but after another 3000 yr, this peat has decayed and is now only 1.5 m thick. Modified from Clymo, R. S. (1984). The limits the peat bog growth. *Philosophical Transactions of the Royal Society of London B Biological Sciences*, 303, 605–54, with permission from the Royal Society, London.

The biological limit to height/mass growth occurs only if the proportional rate of catotelm decay is unaffected by peat age or quality. In reality, peat is likely to become more refractory as it decomposes, so we should expect a decreasing proportion to be lost as the peat ages. Clymo and co-workers (1992; 1998) showed that if the peat becomes more refractory with age, then the bog never reaches a 'limit' – it continues to grow indefinitely, albeit at a steadily decreasing rate. Using a more sophisticated model of peat decay, Bauer (2004) explored the effects of adding differing amounts of litter of differing quality. The results suggest that a wide range of peat accumulation patterns are possible.

A further complication involves the rate at which peat is transferred from the acrotelm to the catotelm. Although many models assume that peat is added to the catotelm at a constant rate, processes occurring in the acrotelm are likely to be the most responsive to environmental change (Belyea and Warner 1996; Wieder 2001) and there may be periodically decreased inputs of material into the catotelm, or periods with increased decay at the top of the catotelm. The discrete change at the acrotelm – catotelm boundary is also an oversimplification, and a gradual decrease in the decay from the acrotelm and down into the catotelm would be more realistic. At some sites, switches in vegetation type or the development of hummock – hollow microtopography have had a strong effect on the rate of peat accumulation (Kilian *et al.* 2000; Belyea and Malmer 2004). Gradual changes in the rate of peat addition may also occur as the peat deposit grows, altering surface hydrological conditions (Yu *et al.* 2003; Belyea and Malmer 2004; Belyea and Baird 2006). In addition exchanges between different layers need to be accounted for. There are roots that penetrate into the catotelm, and Charman *et al.* (1994) found that CH_4 extracted from the catotelm was 500–2000 years younger than the adjacent peat, and the C source may be dissolved organic C transported from younger layers above.

Several advances have been made in vegetation – C – water dynamics modelling. A model by Hilbert *et al.* (2000) demonstrated the sensitivity of peat accumulation to the position of the water table in the peat profile, and a model by Pastor *et al.* (2002) demonstrated the dynamics of plant communities to nutrient cycling. It is significant that both models demonstrated bifurcations from one stable state to another as inputs of either water or nutrients increased. Taken together these two studies emphasize that both water balance and nutrients are critical factors influencing peat development. Surprisingly little work has focused on how lateral expansion interacts with height growth. In early peatland work there was much interest in how high the cupola of a raised bog could develop, and it was hypothesized that with increasing precipitation the cupola would be higher (Granlund 1932). This height should be set by hydrological constraints, such that the higher the precipitation surplus, or the lower the hydraulic conductivity, the higher the bog height can be for a certain width (Ingram

1982). This formula could also be solved for maximum radius given a certain height. These approaches are rather theoretical, however, and more current questions regarding peatland development concern rates of lateral growth in relation to the slope of the adjacent mineral terrain, with the realization that in practice it is likely that the size and shape of a peatland is set by the terrain form (Korhola 1992). Belyea and Baird (2006) show that changes in the size and shape of a bog affect hydrological conditions at the peatland surface, and so may also affect acrotelm structure (i.e. vegetation, water-table depth) and the rate of peat addition to the catotelm.

Measuring the rate of accumulation

There are several ways of measuring the rate of peat accumulation. A very simple measure is, the *long-term apparent rate of carbon accumulation* (LARCA, sometimes referred to as LORCA). This can be calculated by dividing the mass per unit area in the whole peat stack by its age (Fig. 12.8), and then converting the mass value to C (g m^{-2} yr^{-1}). Only one ^{14}C datum is needed, that of the basal peat, which is fortunate considering the cost of such analyses. It is therefore tempting to use this measure to get data from many peatlands for comparative analyses. Although LARCA gives the average rate of peat accumulation, the interpretation is very specific and this limits its usage (Clymo *et al.* 1998). Assume, for example, a peatland with basal age 2000 yr. The LARCA value specifically represents the accumulation from 2000 BP to today, and it specifically represents peat accumulation over the first 2000 yr of the development of this peatland. It cannot be

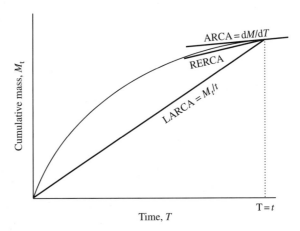

Fig. 12.8   Three measures of peat accumulation rate. LARCA (the long-term average) is simply the total amount accumulated divided by the time since the mire was formed. RERCA is calculated in the same way, but only for a recent period of the peatlands history. ARCA is the current rate, i.e. the slope of the peat accumulation curve today. Since the rate of peat accumulation slows down with age, LARCA > RERCA > ARCA.

unhesitatingly compared with the accumulation from 2000 BP to today in a 5000 yr old peatland, or with the first 2000 yr of a peatland that was initiated in 3000 BP.

A large study in Finland (Turunen 2003) where some of these problems with LARCA were accounted for indicates that bogs in the raised-bog region had LARCA values of 30–35 g m^{-2} yr^{-1}. Lower values were reported for fens and bogs in the more northern aapa mire region (the review by Turunen 2003 gives data also from North America and Asia). The global average value for LARCA is estimated to be 29 g m^{-2} yr^{-1} (Gorham 1991). The LARCA value from all major mire types in the former Soviet Union, for instance, reportedly varies between 12 in northern polygonal mires to 72–80 in fens and marshes, yielding an average LARCA of 30 g m^{-2} yr^{-1} (Botch *et al.* 1995). For Finnish mires, data derived from 1125 peat cores suggested that the average LARCA was 23 g m^{-2} yr^{-1}, with an average for bogs of 24 compared to 15 for fens (Tolonen and Turunen 1996; Turunen and Tolonen 1996). In western boreal Canada, the average over 9000 years in a *Sphagnum fuscum* peat ranged between 14 and 35 g m^{-2} yr^{-1} (Kuhry and Vitt 1996).

A second measure is the *actual rate of carbon accumulation* (ARCA). This is the rate at which a peatland currently accumulates material, and in that sense would be a good measure with which to compare peat accumulation under various scenarios of future climate change. However, according to the model, this parameter decreases with the age of a peatland (Fig. 12.8), so, for instance, a comparison among regions must rest on the assumption that the effect of peatland age must be negligible in relation to the regional differences in peat accumulation. ARCA can be calculated from the fluxes of CO_2 and CH_4 measured by gas exchange techniques in the field (e.g. Bubier *et al.* 2003). Photosynthesis and respiration are measured in chambers as described above, and in addition CH_4 emitted from the peat can be collected in the chambers. There are also techniques by which fluxes of CO_2 and CH_4 can be measured with towers placed on the peatland (e.g. Lafleur *et al.* 2003). The tower measurements are a composite of the gas exchanges for the peatland surface area being sensed.

Given the large inter-annual variation, it may be useful to have a measure of peat accumulation over the last few decades to average out such variation. This measure is the *recent rate of carbon accumulation* (RERCA), which is obtained from the bulk density down to a dated level not far from the surface. Given the recent developments in precise and accurate dating of young peat (see Chapter 6), this is now quite possible.

Rates of accumulation are not uniform over the whole surface, and vary according to microtopography, which in turn is related to differences in vegetation cover, depth to water and depth of oxic layer, temperature, and litter production. In high hummocks and in the surface layers of dry or drained sites, lack of moisture may constrain decomposition, and higher temperatures may enhance the moisture stress (Laiho *et al.* 2004). Mud-bottoms are a special circumstance where *Sphagnum* mosses are absent, peat production is retarded or stopped, and net losses by decomposition (emissions of CO_2 and CH_4) prevail. In some peatlands the proportion of mud-bottoms and pools may be significant enough to be an important influence on the C budget (Karofeld 2004).

It is useful to present a concrete example of how plant communities and rates of C accumulation can vary over time, and can indicate variations in water balance and climate change. Malmer and Wallén (2004) characterized the C balance in Store Mosse, a boreo-nemoral bog in southern Sweden, for the last 1000 years, and compared it with northern subalpine mires. The C/N ratio decreases with depth in the acrotelm (Fig. 12.9), and this was used to measure the loss of C in the acrotelm (the assumption being that the ratio in the litter entering the acrotelm has been constant over time). This was combined with ^{14}C dating to indicate the residence time for litter in the acrotelm (before passing into the catotelm). At Store Mosse a recurrence surface dated to 1000 BP marked an increase in net C balance from <20 g m^{-2} yr^{-1} to 49 in 800 BP, then dropping gradually to the current level of 20 g m^{-2} yr^{-1}. The northern mires had average net C balances decreasing from 20 to 0 for the same period (Table 12.4). The zero net C change indicates that the northern bogs are not growing, compared to the southern bog where there is still a slow net gain in C. In Store Mosse, the increase after 1000 BP was contemporary with a rise of the lake water levels in the region, and can thus be explained as a more humid climate raising the water level in the bog, which reduced decay losses in the acrotelm. The following gradual fall in net balance may be owing to gradual decreases in

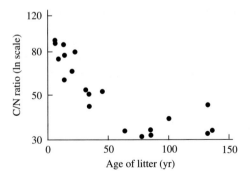

Fig. 12.9 The decrease in C/N ratio as the organic matter decays in the acrotelm. Assuming that the quotient of freshly produced litter has been constant over time, the decrease in C/N ratio indicates the loss of carbon in the acrotelm. From Malmer, N. and Wallén, B. (2004). Input rates, decay losses and accumulation rates of carbon in bogs during the last millenium: internal processes and environmental changes. *Holocene*, 14, 111–17. © 2004 Edward Arnold (Publishers) Ltd.

Table 12.4 Estimated past and present carbon balance (g m^{-2} yr^{-1}) for a boreo-nemoral bog, Store Mosse, southern Sweden, and three subalpine mires, northern Sweden (Malmer and Wallén 2004)

C component	Boreo-nemoral, southern Sweden			Subalpine mires, northern Sweden		
	800 BP	19th century	Present day	800 BP	19th century	Present day
Input to the acrotelm	74	79	29	50	50	17
Decay loss in the acrotelm	21	51	21	27	27	12
Input to the catotelm	53	28	8	23	23	5
Decay loss in the catotelm	4	6	6	3	5	5
Net balance	+49	+22	+2	+20	+17	0

climatic humidity resulting in lower water levels and longer residence times for the litter in the acrotelm. It is also coincident with the cooler climate during the Little Ice Age. There is no evidence of recent change in the climatic humidity in Sweden, but a general rise in temperature has occurred since the end of the nineteenth century. Other changes in southern Sweden that may have led to reductions in C input into the catotelm are extensive drainage in the region, as well as increased deposition of N, causing increased vascular plant cover and decreased *Sphagnum* cover, and maybe also increased decomposition.

Store Mosse data spanning the last 5000 years have been reanalysed with a model of peat accumulation linked to a model of peatland hydrology (Belyea and Malmer 2004). Two distinct patterns of growth were observed: abrupt increases associated with major transitions in vegetation and *Sphagnum* species (*fuscum*, *rubellum–fuscum*, and *magellanicum* stages), and gradual decreases associated with increasing humification of newly formed peat. Vegetation transitions were associated with periods of increasing climatic wetness, during which water tables rose, thus reducing acrotelm decay. Because of the complexity of interactions among climatic humidity and temperature, hydrology, surface structures, and peat formation, the authors cautioned against use of past rates to estimate current or predict future rates of peatland C sequestration.

Carbon store in peatlands

The enormous amount of C stored in peat is of course relevant to the discussion on global change. Several attempts have been made to calculate the volumes and amount of C stored in the northern temperate and boreal peatlands. The estimates differ wildly, largely due to uncertainties for peat depth and peat bulk density (Turunen *et al.* 2002). Mean depths between 1.3 and 2.3 m have been applied and bulk densities between 78 and 112 kg m^{-3}. This means that the lowest and highest estimates differ by a factor of 2.5. We can chose a mean depth of 2 m and bulk density of 91 from a large Finnish survey (Turunen *et al.* 2002), and a C content of 52% (Gorham 1991). The IMCG database (see Chapter 11) gives the area of peatlands in the northern temperate and cold climates as $3.4 \times 10^6 \, km^2$ (about 80% of the global peatland area), indicating that these peatlands store 320 Gt of C. Turunen *et al.* (2002) suggested that the true value is in the range 270–370 Gt, which would then be 37–51% of the 730 Gt C held in atmosphere as CO_2 (IPCC 2001; Fig 3.1). Various calculations suggest that the current annual C sequestering in northern peatlands amounts to 0.096 Gt (Gorham 1991) or 0.07 Gt (Clymo *et al.* 1998). The calculations of peat store for the whole world are even more uncertain, since peat depths and areas are less well known than for the northern peatlands.

Peatlands and climate change

There is no complete agreement about the way the global climate is evolving, but we know definitely that the amount of CO_2 in the atmosphere has been rising steadily since late nineteenth century, and this has been blamed on human activities including use of fossil fuel, agricultural practice, deforestation, fire, and drainage. In essence CO_2 absorbs infrared radiation that would otherwise be lost into space, and by an obvious analogy CO_2 is referred to as a greenhouse gas.

The other greenhouse gases that are important in peatlands are CH_4 and nitrous oxide (N_2O). Like CO_2, CH_4 has been increasing for the last 100 or more years, but during the past decade the rate of increase has declined. Atmospheric CH_4 accounts for 15–20% of the warming effects of these gases (IPCC 2001). Wetlands are the major biogenic source of CH_4, contributing a fifth of the total input of 500–600 Tg CH_4 yr^{-1}, and the contribution from rice fields is of the same order of magnitude. Northern peatlands have been estimated to provide 6–9% of the global emission (review by Pastor *et al.* 2002). However, CH_4 in the atmosphere is gradually oxidized. Nitrous oxide is formed during the anaerobic denitrification, which does not play a large role in virgin peatlands, but will increase somewhat after draining of minerotrophic peatlands, especially nutrient-rich ones, or in combination with N fertilization. In total, northern peatlands are not likely to have any significant climate effect in terms of N_2O emission (Martikainen *et al.* 1993; Augustin *et al.* 1996; Nykänen *et al.* 2002). In general, and historically, peatlands have had a net gain of CO_2 from the atmosphere, and emit CH_4.

Radiative forcing (measured in W m^{-2}) is the perturbation of the Earth's radiation energy budget which forces the global temperature to move towards a new equilibrium. Positive values indicate a potential warming of the atmosphere (i.e. a greenhouse effect) and negative ones a cooling. The *global warming potential* (GWP) or *greenhouse effect* of a gas depends on its ability to absorb infrared radiation, its turnover time in the atmosphere, and the pathways of chemical breakdown. It also depends on the time frame for the calculation. The GWP of a gas is compared with CO_2 (GWP = 1). On a mass basis and for a 100 year horizon the GWP is 23 for CH_4 and 296 for N_2O (IPCC 2001). With a time horizon of 500 years, the values are 7 for CH_4 and 156 for N_2O. These values are higher than earlier published ones, and the reason is that recent calculations give a lower GWP for CO_2, the standard with which other gases are compared. As GWP is expressed on a mass basis, we can also make the comparison that each C atom that is emitted from the peatland as CH_4 has a warming potential about 8 times higher than if it is emitted in the form of CO_2 (on a 100 yr horizon).

As we saw above, there are indications that even today some open northern peatlands have quite low C accumulation rates, in addition to being substantial sources of CH_4 (Waddington and Roulet 2000; Mäkilä *et al.* 2001; Wieder 2001). Whiting and Chanton (2001) compared boreal fens and subtropical *Typha* wetlands and observed that over the 100 yr time horizon the boreal ones have such a high CH_4 emission that they may contribute to global warming despite their positive CO_2 balance. The subtropical ones had higher CO_2 binding per mol of released CH_4, and clearly counteract global warming.

Increasing temperature is likely to have many complicated and interacting effects, but at least for northern peatlands it is possible that direct effects such as increased decomposition (Updegraff *et al.* 2004), increased CH_4 emission (Fowler *et al.* 1995), and increased photosynthesis are overshadowed by the concomitant changes in water level (Gorham 1991; Weltzin *et al.* 2001). After experimentally changing from a constant water table to a fluctuating one, Blodau and Moore (2003) measured a decreased emission of CH_4 and an increased emission of CO_2. Alterations in the hydrological regime will induce many biotic and abiotic changes, and the combined impacts on the C flows in longer term are difficult to predict (Bauer 2004). There are also other uncertainties (Moore *et al.* 1998): shifts in vegetation types and the relative dominance of different types of microforms may have a large effect on C cycling (for example, if pools/hollows expanded, then CH_4 emissions would increase). One of the biggest uncertainties is the effect of permafrost melting.

Effect of drainage and forestry on C balance

Drainage lowers the water table and yields a deeper oxic layer with more aerobic decomposition and more CO_2 release. It also favours the methanotrophs and leads to a strong reduction, or even complete cessation, of CH_4 emission. Future climate warming may lead to lowered water levels in peatlands, and researchers have used the effects of drainage on C balance as a way to mimic the potential effect of climate change (Laine *et al*. 1996). Using data from many drained and virgin Finnish peatlands, Nykänen *et al*. (1998) predicted that a 10 cm lowering of the water level would reduce CH_4 emission by 70% in fens and 45% in bogs. The total effect on radiative forcing depends on the time horizon. Minkkinen *et al*. (2002) suggested that if water levels are lowered by about 30 cm on average there will then be a decrease in radiative forcing of about 0.2 W m^{-2}. However, the total cooling by lowering water levels in all peatlands would be minor compared to the warming caused by anthropogenic greenhouse gas emissions.

Drainage for forestry

Numerous studies have documented CO_2, CH_4, and N_2O flux in drained forested peatlands (e.g. Martikainen *et al*. 1993; Roulet *et al*. 1993). The effect on the C store in peat is very variable, and draining does not lead to higher rates of decomposition in all cases (Laiho *et al*. 2004). Minkkinen and Laine (1998) reported changes in peat C between −50 and +50 kg m^2 60 years after drainage. In some places a rather small increase in decomposition may be balanced by a considerable input of root litter, and wood as the trees grow larger (Laiho *et al*. 2003). The somewhat surprising conclusion is therefore that peatlands drained for forestry may be C sinks, although this may not occur in the early stages of afforestation, and not in all sites. After about a century the C store in trees will saturate, but even after 300 years the radiative forcing may be lower than before draining (Laine *et al*. 1996).

These results may not apply everywhere – the C losses following drainage may be much higher in milder and more oceanic areas (reviewed by Laine *et al*. 1995), but even in Scotland it has been reported that afforested peatlands accumulated more C than they lost from draining over a time horizon of 90–190 years (Hargreaves *et al*. 2003).

Peat harvesting

Drainage followed by peat harvesting will remove peat from the harvested site. When the peat is burned or used as a soil improver it will transfer CO_2 to the atmosphere, and contribute to warming. The effects of drainage and harvesting are to increase the depth of water table, increase aerobic decomposition to CO_2, and decrease CH_4 emissions (except in the water-filled ditches). Hence, overall CH_4 emission will decrease, but for a peat area in Sweden this compensated only for 15% of the radiative forcing caused by the combustion of the peat (Rodhe and Svensson 1995). Drainage for agriculture will usually convert peatlands to C sources, and can also be substantial N_2O sources. Organic matter inputs from crops will not compensate for the increased decomposition rates caused by lowered water levels and fertilization.

Emissions of CH_4 from drained, cutaway peatlands are low, and may stay at a lower level than for pristine mires for a long time even after the sites have been vegetated and colonized fully by mire plants (Tuittila *et al.* 2000). This may be owing to a long recovery period of the methanogens. In a study of a cutaway peatland restored by spreading *Sphagnum angustifolium* fragments, water level and the amount of moss cover were key factors controlling both photosynthesis and the respiration of *Sphagnum*, and water level and temperature were the main factors controlling respiration of the peat (Tuittila *et al.* 2004). A model was developed to show seasonal CO_2 balance and its components. The total C balance was a net loss for dry conditions, and small net gains for optimum and wet conditions (cf. McNeil and Waddington 2003).

13 Uses, functions, and management of peatlands

From prehistoric times humans have utilized wetlands for such basic needs as water, food, and materials. Neolithic and Bronze Age villages have been found in wetlands: pile-dwellings (in Scotland and Ireland known as *crannogs*) resting on rows of piles inserted into the sediment (Coles and Coles 1989). But waterlogged, sodden places were also regarded as wastelands, places to be avoided, even feared. People and their animals could fall into wet holes, become 'mired down', and perish. The fate of the hundreds of 'bog bodies' found in Europe is ghastly: some were buried after execution, but others may have fallen into quagmires and been unable to climb out (Turner and Scaife 1995). Peatlands could have swarms of biting insects, and they were difficult and dangerous to cross. They were a blight on the land, a hindrance that needed to be drained and cleared and converted to useful ends, or as King (1685) expressed it: 'it were good for Ireland, the Bogs were sunk in the Sea'. Linnaeus was decidedly hostile about the mires during his expedition to northern Sweden:

The whole of this Lapp country was bog, which is why I call it the Styx. No priest has ever painted Hell so vile that this does not exceed it, no poet described a Styx so foul that this does not eclipse it (Linnaeus 1732).

If peatlands were wasteland that needed conquering in earlier times, this has changed in the modern world, and nowadays these wetlands are regarded as a treasure of functions and benefits, something to be valued and protected. There are even poetic, artistic, and spiritual values for the peatlands that need to be recognized and defended, and the current idea is one of carefully considered, wise use of a valuable resource (Joosten and Clarke 2002). The prevailing paradigm is well captured in the following:

Wise use of peatlands is essential in order to ensure that sufficient area of peatlands remain on this planet to carry out their vital natural resource functions while satisfying the essential requirements of present and future human generations. This involves evaluation of their functions, uses, impacts and constraints. Through such assessment and reasoning, we must highlight the priorities for their management and use, including mitigation of damage done to them to date (statement from International Mire Conservation Group and International Peat Society, cited in Joosten and Clark 2002).

In this chapter we discuss humanity's relationship to peatlands and cover their use, management, conservation and restoration. References which cover these topics in greater detail are Heathwaite (1993), Mitsch (1994), Moen (1995a), Wheeler *et al.* (1995), Lappalainen (1996), Vasander (1996), Parkyn *et al.* (1997), Keddy (2000), Charman (2002), Joosten and Clarke (2002), Parent and Ilnicki (2003), and Turner *et al.* (2003).

Historical development of peatland use

The large losses of peatland areas to agriculture and forestry were described in Chapter 11. Drainage and agricultural use was well established by the time of the Romans. After this drainage diminished, but was still kept alive through the Middle Ages on a smaller scale. A well-documented case of large-scale drainage during this period is the destruction of peatlands in the Netherlands. Already by the twelfth century the coastal bogs had disappeared after ditching (for agriculture) followed by peat subsidence which made peatlands accessible to sea water. Peat fuel excavation (even below ground water) reached industrial scale in the seventeenth century and peaked in the nineteenth. Much of it was for brick-making and heating lime kilns. Ombrotrophic bogs covered 1800 km^2 in the early seventeenth century, of which about 36 km^2 remains (Wolff 1993).

In the eighteenth century there was an increase in the rate of land drainage to convert peatlands into grain-growing areas; this happened first in Europe, but soon spread to North America as pioneers settled around the Great Lakes. A detailed account of peatland clearing for agriculture was given by Archie McKerracher (1987), descendant of early tenants of Blair Drummond, an estate near Stirling in Scotland. In the 1760s the owner of the estate began a project to drain the 1500 acres (600 ha) of peatland and remove the peat by digging a system of channels and drains and installing a huge hydrological wheel. Tenants were offered 8 acres (3.2 ha) of peatland, timber to build a house, enough oatmeal for one year, and no rent for one year. The tenants – poverty-stricken, dispossessed Highlanders – built their homes as pits in the peat, or on planks floating on top of the peat. Soon over 100 little houses had been built, and the whole area was covered with people toiling to strip off the peat.

In the nineteenth century large undertakings – river regulation, canal construction, and large-scale drainage projects – were conducted. One of these was a huge project in Hungary, begun in 1845, to regulate the course of the

Tisza River and drain almost 800 km² of wetland (Kerner 1863). Another was the Grand Canal which opened the midlands of Ireland, leading to large-scale development of drainage and peat extraction (Feehan and O'Donovan 1996). Other large river and tidal regulation projects were undertaken in many places in Europe and North America, and drainage continued wherever there was agricultural pressure.

Agriculture on peatland

Agriculture on peatlands can be simply the use of existing natural peatland vegetation, such as for a hay crop or grazing of sheep or cattle, or it can be the active preparation of a peat surface for growing crops. Peat extraction might sometimes be the initial reason for removing the surface vegetation, in which case agriculture could be carried out on the surfaces resulting after extraction.

Use of open peatlands for hayfields and pastures

The earliest uses of wetlands were as hayfields – meadow marshes and sedge fens where the fields were cut in summer, hay dried on racks, stored in sheds or barns on site, and then used in winter for the livestock. Because of this use, homesteaders often chose land that contained some of these wetlands. Systems of damming and flooding of these peatlands were developed to provide enrichment to the peats. When dammed, richer waters and silts spread over the surfaces and fertilized the sedges, which then produced more highly for one or more subsequent years. This cultural practice was carried on for hundreds of years, and it is still possible to find old dams, ditch structures, and hay sheds.

Research has been done (e.g. Elveland 1979; Moen 1995b) on ecological impacts of flooding and haying on the fens in Scandinavia. The main species that are hayed in wetter sedge fens are *Carex rostrata, C. lasiocarpa, Equisetum fluviatile*, and *Eriophorum angustifolium*. The haying makes it possible for smaller understorey plants to maintain or increase their cover. Of course, the height of the stubble, the time the hay harvest, the firmness of the peat layer, and the hydrological conditions may influence the outcome of the altered competition introduced by the scything. Some of the smaller species that are favoured are the sedges *Carex chordorrhiza* and *C. dioica* and the herbs *Parnassia palustris* and *Potentilla palustris*, and the rare rich fen orchid *Microstylis monophyllos*. The ground disturbances caused by grazing cattle create seedbed, and favour increases of several species. Conversely, draining is the main cause of extinction of these same species (Elveland 1993).

Present-day agriculture on peatlands

Peatlands for agriculture have traditionally been drained by shallow open ditches with 10–25 m spacing. In the latter half of the twentieth century an increasing proportion of the agricultural drainage was carried out as sub-surface mole and slit drainage, but one still sees open ditches, often with slightly crowned fields to facilitate runoff into the ditches. In most European countries, agricultural use of peatland is advised only for shallow (<1.0 m) or very shallow (<0.5 m) peat deposits (Ilnicki 2003).

The main kinds of agricultural crops are grains (rye, *Secale cereale*; oats, *Avena sativa*; rice, *Oryza sativa*), and grasses for haying, silage, and pasture, for example *Phalaris arundinacea, Festuca* spp., *Phleum pratense*, and *Bromus inermis*. On richer sedge peats, vegetables (carrots, onions, celery, lettuce, potatoes, turnip, etc.) can be grown (Myllys 1996; Okruszko 1996; Ilnicki 2003; Kreshtapova *et al.* 2003). A special kind of cultivation in the USA and Canada is large cranberry (*Vaccinium macrocarpon*) growing on *Sphagnum* mats. Controlled surface flooding is used to float up the cranberries for harvest, and the flooding waters provide slight fertilization with their mineral content. Grass sod for lawns has been grown on organic soils in some areas. An emerging field is non-wood pulp production (for paper) from reed canary grass (*Phalaris arundinacea*). This can also be used as a bio-fuel, together with fescue (*Festuca arundinacea, Festuca pratensis*) and *Salix* shrubs.

The capability of peats for agricultural production depends on climate, mire geomorphology, peat stratigraphy, peat type and humification, the air–water regime, peat physico-chemical properties, and the subsoil properties (see contributions in Parent and Ilnicki 2003). Kreshtapova *et al.* (2003) provide guides for soil quality levels (poor, medium, good, very good) using such parameters as thickness of the residual peat and arable soil layers, carbon/nitrogen (C/N) ratio, pH, bulk density, ash content, decomposition, base saturation, and extractable iron (Fe), phosphorus (P), and potassium (K). Some of the limiting factors that should be taken into account are:

- nutrient imbalance, and possible need for fertilizer
- subsoils high in calcium (Ca), and marl or marl-containing subsoils, may immobilize P and certain trace elements
- subsoils containing pyrite (iron sulfide, FeS_2) may form acid sulfate when exposed to air, and cause injury to plants
- macroclimate and microclimate factors: short vegetation period and low mean annual temperature, large temperature fluctuations during growing period, danger of frost
- an abrupt transition from waterlogging to water deficit owing to impermeable contact horizon between peat and mineral subsoil
- uneven thickness of residual peat layers
- considerable spatial and temporal variations in crop yield and quality.

Rice paddies often occur on peat. In the Sangjiang plains, north-east China, large loess and floodplains areas with marshes and fens have been converted to rice production (Lu and Wang 1994). Recent proposals in Indonesia for massive development schemes to convert tropical forested peatlands into rice-producing fields have been questioned by scientists who point out the numerous problems posed by the magnitude of draining and removing the deep peats, the acidity and low nutrient status of the peat, and potential for development of acid sulfate soils (Driessen 1978; Driessen and Dudal 1991).

Transformations of peat under agricultural use

The transformation of peat soils after drainage and conversion of fens to grasslands has been called the *moorsh-forming process* (Ilnicki and Zeitz 2003; Okruszko and Ilnicki 2003), and the resulting soils are termed *moorsh*. The transformation is caused largely by changes in microbial and invertebrate populations. In moorsh soils, actinomycetes are the dominant microorganisms, but in soils with water control and fertilization they decrease in relation to bacteria. Mites and springtails make up 90% of the moorsh mesofauna, and these are influenced by management regimes (pasture, grassland). Earthworms increase, and earthworm casts show higher numbers of fungi but decreased numbers of actinomycetes. Earthworms reduce the number of microorganisms in soils rich in bacteria, and increase the numbers in soils poor in bacteria (information from several authors, summarized by Ilnicki and Zeitz 2003).

After drainage the peat surface will sink. Natural peat can be almost afloat, and when the water level is lowered the peat will consolidate and the surface may fall as much as 2.5 m over 5–10 years (Berglund 1996). Then aerobic decay sets in, and this continues much longer. Berglund (1996) suggested the following rules of thumb for oxidative peat losses in Sweden (cm yr^{-1}): grazing land, 0.5; hay meadow, 1; cereals 1–2; intensively cultivated crops (carrots, potatoes), 2–3. There are examples of peatlands drained in the late nineteenth century which have subsided 5–6 m (Heathwaite *et al.* 1993). The drainage ditches have to be deepened at intervals, which leads to further loss of peat.

Forestry on peatland

As peatlands were drained for agriculture, it was observed that trees often invaded along ditches, and when ditches extended through forested or wooded peatlands, the trees along the ditches showed improved growth. This observation resulted in drainage to improve forest growth, but also to reduce waterlogging and make the sites more accessible. In some countries drainage campaigns were run as government programmes to provide work during the

A  B

Fig. 13.1   Peatland forestry. (A) Drainage tractor and contoured bucket, making ditch through black spruce swamp in northern Alberta, Canada. (B) Ditch system in the Wally Creek Forest Drainage Experimental Area, Ontario, Canada (photo by Ontario Ministry of Natural Resources). The ditches on the left side of the road range from 20 to 55 m apart; on the right side of the highway are the horse-logging trails created 55 years before in a logging operation.

economic depression in the 1930s. Some of these were rather unsuccessful, because of lack of knowledge about the need for fertilization. Much larger-scale and more effective drainage was carried out from 1950 (Fig. 13.1) and then often in combination with fertilization. The greatest amount of commercial peatland forestry is done in the Nordic countries, the Baltic states, Russia, UK, Ireland, Canada, USA, and south-east Asia (Chapter 11).

The main wave of drainage has now passed in northern Europe, peaking in the 1980s, because it is no longer subsidized by government, and also because conservation has gained more importance. However, considerable efforts are still made to clean and maintain old ditches. Currently South-east Asia, Indonesia, and Malaysia, have the most active peatland forestry, with a large amount of drainage to make peatland forests accessible for clearcutting. Many of these peatlands have been over-drained and water tables are quite deep, with accompanying rapid subsidence and probably much accelerated losses of carbon dioxide (CO_2). They are reforested by a few fast-growing, non-native species, such as *Acacia*, and there is concern about the large biodiversity losses since the original forests have high diversity.

There are several compilations describing the knowledge and practical experience in peatland forestry (Hånell 1988; Jeglum and Overend 1991; Paavilainen and Päivänen 1995; Trettin *et al.* 1997; Jeglum *et al.* 2003). Peatland forestry mostly applies to sites that are already forested or sparsely wooded, and all that is needed is to take away excess water. Afforestation of open peatlands requires drainage, but also additional investment for planting and usually fertilization. Hence, afforestation on open peatlands is only experimental in some countries. However, in the UK and Ireland forestry has occurred on open peatland, largely blanket bogs.

The peat subsidence after drainage for forestry is less than in agricultural use of peatlands. Minkkinen and Laine (1998) showed that peatland forestry in Finland led to an average subsidence of 22 cm within 60 years, most of this taking place during the first 5 years. But there are several other conditions that must be taken into consideration to lessen the impacts of forestry operations on peatland (the effects on carbon (C) balance and emission of CO_2 and methane (CH_4) are dealt with in Chapter 12):

- Water tables are close to the surface, and the root growth and nutrient cycling are restricted to the upper aerated layer. It is necessary to increase the depth of this layer, but not excessively so that losses of C and nutrients are minimized.
- Organic soils and peats are fragile, and easily subjected to compaction. Machines with low ground pressure are necessary, and for the wettest sites harvesting may need to be done when the site is frozen.
- A rise in the water table ('watering up') is a danger after cutting, owing to decreased interception and evapotranspiration, and blockage of micro-drainage lines by machine trails. Partial cutting and low-ground-pressure machines reduce this danger.
- Low-lying peatland sites are subject to cold air drainage and late spring frosts which can damage regeneration. Partial cutting and protection trees may reduce frost hazard.
- Peatland sites are subject to nutrient imbalances and deficiencies. This may be corrected with fertilization, or application of wood ash from wood-burning installations.
- Drainage causes increased erosion and loss of mineral and organic matter, and this needs to be taken care of with sedimentation ponds and filtration strips (Paavilainen and Päivänen 1995).

The numbers of dominant woody species that are harvested commercially vary amongst regions from only one in Siberia and northeast China (*Larix gmelinii*); to two or three species in northwestern Europe (*Betula pubescens, Picea abies, Pinus sylvestris*), North America (*Picea mariana, Larix laricina, Thuja occidentalis*), and the northwestern temperate forests of British Columbia and Washington (*Thuja plicata, Pinus contorta* var. *contorta*, and *Chamaecyparis nootkatensi*) (Asada *et al.* 2003a,b), to 10 or so species in Florida and the Gulf states of the USA (Ewel 1990) to tens of species in the tropical peatlands of south-east Asia (Chapter 11).

Vegetation changes after drainage for forestry

After drainage, as the surface peat becomes drier, there are biological changes related to reduced moisture content and increased aeration. In boreal peatlands, Finnish workers have recognized three classes of change after drainage based on vegetation: recently drained mires, transforming mires, and transformed mires (Paavilainen and Päivänen 1995). The *Sphagnum* cover is reduced to 75–25% in transforming sites, and to less than 25% in transformed sites. *Sphagnum* is replaced by forest floor mosses such as *Polytrichum*

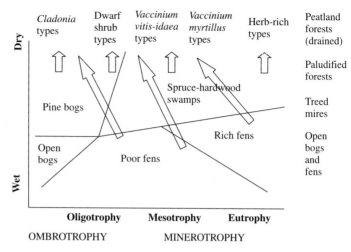

Fig. 13.2   Development of forest types after drainage of peatlands in Finland. The sequence of the resulting vegetation types is in order of decreasing nutrition is: *Herb-rich types* with ferns (*Athryrium filix-femina*) and herbs (*Oxalis acetosella*); *Vaccinium myrtillus types* with *Vaccinium myrtillus* and *V. vitis-idea* dwarf shrubs and herbs typical of mesic upland sites (*Trientalis europaea* and *Dryopteris carthusiana*); *Vaccinium vitis-idaea types* where *V. myrtillus* and *V. vitis-idaea* dominate. Sites which develop from forested mires have dwarf shrubs typical of pine mires present (*Rhododendron tomentosum, Vaccinium uliginosum*), and those from open mires may be dominated by *Betula nana*; *Dwarf-shrub type* are dominated by *Rhododendron tomentosum* and *V. uliginosum*; *Cladonia types* are dominated by *Sphagnum fuscum, Cladonia* lichens, *Calluna vulgaris, Empetrum nigrum*, and *Eriophorum vaginatum*. Redrawn from Paavilainen and Päivänen (1995), with classification based on Laine (1989).

commune, Dicranum spp., and feathermosses (*Pleurozium schreberi, Hylocomium splendens,* and *Ptilium crista-castrensis*). It is worth noting that the main bryophytes in these successions are the same on both sides of the Atlantic, even if the vascular plants differ.

With time, the transformed forests build up a layer of feathermoss and conifer litter over the peat, which takes on upland mor characteristics, and nutrient cycling, soil respiration, and root growth improve. A trophic sequence of drained peatland forests site types has been described in Finland (Fig. 13.2).

Peat extraction

In the following sections, we follow the Irish practice of distinguishing *cutaway peatlands* as those in which most of the peat has almost completely removed, with remaining depths of 1 m to 50 cm or less from the underlying mineral soil, and *cutover peatlands* where there are deeper depths remaining after partial extraction of the peat deposit.

The main uses of peat are for fuel, horticulture, sources of organic compounds, and absorption and treatment systems. Some other uses of peat may be mentioned: *Sphagnum* moss has antiseptic and absorbing properties, and has been commercially developed for use as baby diapers (nappies) and absorbent pads. In peat harvesting, thick layers of the sheaths of *Eriophorum vaginatum* are often encountered and separated in the mill, and these fibres have been mixed with wool, cotton or other fibres to create special textiles used to make clothing and specialty items. Peat is also used as a fuel to dry the grain in whisky production, which adds the characteristic peaty flavour. A final use is balneology – peat baths for therapeutic purposes. For further references see Bélanger *et al.* (1988), Vasander (1996), and Joosten and Clarke (2002).

Peat for fuel

The earliest use for fuel was to heat homes. Originally the peat was dug in sods (Fig. 13.3), stacked to dry, and then carried home to burn in stoves. In Ireland this tradition is retained, and people still own smallholdings at the margins of bogs in which they cut peat sods. Currently, the main commercial harvesting of peat is done by specially constructed machines and methods, aimed at maximum efficiency of extraction (Fig. 13.4).

On a worldwide basis about 50% of the peat extracted is used for energy, although in some countries this is the major use. The proportions of

Fig. 13.3   Blanket bog in Connemara, western Ireland. Peat is extracted by hand digging with the sleán (peat cutting tool) and stacked to dry.

A  B

Fig. 13.4   Modern peat extraction methods: (A) Milled peat being collected from surface of inter-ditch peat strips, Finland; (B) Sod peat being produced with a large excavator and square sod extrusion, Germany.

extracted peat that are burned are 90% in Ireland, over 66% in Finland, 33% in Russia, and 10–15% in Germany (Asplund 1996). Sweden, Russia, Belarus, Ukraine, the Baltic states, China, and Indonesia also use substantial amounts of fuel peat, whereas almost no peat is burned in Canada and the USA. To put this into perspective, the global use of peat for energy is estimated to be 5–6 million tonnes of oil equivalent per year (Joosten and Clarke 2002), which is only 0.1% of energy use globally.

In Ireland, peat represents about 10% of the country's energy use (Asplund 1996). However, there is a growing conservation movement against the burning of peat, and indeed against use of extracted peat for all purposes. A case has been made that the existing extraction areas should continue to be harvested, because they are already prepared and producing. In Finland, about 5% of the country's energy use comes from peat (Asplund 1996), and it will probably continue to be an important energy source.

Peat for horticulture

Sphagnum peat has special properties for growing plants in pots and beds in nurseries: good moisture retention, good aeration, and high cation exchange capacity (CEC). Since it is usually nutrient poor, it may require addition of nutrients for maintaining good plant growth. It can be mixed with other soils or media in order to gain more nutrient supply. Sedge peat and woody peat do not have as good physical qualities as *Sphagnum* peat, but can have higher pH and nutrient content, and therefore be better in terms of nutrition.

Some uses of peat in horticulture are as follows (Bélanger *et al.* 1988):

- *Seedbeds, potting media, and peat mulch:* Often slightly decomposed, surficial peats are used, sometimes mixed with other media (e.g. sand) and fertilizers.
- *Peat pots:* Pots can be made from peat and used to grow seedlings, after which the pot plus seedling may be planted directly.

- *Pre-seeded, fertilized peat blocks or pellets:* Various sizes of blocks and pellets developed for horticulture and forestry. Peat can also be used as a support and water reservoir for plants, as with flower arrangements and wreaths.
- *Casing:* The application of a layer of material on the surface of a bed of compost, promoting fructification in mushroom production.
- *Composting:* Peat can be used as the carrying medium for composting, for example, with manure, with vegetable wastes, and with leaf and grass cuttings.
- *Peat – fertilizer mixes:* Peat can be mixed with fertilizer or lime and applied to crop fields. In Russia, an agriculture use of peat is in combination with ammonium (NH_4^+) fertilizer (Gelfer and Mankinen 1985).

Organic compounds from peat

Peat, being derived from biomass, consists of a great diversity of organic compounds, which may be extracted, hydrolized, or otherwise transformed into useful chemicals. Fuchsman (1980) and Bélanger *et al.* (1988) have given comprehensive treatments of this subject, and here we list only some of the products. Extraction with organic solvents (e.g. acetone) can yield bitumen, peat resins, peat waxes, and peat asphaltenes. By using various reagents such as acids and bases, and different temperatures, extractions can be made of components such as pectin, lignin, cellulose, hemicellulose, monosaccharides, organic acids (humic, fulvic, and other complex and simple acids), and nitrogenous compounds (e.g. proteins, peptides, and amino acids).

Absorption and treatment systems

The adsorption and absorption properties of peat and its high CEC give it high potential as a depolluting agent for industrial, municipal, and domestic wastewater, gas pollutants, and organic solvents. It can act as a natural resin with high capacity to fix several types of pollutants and to act in filtration, and it can act as the support medium for waste substances and for the microorganisms that act to degrade them. Peat beds or bulk peat may be used to adsorb or absorb the following pollutants (e.g. Bélanger *et al.* 1988):

- heavy metals connected in industrial waste
- radioactive substances
- colouring materials, as a substitute for activated charcoal
- pesticides, phenols, and other organic agents
- oil in tanks in oil tankers, and accidental oil spills
- animal urine and manure (bedding), and odours
- municipal wastewater.

In wetlands used as cleaning filters, denitrification could remove some of the N. This requires high pH, anoxic conditions, and that N is available as nitrate (NO_3^-). For these reasons, N removal is more efficient in limnic wetlands where sedimentation can also help to remove it from circulation. In most fens and bogs denitrification is inefficient. One way to achieve N retention is to grow trees on drained peatlands (Lundin 1991). This positive effect should, however, be weighed against the leaching of N occuring for a short period after the ditching and after tree cutting (Nieminen 2004).

Several hundreds of constructed wetlands are in operation in Europe for treatment of municipal wastewater (Brix 1994), and thousands of wetland treatment systems have been built worldwide for treating a variety of wastes, including municipal and industrial wastewaters, mine drainage water, stormwater, and farmland runoff (Kadlec 1994). In some places large natural wetlands/peatlands near a polluting site are used for treatment purposes. One of the first such projects is at Houghton Lake, Michigan (Kadlec 1994) where an 800 ha fen is used to clean municipal wastewater. Several communities in northern Canada utilize small lakes surrounded by peatland vegetation for water polishing before the water drains out through small channels into large lake and river systems.

Functions and values

Peatlands occupy particularly key positions in the landscape, being the ecotones between upland and water. As such they have complex processes and boundary exchanges. They are reservoirs of water that are filled during snow melting and flooding in spring, and during periods of heavy precipitation. In drier periods during summer, they provide runoff and groundwater outflow, and sometimes they dampen the variations in runoff from catchments, and stabilize streamflow (see Chapter 8). They have a large filtering capacity and retain or metabolize organic material, as well as nutrients and polluting agents. However, in some cases, as when drained, they also release certain nutrients, compounds, or heavy metals.

These inherent functions of wetland ecosystems – structure, processes, and regulation – are the basis for all values attached to wetlands by humans. For more in-depth treatments of the functions and ecosystem services provided by peatlands we refer to the works of Joosten and Clarke (2002) and Turner *et al.* (2003). The more we learn about peatlands, the more we come to understand their ecological, social, and economic value. Some peatlands are valued because they are rare or unique, or because they are habitats for threatened species. Other types are important because they have been highly utilized for agriculture or peat extraction, such as in large parts of central Europe, the raised lowland bogs in the northern UK, and large areas of southern Canada and the USA. Here there are few remaining examples to serve as reference types and for peat archives. In order to manage properly, one needs functioning, healthy examples to study and learn how they function.

As noted in Chapter 10, hydromorphology is a fundamental characteristic of peatlands that underlies their functions and processes. Brinson *et al.* (1994) used hydromorphologic classification as the first step to develop an assessment procedure for functions of wetlands. The method consists of the following steps: (1) classify the wetlands according to hydromorphic properties; (2) make connections between the properties of each wetland class and the ecological functions that it performs; (3) develop functional profiles for each wetland class; (4) choose reference wetlands that represent the range of both natural and human-imposed stresses and disturbances, and (5) design the assessment method using indicators calibrated to reference wetlands. This is a subjective process which synthesizes information from many different sources to portray the function of a particular hydromorphological type.

Management of peatlands

Management of peatlands is a complex endeavour which involves communication and participation from natural science, social-economic interests, stakeholders, and decision-makers. The main societal problem is in setting up workable methods to permit optimal solutions for wetland/peatland management questions. Some of the methods for determining management strategies are environmental indicators, economic valuation, and social and deliberative approaches (Turner *et al.* 2003); decision flowcharts, checklists, and codes of conduct (Joosten and Clarke 2002); and spatial and economic modelling (Mitsch 1994; van den Bergh *et al.* 2003).

In their discussion of integrated decision-making (Brouwer *et al.* 2003b) conclude that

In order to deliver the sustainable utilization and management of wetland ecosystem resources, it is necessary to underpin management actions by a scientifically credible but also pragmatic environmental decision-support system. Such a system can have the objective of economic efficiency at its heart, but still recognize other dimensions of wetland resources value and decision-making criteria.

Of course in making the decisions, the critical issue is what weights to apply to the ecological, economic, social, and cultural values.

Sustainability and renewability

The principle of *sustainable development*, brought to prominence by the book *Our Common Future* (World Commission on Environment and Development 1987), has been widely accepted in management of natural resources. The commission called for a global commitment to undertake economic and social development in such a way as to meet the needs of the world's population 'without compromising the ability of future generations to meet their own needs.'

Considerable advances have been made in addressing sustainable use and ecosystem management in forest management. This has led to an approach called criteria and indicators, which has been suggested as a framework for setting objectives and monitoring improvements for forestry in Canada (CCFM 1995) and can equally well be applied to peatlands:

- conserving biological diversity
- maintenance and enhancement of ecosystem health and integrity
- conservation of soil/peat and water resources
- influences on global ecological cycles
- multiple benefits to society
- accepting society's responsibility for sustainable development.

When peat use for fuel is being discussed, questions like 'Is peat a renewable energy source?' and 'Can peat use be sustainable?' often arise. When peat is used, one must bear in mind that it takes thousands of years to form, and in harvesting peat, the first step is to remove the peat-forming vegetation. Peat therefore cannot be considered as a renewable source of energy. This is also the view expressed, for instance, in the Intergovernmental Panel on Climate Change guidelines (IPCC 1997). Whether to extract peat depends on many factors:

- the status of the particular country or region with regard to areas of peatland and reserves of peat
- the needs of society for peatlands for other purposes, such as stabilizing catchment hydrology and biodiversity conservation
- the influence of peat uses on greenhouse gases and global warming (see Chapter 12).

When peat is extracted for fuel or other uses, it is essential to have high-quality wetland inventories as basis for site selection. The inventories are used to prevent the destruction of valuable peatlands, and also as a tool to select sites that have already been damaged by altered hydrology.

In the application of sustainability and ecosystem management, the terms *health* and *integrity* are used to describe the conditions of ecological structure, processes, and functions. Ecosystems are healthy and have integrity when their structure, processes, species composition, and condition are all within or close to the historical range of the successional stage that the ecosystem is in, or for which it is being managed (Kimmins 1997). Two other terms are used to describe how biodiversity and ecosystem functions respond to disturbances: *resistance* is the ability of the system to withstand changes, and *resilience* is the speed of recovery after a change or disturbance (Calow 1998). The longer the time needed to restore a peatland to its original condition, the lower its resilience.

Environmental impact indicators

Indicators of environmental impacts can be applied to peatland management (Brouwer *et al.* 2003a). Environmental indicators that relate to landscape, water regime, and biodiversity are presented in Table 13.1. This is basically a list of attributes that might be considered in assessment of environmental impacts. Land-use changes such as urban development, forestry, and agriculture usually account for the major impacts on wetlands. Water regime, especially the water level and its fluctuations, is the major ecological factor influencing ecosystem structure, functions, and processes. Species richness, including gains and losses of species, is certainly the major indicator of environmental change.

The indicators are not sustainability indicators, they simply reflect certain aspects of a system's state which are considered relevant, according to our scientific understanding of peatland ecosystems, in assessing the impacts of human activities or evaluating management interventions. There is a great deal of interrelatedness amongst indicators, and some are undoubtedly more generally important than others for indicating the state of the ecosystem. In fact, these indicators have considerable similarity to the attributes used for evaluating wetlands for conservation purposes in Ontario (see below). The selection and interpretation of indicators to reflect the status of ecosystems is in the early stages of understanding and development. In order to monitor sustainable use of wetlands, benchmarks or reference points are needed with which to compare the deviations from normal.

Table 13.1 Indicators of environmental impacts or susceptibility to impacts. Modified from Brouwer, R., Crooks, S. and Turner, R. K. (2003). Environmental indicators and sustainable wetland management. In: *Managing wetlands: an ecological economics approach*, (ed. R. K. Turner, J. C. J. M. van den Bergh and R. Brouwer), pp. 41–72, Edward Elgar Publishing, Cheltenham, UK

Landscape indicators	Water regime indicators	Biodiversity indicators
Area size	Water level regime	Species richness
Connectivity	Water balance	Keystone species[a]
Proximity	Sediment load	Umbrella species[b]
Shape	Organic load	Trophic levels of food webs
Location in catchment	Nutrient load	Productivity rates
Hydromorphic type	Toxicant load	
Vegetation type/cover	Oxygen level	
Soil/ peat characteristics	Bacteria, viruses, and	
Land-use changes	water-borne parasites	
	Temperatures	
	Turnover rate	

[a] A keystone species is 'one upon which many other species in an ecosystem depend' (Calow 1998).
[b] An umbrella species has large home range or wide habitat requirements, so that 'if they are adequately protected, other species will automatically be protected as well' (Calow 1998).

Scale and scope of management

Peatland management needs to be defined in the context of the scale of landscape with which one is dealing – international, national, regional, basins, catchments, and specific peatland complexes and bodies. At the international and national scales we are concerned with such broad issues as general wetland policy, conservation, and broad environmental impacts (e.g. global warming and greenhouse gases), and long-distance atmospheric pollution. At the level of large basins and catchments the concerns are such as multiple land-use interactions (e.g. agriculture, clearcutting, drainage), landscape diversity, point-source pollution, and microclimate change. And at the level of peatland complexes and specific peat bodies, the concerns have to do with environmental impacts of peat extraction, drainage, and road-building.

Normally one cannot effectively manage a peatland in isolation, because its hydrology is related to inputs from or outputs to the surroundings; thus the proper level of management is the catchment area to which the peatland belongs. Management may also be concerned with a wide range of disturbance condition, from undisturbed, pristine conditions, as with nature or wildlife reserves, parks, and ecological research sites; to types or areas of peatlands that have been partly disturbed; or to areas that have been so highly disturbed and modified that they no longer fit the definition of peatland.

A particularly useful tool for decision- and policy-makers in scoping environmental problems has been the DPSIR framework: *driving forces – pressure – state – impact – response* (Turner *et al.* 2003). A complete appraisal of peatland projects, programs, or courses of action might involve the following steps (Brouwer *et al.* 2003b):

- Determine the causes of the peatland ecosystem degradation.
- Assess the full ecosystem damage caused.
- Assess the human welfare significance of such changes.
- Formulate practicable indicators of environmental change and sustainable use.
- Evaluate the impacts using monetary and non-monetary indicators.
- Assess alternative peatland uses along with conservation policies.
- Present managers and policy-makers with policy options.

Conservation

In recent years, owing to concern about large-scale and continuing losses of peatlands, conservation has been increasingly emphasized in government and industry, in academia, and by public action groups. Several developed countries have adopted policies for wetland use, for example,

Canada, Finland, Japan, Sweden, the UK, and the USA, and many other countries are in the process of developing policies as their wetlands continue to be disturbed or destroyed. The USA has developed a 'no-net-loss' policy that requires a maintenance of the quantity and quality of wetlands (Tiner 1998). Consequently, it may be an expensive proposition to develop a wetland for other uses, and much effort has gone into defining wetlands and peatlands for regulatory purposes (see references in Chapter 1).

It is important to establish a system of natural, more-or-less undisturbed benchmark sites for the main wetland types, to be able to compare with disturbed sites. Some of these benchmark sites should also serve as long-term ecological research sites, which should provide continuing documentation of processes and dynamics. This information should be a part of continuing research elsewhere into structure and function of peatlands (Keddy 2000; Mitsch and Gosselink 2000; Charman 2002).

This is not the place for a discussion of conservation methods and programmes. However, reasonable aims for a peatland conservation programme are:

- to maintain the complete range of natural variation of peatlands, including vegetational types and species composition, successional history, hydrological systems, and hydromorphic types
- to retain sufficient areas of all peatland types to maintain the function of peatlands within the region of concern
- to maintain the complete range of anthropogenic peatlands, as these relate to sustainable use.

Gap analysis is a method that may prove useful to identify gaps in representation of various components of wetland biodiversity in defined regions or jurisdictions. The biodiversity elements are those that have been mapped in the defined region – vegetation types; species of plants, animals, or invertebrates; peatland types; hydrologic systems or hydromorphic types, and other resource data – and these are entered into a geographic information system (GIS) for distributional and intercorrelative analyses. The goal is to ensure that all, or key, elements of biodiversity are represented adequately throughout the management area of concern (Flather *et al.* 1997).

Peatland evaluation for conservation

However the aims for peatland conservation are formulated, it is essential to have a good inventory, which should be expressed as a comprehensive peatland classification. We give two examples here.

The Ontario Wetland Evaluation System

The Ontario Wetland Evaluation System is a detailed evaluation approach developed primarily to serve the needs of Ontario's planning process (Anon. 1993), and designed to identify and measure values of wetlands, which in Ontario are predominantly peatlands. It was intended to provide a mechanism or framework through which conflicting claims about wetland values and uses could be resolved. Although it is not a complete inventory and certain information is lacking, for example, on rare species and hydrology, the system can also be used to carry out a preliminary or 'first cut' biophysical inventory of a wetland.

In accordance with Ontario's wetland policy to ensure 'no loss of provincially significant wetlands', many wetlands of significance have been rated. The evaluation procedure involves assigning points to the different features of a wetland, based on four components: social, hydrological, biological, and special features. Each component may achieve up to 250 points, so a wetland may score a maximum of 1000 points. Wetlands which achieve 600 or more, or score 200 points or more in either the biological or special features, are considered to be provincially significant.

Because the evaluation system was both costly and time consuming, a more rapid assessment procedure was developed, using only 5–8 key variables that could be determined from aerial photographs, maps, and other documents (Chisholm *et al.* 1995). These provided strong predictive equations for total wetland score using key independent variables, but because of the variability of estimates caution and professional judgement was advised in using the technique. It seems there is no substitute for field site assessment by skilled personnel in obtaining quality evaluations.

The Swedish Wetland Inventory

Many Swedish mires were threatened by drainage for forestry purposes during the 1980s, often with subsidies from the government. After arguments from conservationists the subsidies were stopped, and in 1986 the Nature Conservation Act was changed in a way that permission was required for drainage activities. Only limited special-purpose drainage called 'remedial drainage' is allowed after clear-cutting, to prevent local watering up. The Swedish Environmental Protection Agency has carried out a wetland inventory in cooperation with the County Administration Boards. The objectives were to increase the knowledge base, and also to produce a nature conservation evaluation. All wetlands exceeding 10 ha in the south (2 ha on the calcareous Baltic islands of Gotland and Öland) and 50 ha in the north were included, in addition to a large number of smaller mires. A standard technique was used, involving the interpretation of aerial photographs of all the sites and field surveys of about 10% of the sites. The result is an enormous database with descriptions and classification of wetland types. In total, more than 35 000 wetland sites were included. The wetland sites were classified according to nature conservation value (based on evaluation criteria including naturalness, representativeness, natural richness, and rareness). This yielded about 4000 class 1 sites (very high conservation value), about 8000 in class 2 (high value), about 17 000 in class 3 (some value), and about 4500 in class 4 (no detectable conservation value), and 2600 not yet classified.

Based on the Wetland Inventory a selection of the most valuable mire sites in need of protection was published in 1994 as the 'Mire Protection Plan for Sweden' (Lonnstad and Löfroth 1994). The selection is a representation of mires of all types and from all regions. In addition, priority was given to rare mire types, such as rich fens, important mires for bird life, and important mosaics of mires and natural forests. In total 146 protected and 345 unprotected mires were included, representing an area of almost 0.4 Mha. The objective is that all these sites should be protected by 2010.

Restoration, reclamation, and afteruse

Restoration is the process of returning ecosystems or habitats to their original structure, species composition, function and processes. This requires a detailed knowledge of the original species, ecosystem functions, and interacting processes. Restoration is distinquished from *reclamation*, the latter being planned creation of another ecosystem that is different from the original, and below we mention afteruse of cutaway peatlands as an example. The term restoration is applied to situations where a more-or-less natural peatland is established. More often the result is not exactly the same as the original ecosystem; in these cases the term creation would be better. For example, you may read that a cutaway bog was *restored* to a marsh ecosystem; however, it is more appropriate to say that the marsh was newly *created* where there was once a bog.

Afteruse of cutaway peatland

The main afteruses are agriculture, forestry, and creation of functional wetlands (Selin 1996; Egan 1998; Leinonen *et al.* 1998; Selin and Nyrönen 1998; Farrell and Doyle 2003). A frequent choice in many countries is forestry. For example, at Bellacorick in western Ireland, some 6500 ha of cutaway blanket bog have been planted, the largest such project in the world (Catherine Farrell, pers. comm.). As well, we must mention the creation of multiple-use parklands created from cutaway and cutover peatlands. These can be developed as mixed habitat areas for wildlife, with ponds, wetlands, and uplands, and may be used for hunting, bird watching, fishing, and hiking. An excellent example is the Lough Bora Parkland in Ireland (Egan 1998; 1999). Wildlife may be actively promoted by planting or maintaining vegetation with high values for food or habitat. For example, reed canary grass (*Phalaris arundinacea*) provides excellent nesting and escape cover and the shattered seeds are readily eaten by many bird species (Natural Resource Conservation Service 2002). It should be noted that *P. arundinacea* may become weedy or invasive in some regions or habitats, and may displace desirable vegetation if not properly managed.

Restoration and reclamation

This is currently an active area of research and development, and it is impossible to cover it in detail here. Further references are found in Wheeler and Shaw (1995), Pfadenhauer and Klötzli (1996), Malterer *et al.* (1998), Money and Wheeler (1999), Schouten (2002), Rochefort *et al.* (2003), and Vasander *et al.* (2003).

The critical indicators of restoration of peatlands are hydrological state, vegetation structure and composition, and peat and peat processes. We can visualize restoration as a continuum of kinds of actions, depending on the degree of disturbance (Table 13.2). Hence, with lightly disturbed peatlands the actions will be rather modest, whereas with highly disturbed peatlands the actions will be extreme and expensive. The main steps of natural wetland or mire restoration include the following:

- Determine the pH/nutrient regime for the remaining peat/subsoil, and determine what ecosystem is appropriate to that regime.
- Establish a water level regime appropriate for the ecosystem one has chosen to create.
- Determine if there are enough residual species present for the chosen ecosystem; if necessary add propagules (seeds, vegetative parts, clumps of living vegetation) onto the site.
- Create depressions that will, in the case of dry cutaways, promote carpet or lawn level *Sphagna* (Price *et al.* 2002), or, in the case of flooded basins, provide hummocks or rafts for hummock-forming *Sphagna* and mosses.

The determination of pH/nutrient status is critical, because it will determine what kind of ecosystem to select, for example, rich fen, intermediate fen, poor fen, or ombrotrophic bog. The water level regime is also critical because it should match the water regime for the ecosystem being favoured. A highly fluctuating water level is appropriate to marshes; water levels close to the surface with lower fluctuations to open fens and bogs; and water levels deeper below the surface (20 to 30 cm) will allow for woody plant root development in wooded fens and swamp forests.

Some approaches to restoration and reclamation

The strategy for restoration/creation depends on how much of the original living surface vegetation is remaining (see Table 13.2). If much of the original structure and composition still remains, the job of restoration will be easier. A straightforward solution is to dam up the ditches at appropriate

Table 13.2 The relationship of degree or type of disturbance to ecosystem condition, and restorative actions

Degree/type of disturbance/examples	Hydrological state	Vegetation structure and composition	Peat and peat processes	Restorative actions
Weakly disturbed, no drainage, harvesting biomass. Non-intensive forestry, traditional haying	Water level can rise after harvesting, but becomes normal in a few years	Vegetation partly removed, changes favour some species, disadvantage others	Peat can become mucked up and compacted in tractor trails; some drainage blockage may occur	(1) Ensure forest regeneration (2) Minimize compaction and mucking disturbances
Moderately disturbed, ditched for forestry	Water level lowered, deepest near ditches, reducing further from ditches	Vegetation changes towards drier upland species	Increasing depth of aerated peat; increase in humification, bulk density, mineralization, root production, and CO_2 release; decrease in CH_4 release	(1) Damming or blocking ditches
Highly disturbed, ditched, surface vegetation bladed off, no peat removed. With crops or grassland	Water level lowered very deep	Peatland plants present only in low depressions	As above	As (1) above (2) Spreading peat forming plant inoculum (3) May elect to maintain agricultural use
Highly disturbed, ditched, vegetation removed, peat removed to variable depth. Cutover and cutaway peatlands with peat extraction	Water level lowered to permit efficient extraction without excessive drying	Surface may be bare peat, algal crusts, or variable amounts of pioneer plants and pockets of peatland mosses	As above. High losses of carbon and nutrients	As (1) and (2) above (3) After-use options, including forestry, agriculture, or creation of new wetland ecosystem
Highly disturbed, drained, extraction of peat from deep trenches and ponds. Mostly older methods of peat extraction	Wet slurry extraction. Water levels lowered during peat extraction, but fill again	Ponds often with steep banks, can be colonized by water lilies and floating mosses	Complete change to open water ponds surrounded by peatland, rich in organic matter	(1) Establish rafts of floating brush or clumps of peat as nuclei for floating mosses (2) Accept as new habitat for wildlife

Fig. 13.5 Restoration of fen. The hydrology is restored by blocking in the ditch. The blockings consist of a plank wall and compressed peat a few metres upstream. On the left side of the ditch the trees that encroached after draining (about 40 years earlier) have been removed. Rich fen, eastern Sweden. Photo by Stefani Leupold.

intervals, such that the water levels are raised close to the original levels throughout the area. Ditch blockage is a technique that has been used in many places (Fig. 13.5). Another solution is to fill up the ditches completely, using the bankside spoil originally extracted from the ditches. This might be done where the intent is to cover up the ditch disturbance and establish the original character of the peatland.

The steep-sided ponds created by deep trench extraction are a special feature of some extraction methods, and it may be necessary to accept them as a permanent feature of the peatland. Such trenches re-vegetate very slowly (Soro *et al.* 1999). If extraction ponds are surrounded by remaining peatland vegetation, establishment of floating *Sphagnum* species (notably *S. cuspidatum*) is promoted if a high water level is maintained (Fig. 13.6A). To fully restore a *Sphagnum* mire it is necessary to promote successional series from carpet mosses like *Warnstorfia fluitans* and *Sphagnum cuspidatum*, to lawn species such as *S. fallax* and *S. papillosum*, to low hummock species such as *S. magellanicum* and *S. rubellum*, to higher hummocks with *S. fuscum*. One could try to promote floating mat development by strategically placing clumps of *Sphagnum* peat, logs, or stumps as raised islands for *Sphagnum* development and hummock initiation (Swan and Gill 1970).

If the surface has been bladed and there is only a small amount of vegetation structure and composition remaining, but the site is still deemed to be

Fig. 13.6 Earliest goal of bog restoration is rewetting and *Sphagnum* establishment. (A) Extraction pond with carpet of *Sphagnum cuspidatum*, lawns with *Eriophorum chamissonis* on banks. Burns Bog, British Columbia, Canada. (B) Rewetting a cutaway peatland, Finland, showing clumps of *Eriophorum vaginatum* and *Typha latifolia*. (C) Establishment of *Sphagnum fuscum* from dispersed fragments is augmented by a layer of straw to protect from surface desiccation, Quebec (cf. Rochefort *et al.* 2003).

worth restoring, it may be decided to bring in propagules or clumps of living surface vegetation. Of course, this must be accompanied by an appropriate water level regime (Fig. 13.6B). An approach to restoring cutaway bogs has been developed in Canada (Rochefort *et al.* 2003). Inoculum of living *Sphagnum* fragments is spread onto the bare peat surfaces (Fig. 13.6C). The top parts of the shoots are harvested on nearby peatlands, leaving the source carpet to recover vegetatively. Generally the more diaspores the better, but that will also require larger source areas. A compromise is to use fragments from 1 m² of source for every 10 m² of restoration surface. The sowing material is coarsely milled, and spread evenly with a regular farm manure spreader onto the peat surface. A light covering of straw is spread over the shredded *Sphagnum* layer, in order to give protection against desiccation. Successful establishment of a fragment depends critically on the wetness of the surface. Maintenance of a stable water table is more beneficial to *Sphagnum* growth than repeated wetting and drying events (Rochefort *et al.* 2002). The same approach has been successfully used for fen restoration on a cutaway peatland that had exposed fen peat at

Fig. 13.7   Pond and marsh construction. The Lough Bora Parkland, a multiple-purpose after-use area in a cutaway peatland, central Ireland.

its base (Cobbaert *et al.* 2004). One of the telling conclusions is that further rewetting measures may be necessary to create real functioning ecosystems.

Bogs and poor fens are perhaps the hardest peatland ecosystems to restore. They require low pH and often nutrient-poor water. Restoration will be particularly difficult if the peat is cut deep, because the cutaway site is likely to be strongly influenced by minerogenous water with distinctly higher pH and nutrient content than in a bog. In such circumstances it may be necessary to create a new ecosystem. Marshes are easiest to create, because marsh species have wind-, water-, or bird-dispersed small seeds that are readily transported to sites, but they require fluctuating water levels with both drawdowns (for plant establishment) and flooding (Fig. 13.7).

Peatland societies and organizations

The interest and concern about peatland conservation is growing, as witnessed by the activities of the various conservation groups worldwide. For effective conservation of peatlands it is necessary to be politically proactive, and non-governmental organizations (NGOs) are important movers in promoting conservation. In this final section we list some important societies and organizations that are actively promoting understanding, conservation, and wise management of peatlands.

- The **International Association for Ecology** (INTECOL, *www.intecol.net*) arranges a large International Wetlands Conference every fourth year where the newest developments in wetland science are presented and discussed.
- The **International Mire Conservation Group** (IMCG, *www.imcg.net*) is devoted to conservation and preservation of mires. They publish an electronic newsletter and have elaborated guidelines for wise use of mires, together with the International Peat Society (Joosten and Clarke 2002). IMCG is an international network of specialists which promotes the conservation of mires and enhances the exchange of information and experience relating to mires.
- The **International Peat Society** (IPS, *www.peatsociety.fi*) is an international NGO of scientific, industrial, and regulatory stakeholders. Its main task is the 'advancement, exchange and communication of scientific, technical and social knowledge and understanding for the wise use of peatlands and peat'. IPS is quite strongly linked to the peat industry. They publish an electronic information bulletin, *Peat News*, and the *International Peat Journal*. Every fourth year there is an International Peat Congress.
- The **Ramsar Convention on Wetlands** (*www.ramsar.org*) is an intergovernmental treaty which provides the framework for national action and international cooperation for the conservation and wise use of wetlands and their resources. It was adopted in Ramsar, Iran in 1971 (Ramsar 1987). The convention is managed by the Ramsar Secretariat, located within the IUCN headquarters in Switzerland. Member countries of the Convention commit themselves to designate at least one, but preferably several wetlands that meet the criteria for inclusion in the List of Wetlands of International Importance (the 'Ramsar list') and ensure the maintenance of the ecological character of each of these 'Ramsar sites'. Listed sites do not necessarily require protected legal status, provided their ecological character is maintained through a wise-use approach. Currently there are over 140 member states and the total area of the 1400 or so Ramsar sites is more than 1 million km^2, but most of these wetlands are not peatlands.
- The **Global Environment Centre** (GEC, *www.gecnet.info*), a non-profit organization based in Malaysia, has established a web portal for exchange of information on peatland management, particularly for Asia (*www.peat-portal.net*). There are several projects, for instance the Southeast Asia Peat Network (SEA-PEAT) which is an internet information network.
- **Wetlands International** (*www.wetlands.org*) is an organization dedicated to the work of wetland conservation and sustainable management. The Global Peatland Initiative (GPI) was set up by Wetlands International, IMCG, IPS, and other organizations. It is a platform programme, which promotes the identification and development of projects for the wise use and conservation of peatlands and facilitates their funding.

- Among the various national organizations we can mention the **Finnish Peatland Society** (*www.suoseura.fi*). They publish the journal *Suo*, devoted to all aspects of peatlands, especially forested ones. The **Irish Peatland Conservation Council** (*www.ipcc.ie*) has the mission 'to conserve a representative sample of living intact Irish bogs and peatlands for the benefit of the people of Ireland, and to safeguard their diversity of wildlife'. Based in the USA the **Society of Wetland Scientists** (SWS, *www.sws.org*) is a 'non-profit organization founded in 1980 to promote wetland science and the exchange of information related to wetlands'. SWS publishes the scientific journal *Wetlands* and acts to foster the conservation and understanding of wetlands, and their management.
- Many organizations with a broader scope in nature conservation have activities concerning peatlands, for example the **Worldwide Fund for Nature** (WWF, *www.wwf.org*).

References

Adamec, L. (2002). Leaf absorption of mineral nutrients in carnivorous plants stimulates root nutrient uptake. *New Phytologist*, **155**, 89–100.

Adams, P. W. (1995). The effects of beaver (*Castor canadensis*) on boreal forest peatlands and keys to the *Sphagnum* of Ontario. Ph.D. Thesis, University of Waterloo, Canada.

Aerts, R., Wallén, B., and Malmer, N. (1992). Growth-limiting nutrients in *Sphagnum*-dominated bogs subject to low and high atmospheric nitrogen supply. *Journal of Ecology*, **80**, 131–40.

Aerts, R., Verhoeven, J. T. A., and Whigham, D. F. (1999). Plant-mediated controls on nutrient cycling in temperate fens and bogs. *Ecology*, **80**, 2170–81.

Agriculture Canada Expert Committee on Soil Survey (1987). *The Canadian system of soil classification*, 2nd edn, Research Branch, Agriculture Canada Publication 1646.

Albinsson, C. (1996). Vegetation structure and interactions on mires. Ph.D. thesis, Lund University, Sweden.

Aldous, A. R. (2002). Nitrogen translocation in *Sphagnum* mosses: effects of atmospheric nitrogen deposition. *New Phytologist*, **156**, 241–53.

Alm, J., Tolonen, K., and Vasander, H. (1992). Determination of recent apparent carbon accumulation in peat using dated fire horizons. *Suo*, **43**, 191–94.

Alm, J., Talanov, A., Saarnio, S. *et al.* (1997). Reconstruction of the carbon balance for microsites in a boreal oligotrophic pine fen, Finland. *Oecologia*, **110**, 423–31.

Alm, J., Schulman, L., Walden, J., Nykänen, H., Martikainen, P. J. and Silvola, J. (1999). Carbon balance of a boreal bog during a year with an exceptionally dry summer. *Ecology*, **80**, 161–74.

Almquist-Jacobson, H. and Foster, D. R. (1995). Toward an integrated model for raised-bog development: theory and field evidence. *Ecology*, **76**, 2503–16.

Andersen, S. T. (1973). The differential pollen productivity of trees and its significance for the interpretation of a pollen diagram from a forested region. In: *Quaternary plant ecology*, (ed. H. J. B. Birks and R. G. West), pp. 109–115. Blackwell Scientific Publications, Oxford.

Anderson, J. A. R. (1964). The structure and development of the peat swamps of Sarawak and Brunei. *Journal of Tropical Geography*, **18**, 7–16.

Anderson, J. A. R. (1983). The tropical peat swamps of Western Malesia. In: *Ecosystems of the world. 4B. Mires: swamp, bog, fen and moor. Regional studies*, (ed. A. J. P. Gore), pp. 181–99, Elsevier, Amsterdam.

Anderson, J. P. E. and Domsch, K. H. (1978). A physiological method for the quantitative measurement of microbialbiomass in soils. *Soil Biology and Biochemistry*, **10**, 215–21.

Andrejko, M. J., Fiene, F. and Cohen, A.D. (1983). Comparison of ashing techniques for determination of the inorganic content of peats. In: *Testing of peats and organic soils*, (ed. P.M. Jarrett), pp 5–20. Special Technical Publication 820, American Society for Testing and Materials, West Conshohocken, PA.

Andrus, R. E. (1986). Some aspects of *Sphagnum* ecology. *Canadian Journal of Botany*, **64**, 416–26.

Anon. (1993). *Ontario wetland evaluation system. NEST Technical Manual: TM-001, northern manual; TM-002, southern manual*. Ontario Ministry of Natural Resources, Northeast Science and Technology Unit, South Porcupine.

Anshari, G., Kershaw, A. P., Van der Kaars, S., and Jacobsen, G. (2004). Environmental change and peatland forest dynamics in the Lake Sentarum area, West Kalimantan, Indonesia. *Journal of Quaternary Science*, **19**, 637–55.

Appelo, C. A. J. and Postma, D. (1994). *Geochemistry, groundwater and pollution*. A. A. Balkema, Rotterdam.

Asada, T., Warner, B. G., and Banner, A. (2003a). Growth of mosses in relation to climate factors in a hypermaritime coastal peatland in British Columbia, Canada. *Bryologist*, **106**, 516–27.

Asada, T., Warner, B. G. and Pojar, J. (2003b). Environmental factors responsible for shaping an open peatland – forest complex on the hypermaritime north coast of British Columbia. *Canadian Journal of Forest Research*, **33**, 2380–94.

Ashworth, A. C., Markgraf, V., and Villagran, C. (1991). Late Quaternary climatic history of the Chilean Channels based on fossil pollen and beetle analyses, with an analysis of the modern vegetation and pollen rain. *Journal of Quaternary Science*, **6**, 279–91.

Asplund, D. (1996). Energy use of peat. In: *Peatlands in Finland*, (ed. H. Vasander), pp. 107–13, Finnish Peatland Society, Helsinki.

Auer, V. (1930). *Peats bogs in southeastern Canada*. Canada Department of Mines, Geological Survey Memoir 162.

Auer, V. (1958). The Pleistocene of Fuego-Patagonia. Part II: The history of the Flora and Vegetation. *Annales Academiae Scientiarum Fennicae. Series A III. Geologica-Geographica*, **50**, 7–239.

Augustin, J., Merbach, W., Schmidt, W. and Reining, E. (1996). Effect of changing temperature and water table on trace gas emission from minerotrophic mires. *Angewandte Botanik*, **70**, 45–51.

Backéus, I. (1972). Bog vegetation re-mapped after sixty years. Studies on Skagershultamossen, central Sweden. *Oikos*, **23**, 384–93.

Backéus, I. (1988). Weather variables as predictors of *Sphagnum* growth on a bog. *Holarctic Ecology*, **11**, 146–50.

Backéus, I. (1989). Flarks in the Maloti, Lesotho. *Geografiska Annaler*, **71A**, 105–11.

Backéus, I. (1990a). Production and depth distribution of fine roots in a boreal open bog. *Annales Botanici Fennici*, **27**, 261–65.

Backéus, I. (1990b). The cyclic regeneration of bogs – a hypothesis that became an established truth. *Striae*, **31**, 33–35.

Baird, A. J. and Gaffney, S. W. (1995). A partial explanation of the dependency of hydraulic conductivity on positive pore water pressure in peat soils. *Earth Surface Processes and Landforms*, **20**, 561–66.

Baird, A. J. and Gaffney, S. W. (1996). Discussion: 'Hydrological model of peat-mound form with vertically varying hydraulic conductivity' by Adrian C. Armstrong. *Earth Surface Processes and Landforms*, **21**, 765–67.

Baird, A. J., Beckwith, C. W., and Heathwaite, A. L. (1997). Water movement in undamaged blanket peats. In: *Blanket mire degradation: causes, consequences and challenges*, (ed. J. H. Tallis, R. Meade and P. D. Hulme), pp. 128–39, Macaulay Land Use Institute, Aberdeen.

Barber, K. E. (1981). *Peat stratigraphy and climatic change: a paleoecological test of the theory of cyclic peat bog regeneration.* A. A. Balkema, Rotterdam.

Barber, K. E. (1993). Peatlands as scientific archives of past biodiversity. *Biodiversity and Conservation*, **2**, 474–89.

Barber, K. E., Dumayne-Peaty, L., Hughes, P., Mauquoy, D., and Scaife, R. (1998). Replicability and variability of the recent macrofossil and proxy-climate record from raised bogs: field stratigraphy and macrofossil data from Bolton Fell Moss and Walton Moss, Cumbria, England. *Journal of Quaternary Science*, **13**, 515–28.

Barber, K. E., Battarbee, R. W., Brooks, S. J. *et al.* (1999). Proxy records of climate change in the UK over the last two millenia: documented change and sedimentary records from lakes and bogs. *Journal of the Geological Society, London*, **156**, 369–80.

Bateman, L. E. and Davis, C. C. (1980). The Rotifera of hummock-hollow formations in a poor (mesotrophic) fen in Newfoundland. *Internationale Revue der Gesamten Hydrobiologie*, **65**, 127–53.

Bates, J. W. (2000). Mineral nutrition, substratum ecology, and pollution. In: *Bryophyte biology*, (ed. A. J. Shaw and B. Goffinet), pp. 248–311, Cambridge University Press, Cambridge, UK.

Bauer, I. (2004). Modelling effects of litter quality and environment on peat accumulation over different time-scales. *Journal of Ecology*, **92**, 661–74.

Beckwith, C. W. and Baird, A. J. (2001). Effect of biogenic gas bubbles on water flow through poorly decomposed blanket peat. *Water Resource Research*, **37**, 551–58.

Bélanger, A., Potvin, D., Cloutier, R., Caron, M., and Thériault, G. (1988). *Peat. A resource of the future.* Centre Québécois de Valorisation de la Biomass, Sainte-Foy, Canada.

Bellamy, D. J. and Rieley, J. (1967). Some ecological statistics of a 'miniature bog'. *Oikos*, **18**, 33–40.

Beltman, B., Kooijman, A. M., Rouwenhorst, G., and van Kerkhoven, M. B. (1996). Nutrient availability and plant growth limitation in blanket mires in Ireland. *Biology and Environment*, **96B**, 77–87.

Belyea, L. R. (1999). A novel indicator of reducing conditions and water table depth in mires. *Functional Ecology*, **13**, 431–34.

Belyea, L. R. and Baird, A. J. (2006). Beyond 'the limits to peat bog growth': cross-scale feedback in peatland development. *Ecological Monographs*, **76**, 299–322.

Belyea, L. R. and Clymo, R. S. (2001). Feedback control of the rate of peat formation. *Proceedings of the Royal Society of London B Biological Sciences*, **268**, 1315–21.

Belyea, L. R. and Lancaster, J. (2002). Inferring landscape dynamics of bog pools from scaling relationships and spatial patterns. *Journal of Ecology*, **90**, 223–34.

Belyea, L. R. and Malmer, N. (2004). Carbon sequestration in peatland: patterns and mechanisms of response to climate change. *Global Change Biology*, **10**, 1043–52.

Belyea, L. R. and Warner, B. G. (1994). Dating of the near-surface layer of a peatland in northwestern Ontario, Canada. *Boreas*, **23**, 259–69.

Belyea, L. R. and Warner, B. G. (1996). Temporal scale and the accumulation of peat in a *Sphagnum* bog. *Canadian Journal of Botany*, **74**, 366–77.

Bennett, K. D. and Willis, K. J. (2001). Pollen. In: *Tracking environmental change using lake sediments: terrestrial, algal, and siliceous indicators*, (ed. J. P. Smol, H. J. B. Birks, and W. M. Last), pp. 5–32. Developments in paleoenvironmental research, vol 3. Springer, Berlin.

Bennett, K. D., Haberle, S. G. and Lumley, S. H. (2000). The last Glacial-Holocene transition in southern Chile. *Science*, **290**, 325–28.

Berendse, F., van Breemen, N., Rydin, H. *et al.* (2001). Raised atmospheric CO_2 levels and increased N deposition cause shifts in plant species composition and production in *Sphagnum* bogs. *Global Change Biology*, **7**, 591–98.

Berglund, B. E. with the assistance of Ralska-Jasiewiczowa, M. (eds) (1986). *Handbook of Holocene palaeoecology and palaeohydrology*. John Wiley & Sons, Chichester.

Berglund, B. E., Birks, H. J. B., Ralska-Jasiewiczowa, M and Wright, H. E. (1996). *Palaeoecological events during the last 15 000 years: Regional syntheses of palaeoecological studies of lakes and mires in Europe*. John Wiley & Sons, Chichester.

Berglund, K. (1996). Cultivated organic soils in Sweden: properties and amelioration. PhD thesis, Swedish University of Agricultural Sciences, Uppsala.

Bergman, I., Svensson, B. H., and Nilsson, M. (1998). Regulation of methane production in a Swedish acid mire by pH, temperature and substrate. *Soil Biology and Biochemistry*, **30**, 729–41.

Bergner, K, Bohlin, E. and Albano, Å. (1990). *Vad innehåller torv?* Sveriges lantbruksuniversitet, Umeå.

Berry, A. Q., Gale, F., Daniels, J. L., and Allmark, B. (1996). *Fenn's and Whixall mosses*. Clwyd Archaeology Service, Mold.

Berry, G. J. and Jeglum, J. K. (1991). *Hydrology of drained and undrained black spruce peatlands: Groundwater table profiles and fluctuations*. Forestry Canada, Ontario Region, Sault Ste. Marie, COFRDA Rep. No. 3307.

Biester, H., Kilian, R., Franzen, C., Woda, C., Mangini, A. and Schöler, H. F. (2002). Elevated mercury accumulation in a peat bog of the Magellanic Moorlands, Chile (53° S) – an anthropogenic signal from the Southern Hemisphere. *Earth and Planetary Science Letters*, **201**, 609–20.

Birks, H. H. (2001). Plant macrofossils. In: *Tracking environmental change using lake sediments: terrestrial, algal, and siliceous indicators*, (ed. J. P. Smol, H. J. B. Birks, and W. M. Last), pp. 49–74. Developments in paleoenvironmental research, vol 3. Springer, Berlin.

Birks, H. J. B and Birks, H. H. (1980). *Quaternary palaeoecology*. Edward Arnold, London.

Björkman, L. (1997). The role of human disturbance in the local Late Holocene establishment of *Fagus* and *Picea* forests at Flahult, western Småland, southern Sweden. *Vegetation History and Archaeobotany*, **6**, 79–90.

Björkman, L. (1999). The establishment of *Fagus* sylvatica at the stand-scale in southern Sweden. *The Holocene*, **9**, 237–45.

Blaauw, M., Heuvelink, G.B.M., Mauquoy, D., van der Plicht, J., and van Geel, B. (2003). A numerical approach to ^{14}C wiggle-match dating of organic deposits: best fits and confidence intervals. *Quaternary Science Reviews*, **22**, 1485–1500.

Blades, D. C. A. and Marshall, S. A. (1994). Terrestrial arthropodes of Canadian peatlands: synopsis of pan trap collections at four southern Ontario peatlands. *Memoirs of the Entomological Society of Canada*, **169**, 221–84.

Blodau, C. (2002). Carbon cycling in peatlands – a review of processes and controls. *Environmental Review*, **10**, 111–34.

Blodau, C. and Moore, T. R. (2003). Experimental response of peatland carbon dynamics to a water table fluctuation. *Aquatic Sciences*, **65**, 47–62.

Boatman, D. J., Goade, D. J., and Hulme, P. D. (1981). The Silver Flowe. III. Pattern development on Long Loch B and Craigeazle mires. *Journal of Ecology*, **69**, 897–918.

Bodley, C. L., Jeglum, J. K., and Berry, G. L. (1989). A probe for measuring depth to water surface in wells. *Canadian Journal of Soil Science*, **69**, 683–87.

Boehm, H. V. and Siegert, F. (2002). The impact of logging on land use change in Central Kalimantan, Indonesia. *International Peat Journal*, **11**, 51–58.

Boelter, D. H. (1965). Hydraulic conductivity of peats. *Soil Science*, **100**, 227–31.

Boelter, D. H. (1969). Physical properties of peats as related to degree of decomposition. *Soil Science Society of America Proceedings*, **33**, 606–9.

Boeye, D., Verhagen, B., van Haesebroeck, V., and Verheyen, R. F. (1997). Nutrient limitation in species-rich lowland fens. *Journal of Vegetation Science*, **8**, 415–24.

Bohlin, E. M. (1993). Botanical composition of peat. Ph.D. thesis, Swedish University of Agricultural Sciences, Umeå.

Bohlin, E., Hämäläinen, M. and Sundén, T. (1989). Botanical and chemical characterization of peat using multivariate methods. *Soil Science*, **147**, 252–63.

Børsheim, K. Y., Christensen, B. E., and Painter, T. J. (2001). Preservation of fish embedment in Sphagnum moss, peat or holocellulose: experimental proof of the oxopolysaccharidic nature of the preservative substance and of its antimicrobial and tanning action. *Innovative Food Science and Emerging Technologies*, **2**, 63–74.

Boström, U. and Nilsson, S. G. (1983). Latitudinal gradients and local variations in species richness and structure of bird communities on raised peat-bogs in Sweden. *Oikos*, **14**, 213–26.

Boström, B., Jansson, M., and Forsberg, C. (1982). Phosphorus release from lake sediments. *Archiv für Hydrobiologie, Ergebnisse der Limnologie*, **18**, 5–59.

Botch, M. S. and Masing, V. V. (1983). Mire ecosystems of the U.S.S.R. In: *Ecosystems of the world. 4B. Mires: swamp, bog, fen and moor. Regional studies*, (ed. A. J. P. Gore), pp. 95–152, Elsevier, Amsterdam.

Botch, M. S., Kobak, K. I., Vinson, T. S., and Kolchugina, T. P. (1995). Carbon pools and accumulation in peatlands of the former Soviet Union. *Global Biogeochemical Cycles*, **9**, 37–46.

Bradshaw, R. and Hannon, G. (1992). Climatic change, human influence and disturbance regime in the control of vegetation dynamics within Fiby Forest, Sweden. *Journal of Ecology*, **80**, 625–32.

Bragazza, L., Gerdol, R. and Rydin, H. (2003). Effects of mineral and nutrient input on mire bio-geochemistry in two geographical regions. *Journal of Ecology*, **91**, 417–26.

Bragazza, L., Tahvanainen, T., Kutnar, L. *et al.* (2004). Nutritional constraints in ombrotrophic Sphagnum plants under increasing atmospheric nitrogen deposition in Europe. *New Phytologist*, **163**, 609–16.

Bragazza, L., Limpens, J., Gerdol, R. *et al.* (2005). Nitrogen concentration and ^{15}N signature of ombrotrophic Sphagnum mosses at different N deposition levels in Europe. *Global Change Biology*, **11**, 106–14.

Bragg, O. M., Hulme, P. D., Ingram, H. A. P., and Robertson, R. A. (eds) (1992). *Peatland ecosystems and man: an impact assessment*. British Ecological Society and International Peat Society. Department of Biological Sciences, University of Dundee.

Bridgham, S. D., Pastor, J., Updegraff, K. *et al.* (1999). Ecosystem control over temperature and energy fluxes in northern peatlands. *Ecological Applications,* **9,** 1345–58.

Brinson, M. M., Kruczynski, W., Lee, L.C., Nutter, W. L., Smith, R. D., and Whigham, D.F. (1994). Developing an approach for assessing the functions of wetlands. In: *Global wetlands: Old World and New,* (ed. W. J. Mitsch), pp. 615–24. Elsevier, Amsterdam.

Brix, H. (1994). Constructed wetland for municipal wastewater treatment in Europe. In: *Global wetlands: Old World and New,* (ed. W. J. Mitsch), pp. 325–33, Elsevier, Amsterdam.

Brönmark, C. and Hansson, L.-A. (2005). *The biology of lakes and ponds,* 2nd edn., Oxford University Press, Oxford.

Brooks, K. N. (1992). Surface hydrology. In: *The patterned peatlands of Minnesota* (ed. H. E. Wright Jr., B. A. Coffin, and N. E. P. Asseng), pp. 153–62, University of Minnesota Press, St. Paul.

Brouwer, R., Crooks, S. and Turner, R. K. (2003a). Environmental indicators and sustainable wetland management. In: *Managing wetlands: an ecological economics approach,* (ed. R. K. Turner, J. C. J. M. van den Bergh, and R. Brouwer), pp. 41–72, Edward Elgar Publishing, Cheltenham.

Brouwer, R., Turner, R. K., Georgiou, S. and van den Bergh, J. C. J. M. (2003b). Integrated assessment as a decision support tool. In: *Managing wetlands: an ecological economics approach,* (ed. R. K. Turner, J. C. J. M. van den Bergh, and R. Brouwer), pp. 19–40, Edward Elgar Publishing, Cheltenham.

Brown, R. J. E. (1977). Muskeg [cf. peatland] and permafrost. In: *Muskeg and the northern environment in Canada* (ed. N. W. Radforth and C. O. Brawner), pp.148–163, University of Toronto Press, Toronto.

Bruenig, E. F. (1990). Oligotrophic forested wetlands in Borneo. In: *Forested wetlands* (ed. A. E. Lugo, M. M. Brinson, and S. Brown), pp.299–334, Elsevier, Amsterdam.

Bubier, J., Costello, A., Moore, T. R., Roulet, N. T., and Savage, K. (1993). Microtopography and methane flux in boreal peatlands, northern Ontario, Canada. *Canadian Journal of Botany,* **71,** 1056–63.

Bubier, J., Crill, P., Mosedale, A., and Frolking, S. (2003). Peatland responses to varying interannual moisture conditions as measured by automatic CO_2 chambers. *Global Biogeochemical Cycles,* **17,** 1066, doi:10.1029/2002GB001946.

Bunting, M. J. (2003). Pollen-vegetation relationships in non-arboreal moorland taxa. *Review of Palaeobotany and Palynology,* **125,** 285–98.

Bunting, M. J., Morgan, C. R., van Bakel, M., and Warner, B. G. (1998). Pre-European settlement conditions and human disturbance of a coniferous swamp in southern Ontario. *Canadian Journal of Botany,* **76,** 1770–79.

Cahoon, D. R., Hensel, P. Rybczyk, J., McKee, K. L., Proffitt, E. E., and Perez, B. C. (2003). Mass tree mortality leads to mangrove peat collapse at Bay Islands, Honduras after Hurricane Mitch. *Ecology,* **91,** 1093–1105.

Cajander, A. K. (1913). Studien über die Moore Finnlands. *Acta Forestalia Fennica,* **2,** 1–208.

Calow, P. ed. (1998). *The encyclopedia of ecology and environmental management.* Blackwell Science, Oxford.

Camill, P. and Clark, J. S. (1998). Climate change disequilibrium of boreal permafrost peatlands caused by local processes. *American Naturalist,* **151,** 207–22.

Cantelmo Jr, A. J. and Ehrenfeld, J. G. (1999). Effects of microtopography on mycorrhizal infection in Atlantic white cedar (*Chamaecyparis thyoides* (L.) Mills.). *Mycorrhiza*, **8**, 175–80.

Carter, V. (1986). An overview of the hydrologic concerns related to wetlands in the U. S. *Canadian Journal of Botany*, **64**, 364–74.

CCFM (1995). *Defining sustainable forest management: a Canadian approach to criteria and indicators.* Canadian Council of Forest Ministers, Ottawa, Canada.

Chabot, B. F. and Hicks, D. J. (1982). The ecology of leaf life spans. *Annual Review of Ecology and Systematics*, **1982**, 229–59.

Chapin, C. T., Bridgham, S. D., Pastor, J., and Updegraff, K. (2003). Nitrogen, phosphorus, and carbon mineralisation in response to nutrient and lime additions in peatlands. *Soil Science*, **168**, 409–20.

Chapman, V. J. (ed.) (1977). *Ecosystems of the World 1. Wet coastal ecosystems.* Elsevier, Amsterdam.

Charman, D. (1998). Pool development on patterned fens in Scotland. In: *Patterned mires and mire pools*, (ed. V. Standen, J. H. Tallis, and R. Meade), pp. 39–54, British Ecological Society, London.

Charman, D. (2002). *Peatlands and environmental change.* John Wiley & Sons, Chichester.

Charman, D. J. and Chambers, F. M. (2004). Holocene environmental change: contributions from the peatland archive. *Holocene*, **14**, 1–6.

Charman, D. J. and Warner, B. G. (1997). The ecology of testate amoebae (Protozoa: Rhizopoda) in oceanic peatlands in Newfoundland, Canada: Modelling hydrological relationships for palaeoenvironmental reconstruction. *Écoscience*, **4**, 555–62.

Charman, D. J., Aravena, R., and Warner, B. G. (1994). Carbon dynamics in a forested peatland in north-eastern Ontario, Canada. *Journal of Ecology*, **82**, 55–62.

Chasar, L. I., Chanton, J. P., Glaser, P. H., Siegel, D. I., and Rivers, J. S. (2000). Radiocarbon and stable carbon isotopic evidence for transport and transformation of dissolved organic carbon, dissolved inorganic carbon, and CH_4 in a northern Minnesota peatland. *Global Biogeochemical Cycles*, **14**, 1095–1108.

Chason, D. B. and Siegel, D. I. (1986). Hydraulic conductivity and related physical properties of peat, Lost River Peatland, northern Minnesota. *Soil Science*, **142**, 91–99.

Chisholm, S., Davies, J. C., Mulamoottil, G. and Capatos, D. (1995). *Wetlands evaluation in Ontario: models for predicting wetland score.* Ontario Ministry of Natural Resources, Northeast Science and Technology Unit, Technical Report TR-025, South Porcupine.

Clair, T. R. (1998). Canadian freshwater wetlands and climate change. *Climate Change*, **40**, 163–5.

Clarkson, B. R., Thompson, K., Schipper, L. A., and McLeod, M. (1999). Moanatuatua Bog – proposed restoration of a New Zealand restiad peat bog ecosystem. In: *An international perspective on wetland rehabilitation*, (ed. W. Streever), pp. 127–37, Kluwer, Dordrecht.

Clarkson, B. R., Schipper, L. A., and Clarkson, B. D. (2004a). Vegetation and peat characteristics of restiad bogs on Chatham Island (Rekohu), New Zealand. *New Zealand Journal of Botany*, **42**, 293–312.

Clarkson, B. R., Schipper, L. A. and Lehmann, A. (2004b). Vegetation and peat characteristics in the development of lowland restiad peat bogs, North Island, New Zealand. *Wetlands*, **24**, 133–51.

Clymo, R. S. (1970). The growth of *Sphagnum*: Methods of measurement. *Journal of Ecology*, **58**, 13–49.

Clymo, R. S. (1984). The limits to peat bog growth. *Philosophical Transactions of the Royal Society of London B Biological Sciences*, **303**, 605–54.

Clymo, R. S. (1992). Models of peat growth. *Suo*, **43**, 127–36.

Clymo, R. S. and Duckett, J. G. (1986). Regeneration of *Sphagnum*. *New Phytologist*, **102**, 589–612.

Clymo, R. S. and Hayward, P. M. (1982). The ecology of *Sphagnum*. In: *Bryophyte ecology*, (ed. A. J. E. Smith), pp. 229–89, Chapman & Hall, London.

Clymo, R. S., Turunen, J. and Tolonen, K. (1998). Carbon accumulation in peatland. *Oikos*, **81**, 368–88.

Cobbaert, D., Rochefort, L., and Price, J. S. (2004). Experimental restoration of a fen plant community after peat mining. *Applied Vegetation Science*, **7**, 209–20.

Coles, B. and Coles, J. (1989). *People of the wetlands. Bogs, bodies and lake-dwellers*. Thames & Hudson, London.

Connolly, A., Kelly, L., Lamers, L. *et al.* (2002). Soaks. In: *Conservation and restoration of raised bogs: geological, hydrological and ecological studies*, (ed. M. G. C. Schouten), pp. 170–85. Department of Environment and Local Government, Dublin.

Cronberg, N. (1993). Reproductive biology of *Sphagnum*. *Lindbergia*, **17**, 69–82.

Cross, J. R. (1987). Unusual stands of birch on bogs. *Irish Naturalists' Journal*, **22**, 305–10.

Cross, J. R. (2002). Bog woodlands. Information sheet, Irish Peatland Conservation Council (*www.ipcc.ie/infobogwoodlandsfs.html*).

Crum, H. (1984). *North American flora. Series II, Part 11. Sphagnopsida*. New York Botanical Garden, New York.

Crum, H. (1988). *A focus on peatlands and peat mosses*. University of Michigan Press, Ann Arbor.

Dachnowski, A. (1912). *Peat deposits of Ohio: their origin, formation and uses*. Geological Survey of Ohio, Fourth Series Bulletin 16.

Dachnowski-Stokes, A. P. (1926). *Factors and problems in the selection of peatlands for different uses*. USDA Department Bulletin No 1419.

Dai, T. S. and Sparling, J. H. (1973). Measurement of hydraulic conductivity of peats. *Canadian Journal of Soil Science*, **53**, 21–6.

Daly, D., Johnston, P., and Flynn, R. (1994). The hydrodynamics of raised bogs: an issue for conservation. In: *The balance of water – present and future* (ed. T. Keane and E. Daly), Proceeding of the AGMET Group (Ireland) and Agricultural Group of the Royal Meteorological Society (UK), Trinity College, Dublin, September 7–9, 1994.

Damman, A. W. H. (1965). *Thin iron pans: their occurrence and the conditions leading to their development*. Canadian Forest Service, Newfoundland Forest Research Center, Info. Rep. N-X-2.

Damman, A. W. H. (1988). Spatial and seasonal changes in water chemistry and vegetation in an ombrogenous bog. In: *Vegetation structure in relation to carbon and nutrient economy* (ed. J. T. A. Verhoeven, G. W. Heil, and M. J. A. Werger), pp. 107–19. SPB Academic Publishing, The Hague.

Damman, A. W. H. and Dowhan, J. J. (1981). Vegetation and habitat conditions in Western Head Bog, a southern Nova Scotian plateau bog. *Canadian Journal of Botany*, **59**, 1343–59.

Daniels, R. E. and Eddy, A. (1990). *Handbook of European Sphagna*. 2nd edn., Institute of Terrestrial Ecology, Abbots Ripton.

Danks, H. V. and Rosenberg, D. M. (1987). Aquatic insects of peatlands and marshes in Canada: Synthesis of information and identification of needs for research. *Memoirs of the Entomological Society of Canada*, **140**, 163–74.

Davis, M. B. (1983). Holocene vegetational history of the eastern United States. In: *Late-quaternary environment of the United States, vol 2, the Holocene*, (ed. H. E. Wright Jr.), pp. 166–81, University of Minnesota Press, St. Paul.

Day, J. H., Rennie, P. J., Stanek, W., and Raymond, G. P. (eds) (1979). *Peat testing manual*. National Research Council of Canada, Associate Committee on Geotechnical Research, Technical Memorandum 125.

de Lange, P. R., Heenan, P. B., Clarkson, B. D., and Clarkson, B. R. (1999). *Sporadanthus* in New Zealand. *New Zealand Journal of Botany*, **37**, 413–31.

Deevey, E. S. and Flint, R. F. (1957). Postglacial hypsithermal interval. *Science*, **125**, 182–84.

Dennis, J. V. (1988). *The great cypress swamps*. Louisiana State University Press, Baton Rouge.

Dennis, W. M. and Batson, W. T. (1974). The floating log and stump communities in the Santee Swamp of South Carolina. *Castanea*, **39**, 166–70.

Desrochers, A. (2001). Les oiseaux: diversité et répartition. In: *Écologie des tourbières du Québec-Labrador*, (ed. S. Payette and L. Rochefort), pp. 159–73, Les Presses de l'Université Laval, Saint-Nicolas, Québec.

Devito, K., Waddington, J., and Branfireun, B. (1997). Flow reversals in peatlands influenced by local groundwater systems. *Hydrological Processes*, **11**, 103–10.

Dickinson, C. H. (1983). Micro-organisms in peatlands. In: *Ecosystems of the world. 4A. Mires: swamp, bog, fen and moor. General studies*, (ed. A. J. P. Gore), pp. 225–45, Elsevier, Amsterdam.

Dierssen, K. (1982). *Die wichstigsten Pflanzengesellschaften der Moore NW-Europas*. Conservatoire et Jardin botaniques, Gènève.

Dierssen, K. (1996). *Vegetation Nordeuropas*. Ulmer, Stuttgart.

Dinel, H., A. Larouche and Lévesque, P. E. M. (1983). Evaluation de deux méthodes de quantification de macrofossiles dans les materiaux tourbeux. *Naturaliste Canadien*, **110**, 429–34.

Domenico, P. A. and Schwarz, W. (1997). *Physical and chemical hydrogeology*, 2nd edn. John Wiley & Sons, New York.

Donaldson, G. M. (1996). Oribatida (Acari) associated with three species of *Sphagnum* at Spruce Hole Bog, New Hampshire, U.S.A. *Canadian Journal of Zoology*, **74**, 1706–12.

Dorrepaal, E., Aerts, R., Cornelissen, J. H. C., Callaghan, T. V., and van Logtestijn, R. S. P. (2003). Summer warming and increased winter snow cover affect *Sphagnum fuscum* growth, structure and production in a sub-arctic bog. *Global Change Biology*, **10**, 93–104.

Downie, I. S., Coulson, J. S., Foster, G. N., and Whitfield, D. P. (1998). Distribution of aquatic macroinvertebrates within peatland pool complexes in the Flow Country, Scotland. *Hydrobiologia*, **377**, 95–105.

Driessen, P. M. (1978). Peat soils. In *Soils and rice*, (ed. F. N. Ponnamperuma), pp. 763–79, International Rice Research Institute, Manila.

Driessen, P. M. and Dudal, R. (eds) (1991). *The major soils of the world. Lecture notes on their geography, formation, properties and use*. Wageningen Agricultural University, Wageningen.

Du Rietz, G. E. (1949). Huvudenheter och huvudgränser i svensk myrvegetation. *Svensk Botanisk Tidskrift*, **43**, 274–309 + pl. I–VI.

Du Rietz, G. E. (1954). Die Mineralbodenwasserzeigergrenze als Grundlage einer natürlichen Zweigliederung der nord- und mitteleuropäischen Moore. *Vegetatio*, **5–6**, 571–85.

Efremova, T.T., Efremov, S.P., and Melent'eva, N.V. (1997). The reserves and forms of carbon compounds in bog ecosystems of Russia. *Eurasian Soil Science*, **30**, 1318–25.

Egan, T. (ed.) (1998). *The future use of cutaway bogs. Lough Boora Parklands. Cutaway bogs conference, May 1997, Kilcormac, Ireland.* Brosna Press, Ferbane, Ireland.

Egan, T. (1999). A landscape uncloaked: Lough Boora Parklands – the national centre of cutaway boglands rehabilitation in Ireland. In: *Policies and Priorities for Ireland's Landscape Conference, Tullamore, Co. Offaly, Ireland, April 1999*, pp. 119–32, Heritage Council, Kilkenny.

El-Daoushy, F., Tolonen, K., and Rosenberg, R. (1982). Lead 210 and moss-increment dating of two Finnish *Sphagnum* hummocks. *Nature*, **296**, 429–31.

El-Kahloun, M., Boeye, D., van Haesebroeck, V., and Verhagen, B. (2003). Differential recovery of above- and below-ground rich fen vegetation following fertilization. *Journal of Vegetation Science*, **14**, 451–58.

Elkington, T., Dayton, N., Jackson, D. L., and Strachan, I. M. (2001). *National vegetation classification: Field guide to mires and heaths.* Joint Nature Conservation Committee, Peterborough.

Ellenberg, H. (1979). Zeigerwerte der Gefässpflanzen Mitteleuropas, 2nd edn., *Scripta Geobotanica*, **9**.

Elveland, J. (1979). Dammängar, silängar och raningar – norrländska naturvårdsobjekt [Irrigated and naturally flooded hay-meadows in North Sweden – a nature conservancy problem.] *Statens Naturvårdsverk PM*, **1174**, 1–124.

Elveland, J. (1993). Dynamik hos knottblomster på Storön vid norrbotten skusten [Population dynamics of *Microstylis monophyllos* (Orchidaceae) on the peninsula Storön on the coast of Norbotten, N Sweden.] *Svensk Botanisk Tidskrift*, **87**, 147–67.

Eurola, S. (1968). Über die Ökologie der nordfinnischen Moorvegetation im Herbst, Winter und Frühling. *Annales Botanici Fennici*, **5**, 83–97.

Eurola, S. and Holappa, K. (1985). The Finnish mire type system. *Aquilo, Ser. Botanica*, **21**, 101–10.

Ewel, K. C. (1990). Swamps. In: *Ecosystems of Florida*, (ed. R. L. Myers and J. J. Ewel), pp. 281–323, University of Central Florida Press, Orlando.

Faegri, K. and Iversen, J. (1989). *Textbook of pollen analysis*, 4th edn., John Wiley & Sons, Chichester.

Fagan, B. (2000). *The Little Ice Age – how climate made history 1300–1850.* Basic Books, New York.

Farrell, C. A. and Doyle, G. J. (2003). Rehabilitation of industrial cutaway Atlantic blanket bog in County Mayo, North-West Ireland. *Wetlands Ecology and Management*, **11**, 21–35.

Farrell, C.A. and Doyle, G. (1998). Rehabilitation of Atlantic raised bog industrial cutaway at Bellacorick, north-west Mayo. In: *Towards a conservation strategy for the bogs of Ireland*, (ed. G. O'Leary and F. Gormley), pp. 103–10, Irish Peatland Conservation Council, Dublin.

Faulkner, S. P., Turner Jr., W. H. and Gambrell, R. P. (1989). Field techniques for measuring wetland soil parameters. *Soil Science Society of America Journal*, **53**, 883–90.

Feehan, J. and O'Donovan, G. (1996). *The bogs of Ireland: an introduction to the natural, cultural and industrial heritage of Irish peatlands.* Environmental Institute, University College Dublin.

Fetter, C. W. (2001). *Applied hydrogeology*, 4th edn., Prentice-Hall, Upper Saddle River, NJ.

Finér, L. and Laine, J. (1998). Root dynamics at drained peatland sites of different fertility in southern Finland. *Plant and Soil*, **201**, 27–36.

Flather, C. H., Wilson, K. R., Dean, D. J., and McComb, W. C. (1997). Identifying gaps in conservation networks: of indicators and uncertainty in geographic-based analyses. *Ecological Applications*, **7**, 531–42.

Flensburg, T. (1965). Micro-vegetation of a mire. *Acta Phytogeographica Suecica*, **50**, 159–60.

Foster, D. R. (1984). The dynamics of *Sphagnum* in forest and peatland communities in southeastern Labrador, Canada. *Arctic*, **37**, 133–40.

Foster, D. R. and Fritz, S. C. (1987). Mire development, pool formation, and landscape processes on patterned fens in Dalarna, central Sweden. *Journal of Ecology*, **75**, 409–37.

Foster, D. R. and Wright Jr., H. E. (1990). Role of ecosystem development and climate change in bog formation in central Sweden. *Ecology*, **71**, 450–63.

Fowler, D., Hargreaves, K. J., Macdonald, J. A., and Gardiner, B. (1995). Methane and CO_2 exchange over peatland and the effects of afforestation. *Forestry*, **68**, 327–34.

Främbs, H. (1996). The importance of habitat structure and food supply for carabid beetles (Coleoptera, Carabidae) in peat bogs. *Memoirs of the Entomological Society of Canada*, **169**, 145–59.

Freeze, R. A. and Cheery, J. A. (1979). *Groundwater.* Prentice-Hall, Englewood Cliffs, NJ.

Frenzel, B. (1983). Mires – repositories of climatic information or self-perpetuating ecosystem. In: *Ecosystems of the world. 4A. Mires: swamp, bog, fen and moor. General studies*, (ed. A. J. P. Gore), pp. 35–65, Elsevier, Amsterdam.

Fuchsman, C. H. (1980). *Peat: industrial chemistry and technology.* Academic Press, New York.

Fuchsman, C. H. (1983). The humic acid problem and the prospects of peat utilization. In: *Proceedings of the International Symposium on Peat Utilization* (ed. C. H. Fuchsman and S. A. Spigarelli), pp. 477–93, Bemidji State University, Bemidji, Minnesota.

Gaudet, C. L. and Keddy, P. A. (1988). A comparative approach to predicting competitive ability from plant traits. *Nature*, **334**, 242–43.

Gaudig, G. (2002): Wachstum von Mooren in Kessellage – Gibt es Kesselmoore? *Greifswalder Geographische Arbeiten*, **26** 149–52.

Gelfer, B. Y. and Mankinen, G. W. (1985). USSR experience in the use of peat in agriculture and horticulture. In: *A technical and scientific conference on peat and peatlands, 16–20 June, 1985, Rivière-du-Loup, Québec*, pp. 239–63.

Gibbons, D. (1998). From observation to experiment. In: *Patterned mires and mire pools*, (ed. V. Standen, J. H. Tallis, and R. Meade), pp. 142–46, British Ecological Society, London.

Gignac, L. D. (1992). Niche structure, resource partitioning, and species interactions of mire bryophytes relative to climatic and ecological gradients in western Canada. *Bryologist*, **95**, 406–18.

Gignac, L. D., Vitt, D. H., Zoltai, S. C., and Bayley, S. E. (1991). Bryophyte response surfaces along climatic, chemical, and physical gradients in peatlands of western Canada. *Nova Hedwigia*, **53**, 27–71.

Gignac, L. D., Halsey, L. A., and Vitt, D. H. (2000). A bioclimatic model for the distribution of *Sphagnum*-dominated peatlands in North America under present climatic conditions. *Journal of Biogeography*, **27**, 1139–51.

Giller, P. S. and Malmqvist, B. (1998). *The biology of streams and rivers*. Oxford University Press. Oxford.

Glaser, P. H. (1992a). Ecological development of patterned peatlands. In: *The patterned peatlands of Minnesota*, (ed. D. B. Wright Jr, B. A. Coffin, and N. E. Aaseng), pp. 27–42, University of Minnesota Press, St. Paul.

Glaser, P. H. (1992b). Peat landforms. In: *The patterned peatlands of Minnesota*, (ed. D. B. Wright Jr, B. A. Coffin and N. E. Aaseng), pp. 3–14, University of Minnesota Press, St. Paul.

Glaser, P. H. (1992c). Vegetation and water chemistry. In: *The patterned peatlands of Minnesota* (ed. H. E. Wright Jr., B. A. Coffin and N. E. P. Asseng), pp. 15–26, University of Minnesota Press, St. Paul.

Glaser, P. H. (1998). The distribution and origin of mire pools. In: *Patterned mires and mire pools*, (ed. V. Standen, J. H. Tallis and R. Meade), pp. 4–25, British Ecological Society, London.

Glaser, P. H., Bennett, P. C., Siegel, D. I., and Romanowicz, E. A. (1996). Palaeo-reversals in groundwater flow and peatland development at Lost River, Minnesota, USA. *Holocene*, **6**, 413–21.

Glaser, P. H., Siegel, D. I., Romanowicz, E. A. and Ping Shen, Y. (1997). Regional linkages between raised bogs and the climate, groundwater, and landscape of northwestern Minnesota. *Journal of Ecology*, **85**, 3–16.

Glaser, P. H., Chanton, J. P., Morin, P. *et al.* (2004a). Surface deformations as indicators of deep ebullition fluxes in a large northern peatland. *Global Biogeochemical Cycles*, **18**, doi:10. 1029/2003GB002069.

Glaser, P. H., Hansen, B. C. S., Siegel, D. I., Reeve, A. S., and Morin, P. J. (2004b). Rates, pathways and drivers for peatland development in the Hudson Bay Lowlands, northern Ontario, Canada. *Journal of Ecology*, **92**, 1036–53.

Glob, P. V. (1998). *The bog people: iron-age man preserved*. Translated from Danish. Faber & Faber, London.

Glooschenko, W. A., Martini, I. P. and Clarke-Whistler, K. (1988). Salt marshes of Canada. In: *Wetlands of Canada* (ed. National Wetlands Working Group), pp. 349–77, Environment Canada, Ottawa.

Godley, E. J. (1960). The botany of southern Chile in relation to New Zealand and the Subantarctic. *Proceedings of the Royal Society, London. Series B*, **152**, 457–75.

Godwin, H. (1981). *The archives of the peat bogs*. Cambridge University Press, Cambridge.

Godwin, H. and Conway, V.M. (1939). The ecology of a raised bog near Tregaron, Cardiganshire. *Journal of Ecology*, **27**, 313–63.

Golterman, H. L. (with the assistance of Clymo, R. S.) (1969). *Methods for chemical analysis of fresh waters*. IBP Handbook No. 8. Blackwell, Oxford.

Gore, A. J. P. (1983a). Introduction. In: *Ecosystems of the world. 4A. Mires: swamp, bog, fen and moor. General studies*, (ed. A. J. P. Gore), pp. 1–34, Elsevier, Amsterdam.

Gore, A. J. P. (ed.) (1983b). *Ecosystems of the world 4B. Mires: swamp, bog, fen and moor. Regional studies*. Elsevier, Amsterdam.

Gorham, E. (1957). The development of peat lands. *Quarterly Review of Biology*, **32**, 145–66.

Gorham, E. (1991). Northern peatlands: role in the carbon cycle and probable responses to climatic warming. *Ecological Applications*, **1**, 182–95.

Gorham, E. and Hofstetter, R. H. (1971). Penetration of bog peats and lake sediments by tritium from atmospheric fallout. *Ecology*, **52**, 898–902.

Gorham, E. and Janssens, J. A. (1992). The paleorecord of geochemistry and hydrology in northern peatlands and its relation to global change. *Suo*, **43**, 117–26.

Gorham, E., Eisenreich, S. J., Ford, J., and Santelmann, M. V. (1985). The chemistry of bog waters. In: *Chemical processes in lakes*, (ed. W. Stumm), pp. 339–63, John Wiley & Sons, New York.

Gotelli, N. J. and Ellison, A. M. (2002). Nitrogen deposition and extinction risk in the northern pitcher plant, *Sarracenia purpurea. Ecology*, **83**, 2758–65.

Granberg, G., Mikkelä, C., Sundh, I., Svensson, B. H., and Nilsson, M. (1997). Sources of spatial variation in methane emission from mires in northern Sweden: A mechanistic approach in statistical modeling. *Global Biogeochemical Cycles*, **11**, 135–50.

Granberg, G., Sundh, I., Svensson, B. H., and Nilsson, M. (2001). Effects of temperature, and nitrogen and sulfur deposition, on methane emission from a boreal mire. *Ecology* **82**, 1982–98.

Granberg, K. (1986). The eutrophication of Lake Lestijärvi as a consequence of forest bog ditching. *Aqua Fennica*, **16**, 57–61.

Granhall, U. and Selander, H. (1973). Nitrogen fixation in a subarctic mire. *Oikos*, **24**, 8–15.

Granhall, U. and von Hofsten, A. (1976). Nitrogenase activity in relation to intracellular organisms in *Sphagnum* mosses. *Physiologia Plantarum*, **36**, 88–94.

Granlund, E. (1932). De svenska högmossarnas geologi. *Sveriges Geologiska Undersökning, Avhandlingar och uppsatser, ser C.*, **373**.

Grime, J. P. (2001). *Plant strategies, vegetation processes, and ecosystem properties*, 2nd edn., John Wiley & Sons, Chichester.

Gunnarsson, U. (2005). Global patterns of *Sphagnum* productivity. *Journal of Bryology*, **27**, 267–77.

Gunnarsson, U. and Rydin, H. (1998). Demography and recruitment of Scots pine on raised bogs in eastern Sweden and relationships to microhabitat differentiation. *Wetlands*, **18**, 133–41.

Gunnarsson, U. and Rydin, H. (2000). Nitrogen fertilisation reduces *Sphagnum* production in Swedish bogs. *New Phytologist*, **147**, 527–37.

Gunnarsson, U., Granberg, G., and Nilsson, M. (2004). Growth, production and interspecific competition in *Sphagnum*: effects of temperature, nitrogen and sulphur treatments on a boreal mire. *New Phytologist*, **163**, 349–59.

Hall, D. O. and Rao, K. K. (1999). *Photosynthesis*, 6th edn, Cambridge University Press, Cambridge.

Hallingbäck, T. (2001). Våtmarkens mossor – förlorarna vid kalkning. *Svensk Botanisk Tidskrift*, **95**, 166–79.

Halsey, L. A., Vitt, D. H., and Trew, D. O. (1997). Influence of peatlands on the acidity of lakes in northeastern Alberta, Canada. *Water, Air, and Soil Pollution*, **96**, 17–38.

Hånell, B. (1988). Postdrainage forest productivity of peatlands in Sweden. *Canadian Journal of Forest Research*, **18**, 1443–56.

Haraguchi, A., Iyobe, T., Nishijima, H., and Tomizawa, H. (2003). Acid and sea-salt accumulation in coastal peat mires of a *Picea glehnii* forest in Ochiishi, eastern Hokkaido, Japan. *Wetlands*, **23**, 229–35.

Hargreaves, K. J., Milne, R., and Cannell, M. G. R. (2003). Carbon balance of afforested peatland in Scotland. *Forestry*, **76**, 299–317.

Harris, A. G., McMurray, S. C., Uhlig, P. W. C., Jeglum, J. K., Foster, R. F., and Racey, G. D. (1996). Field guide to the wetland ecosystem classification for northwestern Ontario. *NWST Field Guide*, **FG-01**, 1–74.

Hayward, P. M. and Clymo, R. S. (1982). Profiles of water content and pore size in *Sphagnum* and peat, and their relation to peat bog ecology. *Proceedings of the Royal Society of London B Biological Sciences*, **215**, 299–325.

Hayward, P. M. and Clymo, R. S. (1983). The growth of *Sphagnum*: Experiments on, and simulation of, some effects of light flux and water-table depth. *Journal of Ecology*, **71**, 845–63.

Heathwaite, A. L. (ed.) (1993). *Mires. Process, exploitation, and conservation.* Translated from German *Moor- und Torfkunde* (ed. K. Göttlich), John Wiley & Sons, Chichester.

Heathwaite, A. L., Eggelsmann, R., and Göttlich, K.-H. (1993). Ecohydrology, mire drainage and mire conservation. In: *Mires. Process, exploitation and conservation* (ed. A. L. Heathwaite), Translated from German *Moor- und Torfkunde* (ed. K. Göttlich), pp. 417–84, John Wiley & Sons, Chichester.

Hebda, R. J., Gustavson, K., Golinski, K., and Calder, A. M. (2000). *Burns Bog ecosystem review. Synthesis report for Burns Bog, Fraser River delta, south-western British Columbia, Canada.* Environmental Assessment Office, Victoria BC.

Hedges, R. E. and Gowlett, J. A. J. (1986). Radiocarbon dating by accelerator mass spectrometry. *Scientific American*, January, 100–7.

Heeley, R. and Motts, W. (1976). A model for the evaluation of groundwater resources associated with wetlands. In: *Models for evaluation of freshwater wetlands* (ed. J. Larson), University of Massachusetts, Amherst.

Heery, S. (1993). *The Shannon floodlands: a natural history of the Shannon callows.* Tir Eolas, Newtownlynch.

Heijmans, M. M. P. D., Berendse, F., Arp, W. J. *et al.* (2001). Effects of elevated carbon dioxide and increased nitrogen deposition on bog vegetation in the Netherlands. *Journal of Ecology*, **89**, 268–79.

Heikkilä, R. and Lindholm, T. (1988). Distribution and ecology of *Sphagnum molle* in Finland. *Annales Botanici Fennici*, **25**, 11–19.

Heikkinen, K. (1990a). Transport of organic and inorganic matter in river,brook and peat mining water in the drainage basin of the River Kiiminkijoki. *Aqua Fennica*, **20**, 143–55.

Heikkinen, K. (1990b). Nature of dissolved organic matter in the drainage basin of a boreal humic river in northern Finland. *Journal of Environmental Quality*, **19**, 649–57.

Heikkinen, K. and Visuri, A. (1990): Effect of water quality on bacterioplankton densities in river, brook and peat mining water in the basin of humic River Kiiminkijoki, northern Finland. *Archiv für Hydrobiologie*, **119**, 215–30.

Heilman, P. E. (1966). Change in distribution and availability of nitrogen with forest succession on north slopes in interior Alaska. *Ecology*, **47**, 825–31.

Heilman, P. E. (1968). Relationship of availability of phosphorous and cations to forest succession and bog formation in interior Alaska. *Ecology*, **49**, 331–36.

Heinselman, M. L. (1963). Forest sites, bog processes, and peatland types in the Glacial Lake Agassiz Region, Minnesota. *Ecological Monographs*, **33**, 327–74.

Heinselman, M. L. (1965). String bogs and other patterned organic terrain near Seney, Upper Michigan. *Ecology*, **46**, 185–88.

Heinselman, M. L. (1970). Landscape evolution, peatland types, and the environment in the Lake Agassiz Peatlands Natural Area, Minnesota. *Ecological Monographs*, **40**, 235–61.

Heinselman, M. L. (1975). Boreal peatlands in relation to environment. In: *Coupling of land and water systems* (ed. A. D. Hasler), pp. 93–103, Springer-Verlag, Berlin.

Hellberg, E. Hörnberg, G., Östlund, L., and Zackrisson, O. (2003). Vegetation dynamics and disturbance history in three deciduous forests in boreal Sweden. *Journal of Vegetation Science*, **14**, 267–76.

Helmer, E. H., Urban, N. R., and Eisenreich, S. J. (1990). Aluminum geochemistry in peatland waters. *Biogeochemistry*, **9**, 247–76.

Hemond, H. (1980). Biogeochemistry of Thoreau's Bog, Concord, Massachusetts. *Ecological Monographs*, **50**, 507–26.

Heusser, C. J. (1989). Late Quaternary vegetation and climate of Southern Tierra del Fuego. *Quaternary Research*, **31**, 396–406.

Heusser, C. J. (1995). Palaeoecology of a *Donatia-Astelia* cushion bog, Magellanic Moorland-Subantarctic Evergreen Forest transition, southern Tierra del Fuego, Argentina. *Review of Paleobotany and Palynology*, **89**, 429–40.

Hilbert, D. W., Roulet, N., and Moore, T. (2000). Modelling and analysis of peatlands as dynamical systems. *Journal of Ecology*, **88**, 230–42.

Hingley, M. (1993). *Microscopic life in* Sphagnum. Naturalists' Handbook 20. Richmond Publishing, Slough.

Hogg, E. H. (1993). Decay potential of hummock and hollow *Sphagnum* peats at different depths in a Swedish raised bog. *Oikos*, **66**, 269–78.

Hooijer, A. (1996). Floodplain hydrology – an ecologically oriented study of the Shannon Callows, Ireland. PhD theisis, Vrije Universiteit, Amsterdam.

Hooijer, A. (2005). Hydrology of tropical wetland forests: recent research results from Sarawak peatswamps. In: *Forests – water – people in the humid tropics* (ed. M. Bonell and L. A. Bruijnzeel), pp. 447–61, Cambridge University Press, Cambridge.

Huber, U. M., Markgraf, V. and Schäbitz, F. (2004). Geographical and temporal trends in Late Quaternary fire histories of Fuego-Patagonia, South America. *Quaternary Science Reviews*, **23**, 1079–97.

Hughes, J. and Heathwaite, A. L. (1995). *Hydrology and hydrochemistry of British wetlands*. John Wiley & Sons, Chichester.

Hughes, P. D. M. and Barber, K. E. (2004). Contrasting pathways to ombrotrophy in three raised bogs from Ireland and Cumbria, England. *Holocene*, **14**, 65–77.

Hulme, P. D. (1994). A paleobotanical study of paludifying pine forest on the island Hailuoto, northern Finland. *New Phytologist*, **126**, 153–62.

Huntley, B. and Birks, H. J. B. (1983). *An atlas of past and present pollen maps for Europe 0–13,000 years ago*. Cambridge University Press, Cambridge.

Huntley, B., Cramer, W., Morgan, A.V., Prentice, H.C., and Allen, J.R.M. (1997). *Past and future rapid environmental changes: the spatial and evolutionary responses of terrestrial biota*. Springer-Verlag, Berlin.

Huston, M. and Smith, T. (1987). Plant succession: life history and competition. *American Naturalist*, **130**, 168–98.

Hutchinson, G. E. (1975). *A treatise on limnology. III Limnological botany*. John Wiley & Sons, New York.

Ilnicki, P. (2003). Agricultural production systems for organic soil conservation. In: *Organic soils and peat materials*, (ed. L.-E. Parent and P. Ilnicki), pp. 187–99, CRC Press, Boca Raton.

Ilnicki, P. and Zeitz, J. (2003). Irreversible loss of organic soil functions after reclamation. In: *Organic soils and peat materials*, (ed. L.-E. Parent and P. Ilnicki), pp. 15–32, CRC Press, Boca Raton.

Ingram, H. A. P. (1982). Size and shape in raised mire ecosystems: a geophysical model. *Nature*, **297**, 300–3.

Ingram, H. A. P. (1983). Hydrology. In: *Ecosystems of the world. 4A. Mires: swamp, bog, fen and moor. General studies*, (ed. A. J. P. Gore), pp. 67–158, Elsevier, Amsterdam.

Ingram, H. A. P. (1987). Ecohydrology of Scottish peatlands. *Transactions of the Royal Society Edinburgh: Earth Sciences*, **78**, 287–96.

IPCC (1997). *Revised 1996 IPCC guidelines for national greenhouse gas inventories*. Intergovernmental Panel on Climate Change, Geneva.

IPCC (2001). *Climate change 2001: The scientific basis*. Cambridge University Press, Cambridge, UK.

IPS (1984). *Russian–English–German–Finnish–Swedish peat dictionary*. International Peat Society, Helsinki.

Iqbal, R., Hotes, S., and Tachibana, H. (2005). Water quality restoration after damming and its relevance to vegetation succession in a degraded mire. *Journal of Environmental Systems and Engineering*, JSCE **790/VII-35**, 56–69.

Itämies, J. and Jarva-Kärenlampi, M.-L. (1989). Wolf spiders (Araneae, Lycosidae) on the bog at Pulkkila, Central Finland. *Memoranda Societatis pro Fauna et Flora Fennica*, **65**, 103–8.

Ivanov, K. E. (1981). *Water movements in mirelands*. Academic Press, London.

Janssen, C. R. (1968). Myrtle Lake: a late- and post-glacial pollen diagram from northern Minnesota. *Canadian Journal of Botany*, **46**, 1397–1408.

Janssen, C. R. (1992). The Myrtle Lake peatland. In: *The patterned peatlands of Minnesota* (ed. H. E. Wright Jr., B. A. Coffin, and N. E. P. Asseng), pp. 223–35, University of Minnesota Press, St. Paul.

Janssens, J. A. (1983). A quantitative method for stratigraphic analysis of bryophytes in Holocene peat. *Journal of Ecology*, **71**, 189–96.

Janssens, J. A. (1990). Methods in quaternary ecology. 11. Bryophytes. *Geoscience Canada*, **17**, 13–24.

Janssens, J.A., Hansen, B.C.S., Glaser, P.A., and Whitlock, C. (1992). Development of a raised-bog complex. In: *The patterned peatlands of Minnesota* (ed. H. E. Wright Jr., B. A. Coffin, and N. E. P. Asseng), pp. 189–221, University of Minnesota Press, St. Paul.

Jäppinen, J-P. and Hotanen, J-P. (1990). Effect of fertilization on the abundance of bryophytes in two drained peatland forests in Eastern Finland. *Annales Botanici Fennici*, **27**, 93–108.

Järvet, A. and Lode, E. (eds) (2003). *Ecohydrological processes in northern wetlands*. Tartu University Press, Tartu.

Järvinen, O. and Väisänen, R. A. (1978). Ecological zoo geography of north European waders, or why do so many waders breed in the north? *Oikos*, **30**, 496–507.

Jeglum, J. K. (1971). Plant indicators of pH and water level in peatlands at Candle Lake, Saskatchewan. *Canadian Journal of Botany*, **49**, 1661–76.

Jeglum, J. K. (1991). Definition of trophic classes in wooded peatlands by means of vegetational types and plant indicators. *Annales Botanici Fennici*, **28**, 175–92.

Jeglum, J. K. and Overend, R. P. (eds) (1991). *Peat and peatlands diversification and innovation: Volume I – Peatland forestry, 6–10 August 1989*. Canadian Society for Peat and Peatlands, Québec City.

Jeglum, J. K., Rothwell, R. L., Berry, G. J., and Smith, G. K. M. (1992). *A peat sampler for rapid survey*. Frontline Technical Note, Canadian Forest Service, Sault Ste. Marie, Ontario.

Jeglum, J. K., Kershaw, H. M., Morris, D. M., and Cameron, D. A. (2003). *Best forestry practices: a guide for the boreal forest in Ontario*. Canadian Forest Service, Sault Ste Marie, Ontario.

Jeschke, L., Knapp, H. D. and Succow, M. (2001). Mooregionen Europas. In: *Lantschaftsökologische Moorkunde*, (ed. M. Succow and H. Joosten), pp. 256–316, 2nd edn., E. Schweizerbart'sche Verlagsbuchhandlung, Stuttgart.

Joensuu, S., Ahti, E., and Vuollekoski, M. (2002) Effects of ditch network maintenance on the chemistry of run-off water from peatland forests. *Scandinavian Journal of Forest Research*, **17**, 238–47.

Johnson, L. C. and Damman, A. W. H. (1991). Species-controlled *Sphagnum* decay on a South Swedish raised bog. *Oikos*, **61**, 234–42.

Johnson, L. C. and Damman, A. W. H. (1993). Decay and its regulation in *Sphagnum* peatlands. *Advances in Bryology*, **5**, 249–96.

Jones, R. I. (1992). The influence of humic substances on lacustrine planktonic food chains. *Hydrobiologia*, **229**, 73–91.

Jongman, R. H. G., ter Braak, C. J. F., and van Tongeren, O. F. R. (1995). *Data analysis in community and landscape ecology*, 2nd edn, Cambridge University Press, Cambridge.

Joosten, H. and Clarke, D. (2002). *Wise use of mires and peatlands*. International Mire Conservation Group and International Peat Society.

Joosten, H. and Succow, M. (2001). Hydrogenetische Moortypen. In: *Lantschaftsökologische Moorkunde*, (ed. M. Succow and H. Joosten), pp. 234–40, 2nd edn., E. Schweizerbart'sche Verlagsbuchhandlung, Stuttgart.

Junk, W. J. (1983). Ecology of swamps on the middle Amazon. In: *Ecosystems of the world. 4B. Mires: swamp, bog, fen and moor. Regional studies*, (ed. A. J. P. Gore), pp. 269–94, Elsevier, Amsterdam.

Kadlec, R. H. (1994). Wetlands for water polishing: free water surface wetlands. In: *Global wetlands: Old World and New*, (ed. W. J. Mitsch), pp. 335–49, Elsevier, Amsterdam.

Kalaitzidis, S. and Christanis, K. (2002). Mineral matter in the Philippi peat in relation to peat/lignite-forming conditions in Greece. *Energy Sources*, **24**, 69–81.

Kalra, Y. P. and Maynard, D. G. (1991). *Methods manual for forest soil and plant analysis*. Forestry Canada, NW Region, Report NOR-X-319.

Karlsson, P. S., Svensson, B. M., and Carlsson, B. Å. (1996). The significance of carnivory for three *Pinguicula* species in a subarctic environment. *Ecological Bulletins*, **45**, 115–20.

Karofeld, E. (2004). Mud-bottom hollows: exceptional features in carbon-accumulating bogs? *Holocene*, **14**, 119–24.

Karofeld, E. and Pajula, R. (2003). Regularities in the formation and distribution of necrotic *Sphagnum* patches in raised bogs. In: *Ecohydrological processes in northern wetlands*, (ed. A. Järvet and E. Lode), pp. 149–54, Tartu University Press, Tartu.

Kaunisto, S. (1987). Effect of refertilization on the development and foliar nutrient contents of young Scots pine stands on drained mires of different nitrogen status. *Communicationes Instituti Forestalis Fenniae*, **140**, 1–58.

Kaunisto, S. (1997). Peatland forestry in Finland: problems and possibilities from the nutritional point of view. In: *Northern forested wetlands: ecology and management*, (ed. C. C. Trettin, M. F. Jurgensen, D. F. Grigal, M. R. Gale, and J. K. Jeglum), pp. 387–401, CRC Press, Boca Raton.

Kearns, F. L., Autin, W. J., and Gerdes, R.G. (1982). Occurrence and stratigraphy of organic deposits, St. Mary Parish, Louisiana. *Geological Society of America Meeting of NE and SE Sections, Abstracts with programs*, **14**, 30.

Keddy, P. A. (2000). *Wetland ecology. Principles and conservation*. Cambridge University Press, Cambridge.

Kellner, E. (2001a). Surface energy exchange and hydrology of a poor *Sphagnum* mire. PhD thesis, Uppsala University.

Kellner, E. (2001b). Surface energy fluxes and control of evapotranspiration from a Swedish *Sphagnum* mire. *Agricultural and Forest Meteorology*, **110**, 101–23.

Kellner, E. and Lundin, L.-C. (2001). Calibration of time domain reflectometry for water content in peat soil. *Nordic Hydrology*, **32**, 315–32.

Kellogg, L. E. and Bridgham, S. D. (2003). Phosphorus retention and movement across an ombrotrophic-minerotrophic peatland gradient. *Biogeochemistry*, **63**, 299–315.

Kenkel, N. C. (1988). Spectral analysis of hummock-hollow pattern in a weakly minerotrophic mire. *Vegetatio*, **78**, 45–52.

Kerner, A. (1863). *The plant life of the Danube Basin*. Translated from the German by H. S. Conard, 1951. Iowa State College Press, Ames.

Keys, D. (1983). Summary of the techniques used in the inventory of the peatlands of New Brunswick. In: *Proceedings of Peatland Inventory Methodology Workshop* (ed. S. M. Morgan and F. C. P. Pollett), pp. 75–87, Agriculture Canada, Ottawa and Environment Canada, St. John's.

Kilian, M. R., van Geel, B. and van der Plicht, J. (2000). [14]C AMS wiggle matching of raised bog deposits and models of peat accumulation. *Quaternary Science Reviews*, **19**, 1011–33.

Kilian, R., Hohner, M., Biester, H., Wallrabe-Adams, H. J., and Stern, C. R. (2003). Holocene peat and lake sediment tephra record from the southernmost Chilean Andes (53–55° S). *Revista Geológica de Chile*, **30**, 23–37.

Kimmins, J. P. (1997). Biodiversity and its relationship to ecosystem health and integrity. *Forest Chronicle*, **73**, 229–32.

King, W. (1685). Of the bogs, and loughs of Ireland. *Philosophical Transactions of the Dublin Society*, **15**, 947–60.

Kivinen, E. and Pakarinen, P. (1981). Geographical distribution of peat resources and major peatland complex types in the world. *Annales Academiae Scientiarum Fennicae, Series A III*, **132**, 1–28.

Klarqvist, M. (2001). Peat growth and carbon accumulation rates during the Holocene in Boreal mires. Ph.D. Thesis, Swedish University of Agricultural Sciences, Umeå.

Knott, J. F., Nuttle, W. K., and Hemond, H. F. (1987). Hydrologic parameters of salt marsh peat. *Hydrological Processes*, **1**, 211–20.

Kolari, K. K. (ed). (1983). Growth disturbances of forest trees. *Communicationes Instituti Forestalis Fenniae*, **116**, 1–208.

Kooijman, A. and Hedenäs, L. (1991). Differentiation in habitat requirements within the genus *Scorpidium*, especially between *S. revolvens* and *S. cossonii*. *Journal of Bryology*, **16**, 619–27.

Kooijman, A. M. and Bakker, C. (1995). Species replacement in the bryophyte layer in mires: the role of water type, nutrient supply and interspecific interactions. *Journal of Ecology*, **83**, 1–8.

Korhola, A. (1992). Mire induction, ecosystem dynamics and lateral extension on raised bogs in the southern coastal area of Finland. *Fennia*, **170**, 25–94.

Korhola, A., Virkanen, J., Tikkanen, M., and Blom, T. (1996). Fire-induced pH rise in a naturally acid hill-top lake, southern Finland: A palaeoecological survey. *Journal of Ecology*, **84**, 257–65.

Kotowski, W., van Andel, J., van Diggelen, R., and Hogendorf, J. (2001). Responses of fen plant species to groundwater level and light intensity. *Plant Ecology*, **155**, 147–56.

Kotowski, W. and van Diggelen, R. (2004). Light as an environmental filter in fen vegetation. *Journal of Vegetation Science*, **15**, 583–94.

Koutaniemi, L. (1999). Twenty-one years of string movements on the Liippasuo aapa mire, Finland. *Boreas*, **28**, 521–30.

Kreshtapova, V. N., Krupnov, R. A., and Uspenskayja, O. N. (2003). Quality of organic soils for agricultural use of cutover peatlands in Russia. In: *Organic soils and peat materials for sustainable agriculture*, (ed. L.-E. Parent and P. Ilnicki), pp. 175–86, CRC Press, Boca Raton.

Kuhry, P. and Vitt, D. H. (1996). Fossil carbon/nitrogen ratios as a measure of peat decomposition. *Ecology*, **77**, 271–75.

Kuhry, P., Nicholson, B. J., Gignac, L. D., Vitt, D. H., and Bayley, S. E. (1993). Development of *Sphagnum* dominated peatlands in boreal continental Canada. *Canadian Journal of Botany*, **71**, 10–22.

Kulczyński, S. (1949). Peat bogs of Polesie. *Mémoires de l'Académie Polonaise des Sciences et des Lettres*, **B15**.

Lafleur, P. M. (1990). Evapotranspiration from sedge-dominated wetland surface. *Aquatic Botany*, **37**, 341–53.

Lafleur, P. M., McCaugey, J. H., Joiner, D. W., Bartlett, P. A., and Jelinski, D. E. (1997). Seasonal trends in energy, water, and carbon dioxide fluxes at a northern boreal wetland. *Journal of Geophysical Research, Atmosphere*, **102**, 29009–20.

Lafleur, P. M., Roulet, N. T., Bubier, J. L., Frolking, S., and Moore, T. R. (2003). Interannual variability in the peatland-atmosphere carbon dioxide exchange at an ombrotrophic bog. *Global Biogeochemical Cycles*, **17**, 1036, doi:10.1029/2002GB001983, 2003.

Laiho, R. and Laine, J. (1994). Nitrogen and phosphorus stores in peatlands drained for forestry in Finland. *Scandinavian Journal of Forest Research*, **9**, 251–60.

Laiho, R. and Laine, J. (1997). Tree stand biomass and carbon content in an age sequence of drained pine mires in southern Finland. *Forest Ecology and Management*, **93**, 161–9.

Laiho, R., Vasander, H., Penttilä, T. and Laine, J. (2003). Dynamics of plant-mediated organic matter and nutrient cycling following water-level drawdowns in boreal peatlands. *Global Biogeochemical Cycles*, **17**, 1053, doi:10.1029/2002GB002015.

Laiho, R., Laine, J., Trettin, C. C. and Finér, L. (2004). Scots pine litter decomposition along drainage succession and soil nutrient gradients in peatland forests, and the effects of inter-annual weather variation. *Soil Biology and Biochemistry*, **36**, 1095–109.

Laine, J. (1989). Metsäojitettujen soiden luokittelu [Classification of peatlands drained for forestry]. *Suo*, **40**, 37–51.

Laine, J. and Vasander, H. (1996). Ecology and vegetation gradients of peatlands. In: *Peatlands in Finland*, (ed. H. Vasander), pp. 10–19, Finnish Peatland Society, Helsinki.

Laine, J., Vasander, H., and Sallantaus, T. (1995). Ecological effects of peatland drainage for forestry. *Environmental Reviews*, **3**, 286–303.

Laine, J., Silvola, J., Tolonen, K. *et al.* (1996). Effect of water-level drawdown on global climatic warming: Northern peatlands. *Ambio*, **25**, 179–84.

Laine, J., Komulainen, V.-M., Laiho, R. *et al.* (2004). *Lakkasuo – a guide to mire ecosystem*. Publication 26, Department of Forest Ecology, University of Helsinki.

Laitinen, J., Rehell, S., and Huttunen, A. (2005). Vegetation-related hydrotopographic and hydrologic classification for aapa mires (Hirvisuo, Finland). *Annales Botanici Fennici*, **42**, 107–21.

Landva, A. O., Korpijaakko, E. O., and Pheeney, P. E. (1983a). Geotechnical classification of peats and organic soils. In: *Testing of peats and organic soils*, (ed. P.M. Jarrett), pp. 37–51, Special Technical Publication 820, American Society for Testing and Materials, West Conshohocken, PA.

Landva, A. O., Pheeney, P. E., and Mersereau, D. E. (1983b). Undisturbed sampling of peat. In: *Testing of peats and organic soils*, (ed. P.M. Jarrett), pp. 141–56, Special Technical Publication 820, American Society for Testing and Materials, West Conshohocken, PA.

Lapen, D.R., Price, J.S. and Gilbert, R. (2000). Soil water storage dynamics in peatlands with shallow water tables. *Canadian Journal of Soil Science*, **80**, 43–52.

Lapen, D. R., Price, J. S. and Gilbert, R. (2005). Modelling two-dimensional steady-state groundwater flow and flow sensitivity to boundary conditions in blanket peat complexes. *Hydrological Processes*, **19**, 371–86.

Lappalainen, E. (1996). *Global peat resources*. International Peat Society, Jyskä, Finland.

Larson, D. J. and House, N. L. (1990). Insect communities of Newfoundland bog pools with emphasis on the Odonata. *Canadian Entomologist*, **122**, 469–501.

Lee, G. F., Bentley, E., and Amundson, R. (1975). Effects of marshes on water quality. In: *Coupling of land and water systems*, (ed. A. D. Hasler), pp.105–27, Springer-Verlag, Berlin.

Lee, J. A. and Studholme, C. J. (1992). Responses of *Sphagnum* species to polluted environments. In: *Bryophytes and lichens in a changing environment*, (ed. J. W. Bates and A. M. Farmer), pp. 314–32, Oxford University Press, Oxford.

Leinonen, A., Lindh, T., Paappanen, T. *et al.* (1998). Cultivation and production of reed canary grass for mixed fuel as a method for reclamation of a peat production area. In: *Peatland restoration and reclamation. Techniques and regulatory considerations*, (ed. T. Malterer, K. Johnson and J. Stewart), pp. 120–24, International Peat Society, Jyväskylä, Finland.

Lemasters, G. S., Bartelli, L. J., and Smith, M. R. (1983). Characterization of organic soils as energy sources. In: *Testing of peats and organic soils*, (ed. P.M. Jarrett), pp. 122–37, Special Technical Publication 820, American Society for Testing and Materials, West Conshohocken, PA.

Lesica, P. and Kannowski, P. B. (1998). Ants create hummocks and alter structure and vegetation of a Montana fen. *American Midland Naturalist*, **139**, 58–68.

Lévesque, M., Dinel, H., and Larouchc, A. (1988). *Guide to the identification of plant macrofossils in Canadian peatlands.* Land Resource Research Institute Publication No 1817, Agriculture Canada, Ottawa.

Lévesque, M., Morita, H., Schnitzer, M., and Mathur, S. P. (1980). *The physical, chemical and morphological features of some Quebec and Ontario peats.* Land Resource Research Institute Contribution 62, Agriculture Canada, Ottawa.

Limpens, J., Raymakers, J. T. A. G., Baar, J., Berendse, F., and Zijlstra, J. D. (2003). The interaction between epiphytic algae, a parasitic fungus and Sphagnum as affected by N and P. *Oikos*, **103**, 59–68.

Lindsay, R. A., Rigall, J., and Burd, F. (1985). The use of small-scale surface patterns in the classification of British peatlands. *Aquilo, Ser Botanica*, **21**, 69–79.

Linnaeus, C. (1732). *Iter Lapponicum. The Lapland journey.* Edited and translated by P. Graves, 1995. Lockharton Press, Edinburgh.

Lissey, A. (1971). Depression-focussed transient groundwater flow patterns in Manitoba. *Geological Association of Canada Special Papers*, **9**, 333–41.

Lonnstad, J. and Löfroth, M. (1994). *Myrskyddsplan för Sverige [Mire protection plan for Sweden.]* Naturvårdsverket, Solna.

Lu, X. and Wang, R. (eds) (1994). *Wetland environment and peatland utilization. International Conference 1994*, Changchun, China. Jilin People's Publishing House, Changchun.

Lucas, R.E. (1982). *Organic soils (histosols). Formation, distribution, physical and chemical properties and management for crop production.* Research report 435, FarScience, Michigan State University.

Lucas, R.E. and Davis J.F. (1961). Relationships between pH values of organic soils and availabilities of 12 plant nutrients. *Soil Science*, **92**, 177–81.

Luken, J. O. (1985). Zonation of *Sphagnum* mosses: Interactions among shoot growth, growth form, and water balance. *Bryologist*, **88**, 374–79.

Lundin, L. (1991). Retention or loss of nitrogen in forest wetlands. *Vatten*, **47**, 301–04.

Lundin, L. and Bergquist, B. (1990). Effects on water chemistry after drainage of a bog for forestry. *Hydrobiologia*, **196**, 167–81.

Lundqvist, G. (1951). En palsmyr sydost om Kebnekajse [In Swedish with English summary: A palsa mire southeast of Kebnekajse.]. *Geologiska Föreningens i Stockholm Förhandlingar*, **73**, 209–35.

Lundqvist, G. (1955). Myrar. *Atlas över Sverige (Vegetation)*, **41–42**, 1–6.

Mäkilä, M., Saarnisto, M., and Kankainen, T. (2001). Aapa mires as a carbon sink and source during the Holocene. *Journal of Ecology*, **89**, 589–99.

Malmer, N. (1962a). Studies on mire vegetation in the archaean area of Southwestern Götaland (South Sweden). I. Vegetation and habitat conditions on the Åkhult mire. *Opera Botanica*, **7(1)**, 1–322.

Malmer, N. (1962b). Studies on mire vegetation in the archaean area of southwestern Götaland (South Sweden). II. Distribution and seasonal variation in elementary constituents of some mire sites. *Opera Botanica*, **7(2)**, 1–67.

Malmer, N. and Wallén, B. (2004). Input rates, decay losses and accumulation rates of carbon in bogs during the last millenium: internal processes and environmental changes. *Holocene*, **14**, 111–17.

Malmer, N., Svensson, B. M., and Wallén, B. (1994). Interactions between *Sphagnum* mosses and field layer vascular plants in the development of peat-forming systems. *Folia Geobotanica et Phytotaxonomica*, **29**, 483–96.

Malmström, C. (1923). Degerö stormyr. *Meddelanden från Statens Skogsförsöksanstalt*, **20**, 1–176.

Malmström, C. (1931). Om faran för skogsmarkens försumpning i Norrland. En studie från Kulbäckslidens och Roklidens försöksfält, *Meddelanden från Statens Skogsförsöksanstalt*, **h 26**.

Malterer, T., Johnson, K., and Stewart, J. (eds) (1998). *Peatland restoration and reclamation: techniques and regulatory considerations*. International Peat Society, Jyväskylä.

Malterer, T. J., Verry, E. S., and Erjavec, J. (1992). Fiber content and degree of decomposition in peats: review of national methods. *Soil Science Society of America Journal*, **56**, 1200–11.

Mangerud, J., Andersen, S. T., Berglund, B. E., and Donner, J. J. (1974). Quaternary stratigraphy of Norden, a proposal for terminology and classification. *Boreas*, **3**, 109–28.

Mankinen, G. W. and Gelfer, B. (1982). *Comprehensive use of peat in the U.S.S.R.* US Department of Environment, Fifth Technical Conference on Peat, Bethesda, MD.

Marino, P. C. (1991). Dispersal and coexistence of mosses (Splachnaceae) in patchy habitats. *Journal of Ecology*, **79**, 1047–60.

Mark, A. F., Johnson, P. N., Dickinson, K. J. M., and Mcglone, M. S. (1995). Southern hemisphere patterned mires, with emphasis on southern New Zealand. *Journal of the Royal Society of New Zealand*, **25**, 23–54.

Markgraf, V. (1993). Younger Dryas in southernmost South America – an update. *Quaternary Science Reviews*, **12**, 351–55.

Martikainen, P. J., Nykänen, H., Crill, P., and Silvola, J. (1993). Effect of a lowered water table on carbon dioxide, methane and nitrous oxide due to forest drainage or mire sites of different trophy. *Plant and Soil*, **168–169**, 571–77.

Mataloni, G. (1999). Ecological studies on algal communities from Tierra del Fuego peat bogs. *Hydrobiologia*, **391**, 157–71.

Mathur, S. P. and Farnham, R. S. (1985). Geochemistry of humic substances in natural and cultivated peatlands. In: *Humic substances in soil, sediment, and water: Geochemistry, isolation, and characterization*, (ed. G. R. Aiken, D. M. McKnight, R. L. Wershaw, and P. MacCarthy), pp. 53–85, John Wiley & Sons, New York.

Mauquoy, D. and Barber, K. (1999). Evidence for climatic deteriorations associated with the decline of *Sphagnum imbricatum* Hornsch. ex Russ. in six ombrotrophic mires from northern England and the Scottish Borders. *Holocene*, **9**, 423–37.

Mauquoy, D., Blaauw, M., van Geel, B. *et al.* (2004). Late Holocene climatic changes in Tierra del Fuego based on multi-proxy analyses of peat deposits. *Quaternary Research*, **61**, 148–58.

McAndrews, J. H. (1966). Postglacial history of prairie, savanna and forest in northwestern Minnesota. *Torrey Botanical Club Memoirs*, **22**, 1–72.

McBride, M. B. (1994). *Environmental chemistry of soils*. Oxford University Press, Oxford.

McCulloch, R. D. and Davies, S. J. (2001). Late-glacial and Holocene palaeoenvironmental change in the central Strait of Magellan, southern Patagonia. *Palaeogeography, Palaeoclimatology, Palaeoecology*, **173**, 143–73.

McKee, K. L. and Faulkner, P. L. (2000). Mangrove peat analysis and the reconstruction of vegetation history at the Pelican Cays, Belize. *Atoll Research Bulletin*, **468**, 45–58.

McKerracher, A. (1987). The moss lairds. *Scots Magazine*, **127**, 425–33.

McLaren, B. E. and Jeglum, J. K. (1998). Black spruce growth and foliar nutrient responses to drainage and fertilization: Wally Creek, Ontario. *Forest Chronicle*, **74**, 106–15.

McNeil, P. and Waddington, J. M. (2003). Moisture controls on *Sphagnum fuscum* growth and CO_2 exchange on a cutover bog. *Journal of Applied Ecology*, **40**, 354–67.

McQueen, C. B. (1990). *Field guide to the peat mosses of boreal North America*, University Press of New England, Hanover, NH.

Mellars, P. and Dark, P. (1998). *Star Carr in context*. McDonald Institute for Archaeological Research, Cambridge.

Melloh, R. A. and Crill, P. M. (1995). Winter methane dynamics beneath ice and in snow in a temperate poor fen. *Hydrological Processes*, **9**, 947–56.

Mengel, K. and Kirkby, E. A. (1982). *Principles of plant nutrition*, 3rd edn., International Potash Institute, Bern.

Metsävainio, K. (1931). Untersuchungen über das Wurzelsystem der Moorpflanzen. *Annales Botanici Societatis Zoologicae Botanici Fennici 'Vanamo'*, **1(1)**, 1–418.

Miller, R. M., Smith, C. I., Jastrow, J. D., and Bever, J. D. (1999). Mycorrhizal status of the genus *Carex* (Cyperaceae). *American Journal of Botany*, **86**, 547–53.

Minkkinen, K., Korhonen, R., Savolainen, I., and Laine, J. (2002). Carbon balance and radiative forcing of Finnish peatlands 1900–2100 – the impact of forestry drainage. *Global Change Biology*, **8**, 785–99.

Minkkinen, K. and Laine, J. (1998). Long-term effect of forest drainage on the peat carbon stores of pine mires in Finland. *Canadian Journal of Forest Research*, **28**, 1267–75.

Mitchell, E. A. D., Buttler, A., Grosvernier, P. *et al*. (2000). Relationships among testate amoebae (Protozoa), vegetation and water chemistry in five *Sphagnum*-dominated peatlands in Europe. *New Phytologist*, **145**, 95–106.

Mitsch, W. J. (ed.) (1994). *Global wetlands: Old World and New*. Elsevier, Amsterdam.

Mitsch, W. J. and Gosselink, J. G. (2000). *Wetlands*, 3rd edn, John Wiley & Sons, New York.

Moen, A. (1995a). Regional variation and conservation of mire ecosystems. *Gunneria*, **70**, 1–344.

Moen, A. (1995b). Vegetational changes in boreal rich fens induced by haymaking: management plan for the Sølendet nature reserve. In: *Restoration of temperate wetlands*, (ed. B. D. Wheeler, S. C. Shaw, W. J. Fojt, and R. A. Robertson), pp. 167–81, John Wiley & Sons, Chichester.

Moen, A. (1999). *National atlas of Norway: Vegetation*. Norwegian Mapping Authority, Hønefoss.

Moen, A. (2002). Mires and peatlands in Norway: status, distribution, and nature conservation. In: *Third interntional symposium on the biology of Sphagnum, Uppsala-Trondheim August 2002: excursion guide*, (ed. K. Thingsgaard and K. I. Flatberg), pp. 41–60, Norges teknisk-naturvitenskaplige universitet, Vitenskapsmuseet, Trondheim.

Money, R. P. and Wheeler, B. D. (1999). Some critical questions concerning the restorability of damaged raised bogs. *Applied Vegetation Science*, **2**, 107–16.

Moore, D. M. (1979). Southern oceanic wet-heathlands (including Magellanic Moorland). In: *Heathlands and related shrublands. Descriptive studies* (ed R. L. Specht), pp. 489–497, Ecosystems of the World 9A, Elsevier, Amsterdam.

Moore, D. M. (1983). *Flora of Tierra del Fuego*. Anthony Nelson, Oswestry, UK.

Moore, P. D. (1993). The origin of blanket mire, revisited. In: *Climate change and human impact on the landscape* (ed. F. M. Chambers), pp. 217–24, Chapman & Hall, London.

Moore, P. D., Webb, J. A., and Collinson, M. E. (1991). *Pollen analysis*, 2nd edn, Blackwell Scentific, Oxford.

Moore, T. R. (1989). Growth and net production of *Sphagnum* at five fen sites, subarctic eastern Canada. *Canadian Journal of Botany*, **67**, 1203–7.

Moore, T. R. (2003). Dissolved organic carbon in a northern boreal landscape. *Global Biogeochemical Cycles*, **17**, 1109, doi:10.1029/2003GB002050, 2003.

Moore, T. R., Bubier, J. L., Frolking, S. E., Lafleur, P. M., and Roulet, N. T. (2002). Plant biomass and production and CO_2 exchange in an ombrotrophic bog. *Journal of Ecology*, **90**, 25–36.

Moore, T. R., Roulet, N. T., and Waddington, J. M. (1998). Uncertainty in predicting the effect of climatic change on the carbon cycling of Canadian peatlands. *Climatic Change*, **40**, 229–45.

Morrison, I. K. (1974). *Mineral nutrition of conifers with special reference to nutrient status interpretation: a review of literature.* Publication 1343, Department of Environment, Canadian Forest Service, Ottawa.

Myers, R. L. and Ewel, J. J. (1990). *Ecosystems of Florida*. University of Central Florida Press, Orlando.

Myllys, M. (1996). Agriculture on peatlands. In: *Peatlands in Finland*, (ed. H. Vasander), pp. 64–71, Finnish Peatland Society, Helsinki.

Näsholm, T., Ekblad, A., Nordin, A., Giesler, R., Högberg, M., and Högberg, P. (1998). Boreal forest plants take up organic nitrogen. *Nature*, **392**, 914–16.

National Wetlands Working Group (1988). *Wetlands of Canada*. Environment Canada, Ottawa.

National Wetlands Working Group (1997). *The Canadian wetland classification system*, 2nd edn., University of Waterloo, Canada.

Natural Resource Conservation Service. (2002). *Reed canary grass – fact sheet.* United States Department of Agriculture.

Neiland, B. J. (1971). The forest-bog complex in southeast Alaska. *Vegetatio*, **22**, 1–64.

Nichols, D. S. (1998). Temperature of upland and peatland soils in a north central Minnesota forest. *Canadian Journal of Soil Science*, **78**, 493–509.

Nichols, D. S. and Brown, J. M. (1980). Evaporation from a *Sphagnum* moss surface. *Journal of Hydrology*, **48**, 289–302.

Nicholson, B. J. and Gignac, L. D. (1995). Ecotope dimensions of peatland bryophyte indicator species along gradients in the Mackenzie River Basin, Canada. *Bryologist*, **98**, 437–51.

Nieminen, M. (2004). Export of dissolved organic carbon, nitrogen and phosphorus following clear-cutting of three Norway spruce forests growing on drained peatlands in southern Finland. *Silva Fennica*, **38**, 123–32.

Niering, W. A. and Warren, R. S. (1980). Vegetation patterns and processes in New England salt marshes. *BioScience*, **30**, 301–07.

Nilsson, M., Bååth, M., and Söderström, B. (1992). The microfungal communities of a mixed mire in northern Sweden. *Canadian Journal of Botany*, **70**, 272–76.

Nilsson, M., Klarqvist, M., Bohlin, E., and Possnert, G. (2001a). Variation in ^{14}C age of macrofossils and different fractions of minute peat samples dated by AMS. *Holocene*, **11**, 579–86.

Nilsson, M., Mikkelä, C., Sundh, I., Granberg, G., Svensson, B. H. and Ranneby, B. (2001b). Methane emission from Swedish mires: National and regional budgets and dependence on mire vegetation. *Journal of Geophysical Research*, **106**, 847–60.

Noble, M. G., Lawrence, D. B., and Streveler, G. P. (1984). *Sphagnum* invasion beneath an evergreen forest canopy in southeastern Alaska. *Bryologist*, **87**, 119–27.

Nömmik, H. (1974). Ammonium chloride-imidazole extraction procedure for determining titratable acidity, exchangeable base cations, and cation exchange capacity in soils. *Soil Science*, **118**, 254–62.

Nordbakken, J.-F. (2001). Fine-scale five-year vegetation change in boreal bog vegetation. *Journal of Vegetation Science*, **12**, 771–78.

Nordbakken, J.-F., Ohlson, M., and Högberg, P. (2003). Boreal bog plants: nitrogen sources and uptake of recently deposited nitrogen. *Environmental Pollution*, **126**, 191–200.

Nykänen, H., Alm, J., Silvola, J., Tolonen, K., and Martikainen, P. J. (1998). Methane fluxes on boreal peatlands of different fertility and the effect of long-term experimental lowering of the water table on flux rates. *Global Biogeochemical Cycles*, **12**, 53–69.

Nykänen, H., Vasander, H., Huttunen, J. T., and Martikainen, P. J. (2002). Effect of experimental nitrogen load on methane and nitrous oxide fluxes on ombrotrophic boreal peatlands. *Plant and Soil*, **242**, 147–55.

Ohlson, M. and Dahlberg, B. (1991). Rate of peat increment in hummock and lawn communities on Swedish mires during the last 150 years. *Oikos*, **61**, 369–78.

Ohlson, M., Økland, R. H., Nordbakken, J.-F., and Dahlberg, B. (2001). Fatal interactions between Scots pine and *Sphagnum* mosses in bog ecosystems. *Oikos*, **94**, 425–32.

Øien, D. I. (2004). Nutrient limitation in boreal rich-fen vegetation: A fertilization experiment. *Journal of Vegetation Science*, **7**, 119–32.

Øien, D. I. and Moen, A. (2001). Nutrient limitations in boreal plant communities and species influenced by scything. *Applied Vegetation Science*, **4**, 197–206.

Økland, R. H. (1990a). A phytoecological study of the mire Northern Kisselbergmosen, SE Norway. II. Identification of gradients by detrended (canonical) correspondence analysis. *Nordic Journal of Botany*, **10**, 79–108.

Økland, R. H. (1990b). A phytoecological study of the mire Northern Kisselbergmosen, SE Norway. III. Diversity and habitat niche relationships. *Nordic Journal of Botany*, **10**, 191–220.

Økland, R. H. (1990c). Regional variation in SE Fennoscandian mire vegetation. *Nordic Journal of Botany*, **10**, 285–310.

Økland, R. H., Økland, T., and Rydgren, K. (2001a). A Scandinavian perspective on ecological gradients in north-west European mires: reply to Wheeler and Proctor. *Journal of Ecology*, **89**, 481–86.

Økland, R. H., Økland, T., and Rydgren, K. (2001b). Vegetation-environment relationships of boreal spruce swamp forests in Østmarka Nature Reserve, SE Norway. *Sommerfeltia*, **29**, 1–190.

Okruszko, H. (1993). Transformation of fen-peat soil under the impact of draining. Polish Academy of Sciences Publication **406**, 3–75.

Okruszko, H. (1996). Agricultural use of peatlands. In: *Global peat resources*, (ed. E. Lappalainen), pp. 303–9, International Peat Society, Jyskä.

Okruszko, H. and Ilnicki, P. (2003). The moorsh horizons as quality indicators of reclaimed organic soils. In: *Organic soils and peat materials*, (ed. L.-E. Parent and P. Ilnicki), pp. 1–14, CRC Press, Boca Raton.

Oldfield, F., Appelby, P., Cambray, R. *et al.* (1979). ^{210}Pb, ^{137}Cs, and ^{239}Pu profile in ombrotrophic peat. *Oikos*, **33**, 40–5.

Oldfield, F., Thompson, R., and Barber, K. (1978). Changing atmospheric fallout of magnetic particule in recent ombrotrophic peat sections. *Science*, **199**, 679–80.

Olsson, I. U. (1986). Radiometric dating. In: *Handbook of Holocene palaeoecology and palaeohydrology*, (ed. B. E. Berglund. with the assistance of M. Ralska-Jasiewiczowa), pp. 273–312, John Wiley & Sons, Chichester.

Osvald, H. (1923). Die Vegetation des Hochmoores Komosse. *Svenska Växtsociologiska Sällskapets Handlingar*, **1**, 1–436.

Osvald, H. (1925). Die Hochmoortypen Europas. *Veröffentlichungen des Geobotanischen Institutes Rübel in Zürich*, **3**, 707–23.

Osvald, H. (1949). Notes on the vegetation of British and Irish mosses. *Acta Phytogeographica Suecica*, **26**, 1–62.

Paarlahti K., Reinikainen A., and Veijalainen H. (1971). Nutritional diagnosis of Scots pine stands by needle and peat analysis. *Communications Instituti Forestalis Fenniae*, **61**(1), 1–110.

Paavilainen, E. and Päivänen, J. (1995). *Peatland forestry. Ecology and principles.* Springer-Verlag, Berlin.

Page, S. E., Siegert, F., Rieley, J. O., Boehm, H-D. V., Jaya, A., and Limin, S. (2002). The amount of carbon released from peat and forest fires in Indonesia during 1997. *Nature*, **420**, 61–65.

Painter, T. J. (1995). Chemical and microbiological aspects of the preservation process in *Sphagnum* peat. In: *Bog bodies: new discoveries and new perspectives*, (ed. R. C. Turner and R. G. Scaife), pp. 88–99, British Museum Press, London.

Painter, T. J. (1998). Carbohydrate polymers in food preservation: an integrated view of the Maillard reaction with special reference to discoveries of preserved foods in *Sphagnum*-dominated peat bogs. *Carbohydrate Polymers*, **36**, 335–47.

Päivänen, J. (1966). The distribution of rainfall in different types of forest stands (In Finnish with English summary.] *Silva Fennica*, **119**(3).

Päivänen, J. (1973). Hydrologic conductivity and water retention in peat soils. *Acta Forestalia Fennica*, **129**, 1–70.

Päivänen, J. (1982). Main physical properties of peat soils. In: *Peatlands and their utilization in Finland*, (ed. J. Laine), pp. 33–36, Finnish Peatland Socity, Helsinki.

Pakarinen, P. (1984). Definitions of peats and organic sediments. *Bulletin of the International Peat Society*, **15**, 1–7.

Pakarinen, P. (1995). Classification of boreal mires in Finland and Scandinavia: A review. *Vegetatio*, **118**, 29–38.

Parent, L.-E. and Ilnicki, P. (eds) (2003). *Organic soils and peat materials for sustainable agriculture.* CRC Press, Boca Raton.

Parkyn, L., Stoneman, R. E., and Ingram, H. A. P. (eds) (1997). *Conserving peatlands.* CAB International, Wallingford.

Pastor, J., Peckham, B., Bridgham, S. D., Weltzin, J., and Chen, J. (2002). Plant community dynamics, nutrient cycling, and alternative stable equilibria in peatlands. *American Naturalist*, **160**, 553–68.

Patterson, W. A., Edwards, K. J. and Maguire, D. J. (1987). Macroscopic charcoal as fossil indicator of fire. *Quaternary Science Review*, **6**, 3–23.

Paulissen, M. (2004). Effects of nitrogen enrichment on bryophytes in fens. PhD thesis, Utrecht University.

Peat, H. J. and Fitter, A. H. (1993). The distribution of arbuscular mycorrhizas in the British flora. *New Phytologist*, **125**, 845–54.

Pejler, B. and Berzins, B. (1993). On the ecology of mire rotifers. *Limnologica*, **23**, 295–300.

Pendall, E., Markgraf, V., White, J. W. C., and Dreier, M. (2001). Multiproxy record of late Pleistocene-Holocene climate and vegetation changes from a peat bog in Patagonia. *Quaternary Research*, **55**, 168–78.

Pfadenhauer, J. and Klötzli, F. (1996). Restoration experiments in middle European wet terrestrial ecosystems: An overview. *Vegetatio*, **126**, 101–15.

Pielou, E. C. (1998). *Fresh water*. University of Chicago Press, Chicago.

Pisano, E. (1983). The Magellanic tundra complex. In: *Ecosystems of the world. 4B. Mires: swamp, bog, fen and moor. Regional studies*, (ed. A. J. P. Gore), pp. 295–329, Elsevier, Amsterdam.

Porter, S. C. (ed.) (1983). *Late-quaternary environments of the United States. Vol 1, The late Pleistocene*, University of Minnesota Press, St. Paul.

Prevost, M., Plamondon, A. P., and Belleau, P. (1999). Effects of drainage of a forested peatland on water quality and quantity. *Journal of Hydrology*, **214**, 130–43.

Price, J. S. (1991). Evaporation from a blanket bog in a foggy coastal environment. *Boundary-Layer Meteorology*, **51**, 391–406.

Price, J. S. (2003). Role and character of seasonal peat soil deformation on the hydrology of undisturbed and cutover peatlands. *Water Resource Research*, **39**, 1241, doi:10. 1029/2002WR001302.

Price, J. S. and FitzGibbon, J. E. (1987). Groundwater analysis and streamflow in a subarctic wetland, Saskatchewan. *Canadian Journal of Earth Sciences*, **24**, 2074–81.

Price, J. S. and Maloney, D. A. (1994). Hydrology of a patterned bog-fen complex in southeastern Labrador, Canada. *Nordic Hydrology*, **25**, 313–30.

Price, J. S. and Schlotzhauer, S. M. (1999). Importance of shrinkage and compression in determining water analyze changes in peat: the case of a mined peatland. *Hydrological Processes*, **13**, 2591–2601.

Price, J. S. and Woo, M. K. (1990). Studies of a subarctic coastal marsh. III. Modelling the subsurface water fluxes and chloride distribution. *Journal of Hydrology*, **120**, 1–13.

Price, J. S., Rochefort, L., and Campeau, S. (2002). On the use of shallow basins to restore cutover peatlands. *Restoration Ecology*, **10**, 259–66.

Price, J. S., Heathwaite, A. L., and Baird, A. J. (2003) Hydrological processes in abandoned and restored peatlands: an overview of management approaches. *Wetlands Ecology and Management*, **11**, 65–83.

Price, J. S., Branfireun, B. A., Waddington, J. M., and Devito, K. J. (2005). Advances in Canadian wetland hydrology, 1999–2003. *Hydrological Processes*, **19**, 201–14.

Proctor, M. C. F. (1992). Regional and local variation in the chemical composition of ombrogenous mire waters in Britain and Ireland. *Journal of Ecology*, **80**, 719–36.

Punt, W., Blackmore, S., Hoen, P. P., and Stafford, P. J. (2003). *The northwest European pollen flora*, vol 8. Elsevier, Amsterdam.

Quinton, W. L. and Roulet, N. T. (1998). Spring and summer runoff hydrology of a subarctic patterned wetland. *Arctic and Alpine Research*, **30**, 285–94.

Rabassa J., Heusser, C. J., and Coronato, A. (1989). Peat-bog accumulation rate in the Andes of Tierra del Fuego and Patagonia (Argentina and Chile) during the last 43000 years. *Pirineos*, **133**, 113–22.

Rabassa, J., Coronato, A., and Roig, C. (1996). The peat bogs of Tierra del Fuego, Argentina. In: *Global peat resources* (ed. E. Lappalainen), pp. 261–266. International Peat Society, Jyskä, Finland.

Rabassa, J., Coronato, A., Bujalesky, G. *et al.* (2000). Quaternary of Tierra del Fuego, Southernmost South America: an updated review. *Quaternary International*, **68–71**, 217–40.

Racey, G. D., Harris, A. G., Jeglum, J. K., Foster, R. F., and Wickware, G. M. (1996). Terrestrial and wetland ecosites of northwestern Ontario. *NWST Field Guide*, **FG-02**, 1–94.

Ramsar (1987). *Convention on wetlands (Ramsar, Iran, 1971), as amended in 1982 and 1987*. Ramsar Convention Bureau, Gland.

Ratcliffe, D. A. and Walker, D. (1958). The Silver Flowe, Galloway, Scotland. *Journal of Ecology*, **46**, 407–45.

Ravazzi, C. (2003). An overview of the Quaternary continental stratigraphic units based on biological and climatic events in Italy. *Italian Journal of Quaternary Science*, **16**, 11–18.

Raven, P. H., Evert, R. F., and Eichhorn, S. E. (1999). *Biology of plants*, 6th edn, W. H. Freeman, New York.

Reader, R. J. (1975). Competitive relationships of some bog ericads for major insect pollinators. *Canadian Journal of Botany*, **53**, 1300–05.

Redbo-Torstensson, P. (1994). The demographic consequences of nitrogen fertilization of a population of sundew, *Drosera rotundifolia. Acta Botanica Neerlandica*, **43**, 175–88.

Reimer, P. J., Baillie, M.G.L., Bard, E. *et al.* (2004). IntCal04 Terrestrial radiocarbon age calibration, 0–26 cal kyr BP. *Radiocarbon*, **46**, 1029–58.

Reinikainen, A., Vasander, H., and Lindholm, T. (1984). Plant biomass and primary production of southern boreal mire-ecosystems in Finland. *Proceedins of the 7th International Peat Congress, Dublin*, **4**, 1–20.

Richardson, C. J. (1985). Mechanisms controlling phosphorus retention capacity in freshwater wetlands. *Science*, **228**, 1424–27.

Richardson, J. L. and Vepraskas, M. J. (2001). *Wetland soils: genesis, hydrology, landscapes, and classification*. Lewis Publishers, Boca Raton.

Richter, C. and Dainty, J. (1989). Ion behavior in plant cell walls. I. Characterization of the *Sphagnum russowii* cell wall ion exchanger. *Canadian Journal of Botany*, **67**, 451–59.

Rieley, J. O., Page, S. E., and Shepherd, P. A. (1997). Tropical bog forests of South East Asia. In: *Conserving peatlands* (ed. L. Parkyn, R. E. Stoneman, and H. A. P Ingram), pp. 35–41, CAB International, Wallingford.

Riley, J. L. (1989). *Ontario peatland inventory project: laboratory methods for testing peat*. Miscellaneous Papers 145, Ontario Geological Survey, Sudbury.

Riley, J. L. and Michaud, L. (1994). *Ontario peatland inventory: field work methods*. Miscellaneous Papers 155, Ontario Geological Survey, Sudbury.

Rochefort, L., Campeau, S., and Bugnon, J-L. (2002). Does prolonged flooding prevent or enhance regeneration and growth of *Sphagnum. Aquatic Botany*, **74**, 327–41.

Rochefort, L., Quinty, F., Campeau, S., Johnson, K., and Malterer, T. J. (2003). North American approach to the restoration of *Sphagnum* dominated peatlands. *Wetlands Ecology and Management*, **11**, 3–20.

Rodhe, H. and Svensson, B. (1995). Impact on the greenhouse effect of peat mining and combustion. *Ambio*, **24**, 221–25.

Rodwell, J. S. (ed.) (1991). *British plant communities. Volume 2. Mires and heaths.* Cambridge University Press, Cambridge.

Roig Jr., F., Roig, C., Rabassa, J., and Boninsegna, J. (1996). Fuegian floating tree-ring chronology from subfossil *Nothofagus* wood. *Holocene* **6**, 469–476.

Romanov, V. V. (1968a). *Hydrophysics of bogs.* Israel Program for Scientific Translations, Jerusalem.

Romanov, V. V. (1968b). *Evaporation from bogs in the European territory of the USSR.* Israel Program for Scientific Translations, Jerusalem.

Rosenberg, D. M. and Danks, H. V. (eds) (1987). *Aquatic insects of peatland and marshes in Canada.* Memoirs of the Entomological Society of Canada, vol. 140.

Rosenberg, D. M., Wiens, A. P., and Bilyj, B. (1988). Chironomidae (Diptera) of peatlands in northwestern Ontario Canada. *Holarctic Ecology*, **11**, 19–31.

Rosenberg, N. J., Blad, B. L. and Verma, S. B. (1983). *Microclimate – the biological environment*, 2nd edn, John Wiley & Sons, Chichester.

Rothwell, R. L. (1991). Substrate environments on drained and undrained peatlands, Wally Creek Experimental Drainage Area, Cochrane, Ontario. In: *Symposium '89, Peat and Peatlands: Diversification and Innovation, Vol I – Peatland Forestry, Quebec City, 7–11, 1989* (ed. J. K. Jeglum and R. P. Overend), pp. 103–14, Canadian Society for Peat and Peatlands.

Rothwell, R. L., Silins, U., and Hillman, G. R. (1996). The effects of drainage on substrate water content at several forested Alberta peatlands. *Canadian Journal of Forest Research*, **26**, 53–62.

Roulet, N. T. (1991). Surface level and water table fluctuations in a subarctic fen. *Arctic and Alpine Research*, **23**, 303–10.

Roulet, N. T, Ash, R., Quinton, W., and Moore, T. R. (1993). Methane flux from drained northern peatlands: effect of a persistent water table lowering on flux. *Global Biogeochemical Cycles*, **7**, 749–69.

Rouse, W. R. (1998). A water balance model for a subarctic sedge fen and its application to climate change. *Climate Change*, **38**, 207–34.

Rull, V. (2004). Is the 'Lost World' really lost? Palaeoecological insights into the origin of the peculiar flora of the Guayana Highlands. *Naturwissenschaften*, **91**, 139–42

Ruthsatz, B. and Villagran, C. (1991). Vegetation pattern and soil nutrients of a Magellanic moorland on the Cordillera de Piuchué, Chiloé, Chile. *Revista Chilena de Historia Natural*, **64**, 461–78.

Ruuhijärvi, R. (1960). Über die regionale Einteilung der Nordfinnischen Moore. *Annales Botanici Societatis Zoologicae Botanici Fennici 'Vanamo'*, **31**.

Ruuhijärvi, R. (1983). The Finnish mire types and their regional distribution. In: *Ecosystems of the world. 4B. Mires: swamp, bog, fen and moor. Regional studies*, (ed. A. J. P. Gore), pp. 47–67, Elsevier, Amsterdam.

Ruuhijärvi, R. and Reinikainen, A. (1981). Research program of the project 'Comparative analysis of virgin and forest-improved mire-ecosystems'. *Suo*, **32**, 86–91.

Rydin, H. (1984). Some factors affecting temperature in *Sphagnum* vegetation. – An experimental analysis. *Cryptogamie, Bryologie et Lichénologie*, **5**, 361–72.

Rydin, H. (1985). Effect of water level on desiccation of *Sphagnum* in relation to surrounding *Sphagna*. *Oikos*, **45**, 374–79.

Rydin, H. (1986). Competition and niche separation in *Sphagnum*. *Canadian Journal of Botany*, **64**, 1817–24.

Rydin, H. (1993a). Interspecific competition among *Sphagnum* mosses on a raised bog. *Oikos*, **66**, 413–23.

Rydin, H. (1993b). Mechanisms of interactions among *Sphagnum* species along water-level gradients. *Advances in Bryology*, **5**, 153–85.

Rydin, H. (1995). Effects of density and water level on recruitment, mortality and shoot size in *Sphagnum* populations. *Journal of Bryology*, **18**, 439–53.

Rydin, H. and Barber, K. E. (2001). Long-term and fine-scale co-existence of closely related species. *Folia Geobotanica*, **36**, 53–62.

Rydin, H. and Clymo, R. S. (1989). Transport of carbon and phosphorus compounds about *Sphagnum*. *Proceedings of the Royal Society of London B Biological Sciences*, **237**, 63–84.

Rydin, H., Diekmann, M., and Hallingbäck, T. (1997). Biological characteristics, habitat associations, and distribution of macrofungi in Sweden. *Conservation Biology*, **11**, 628–40.

Rydin, H., Sjörs, H., and Löfroth, M. (1999). Mires. *Acta Phytogeographica Suecica*, **84**, 91–112.

Saarinen, T. (1996). Biomass and production of two vascular plants in a boreal mesotrophic fen. *Canadian Journal of Botany*, **74**, 934–38.

Saarinen, T. (1998). Internal C:N balance and biomass partitioning of *Carex rostrata* grown at three levels of nitrogen supply. *Canadian Journal of Botany*, **76**, 762–68.

Saarnio, S., Alm, J., Silvola, J., Lohila, A., Nykänen, H., and Martikainen, P. J. (1997). Seasonal variation in CH_4 emissions and production and oxidation potentials at microsites on an oligotrophic pine fen. *Oecologia*, **110**, 414–22.

Saito, S. (2004). Nakaikemi, a miraculous Japanese peatland – how has it been saved? *Peatlands International*, **2004 (1)**, 36–40.

Salo, K. (1993). The composition and structure of macrofungus communities in boreal upland type forests and peatlands in North Karelia, Finland. *Karstenia*, **33**, 61–99.

Såstad, S. M., Stenøien, H. K., Flatberg, K. I., and Bakken, S. (2001). The narrow endemic *Sphagnum troendelagicum* is an allopolyploid derivative of the widespread *S. balticum* and *S. tenellum*. *Systematic Botany*, **26**, 66–74.

Schimper, W. P. (1858). *Versuch einer Entwickelungs-Geschichte der Torfmoose (Sphagnum) und einer Monographie der in Europa vorkommenden Arten dieser Gattung*. E. Schweizerbart's Verlagsbuchhandlung, Stuttgart.

Schipper, L. A., Clarkson, B. R., Vojvodic-Vukovic, M., and Webster R. (2002). Restoring cut-over peat bogs: a factorial experiment of nutrients, seeds and cultivation. *Ecological Engineering*, **19**, 29–44.

Schlotzhauer, S. M. and Price, J. S. (1999). Soil water flow dynamics in a managed cutover peat field, Quebec: field and laboratory investigations. *Water Resources Research*, **35**, 3675–83.

Schouten, M. G. C. (ed.) (2002). *Conservation and restoration of raised bogs: geological, hydrological and ecological studies*. Department of Environment and Local Government, Dublin.

Segerström, U., Bradshaw, R., and Bohlin, E. (1994). Disturbance history of a swamp forest refuge in northern Sweden. *Biological Conservation*, **68**, 189–96.

Selin, P. (1996). Many uses for peatland cut-away areas. In: *Peatlands in Finland*, (ed. H. Vasander), pp. 128–29, Finnish Peatland Society, Helsinki.

Selin, P. and Nyrönen, T. (1998). The use of cutaway areas in Finland. In: *Peatland restoration and reclamation. Techniques and regulatory considerations*, (ed. T. Malterer, K. Johnson, and J. Stewart), pp. 18–22, International Peat Society, Jyväskylä.

Seppä, H. (1998). Suomen soiden pinnanmuodot. In: *Suomen suot*, (ed. H. Vasander), pp. 27–33, Finnish Peatland Society, Helsinki.

Seppälä, M. and Koutaniemi, L. (1985). Formation of a string and pool topography as expressed by morphology, stratigraphy and current processes on a mire in Kuusamo, Finland. *Boreas*, **14**, 287–309.

Shaw, A. J., Cox, C. J. and Boles, S. B. (2003). Global patterns in peatmoss biodiversity. *Molecular Ecology*, **12**, 2553–70.

Siegel, D. I. (1992). Groundwater hydrology. In: *The patterned peatlands of Minnesota* (ed. H. E. Wright Jr., B. A. Coffin, and N. E. P. Asseng), pp. 163–72, University of Minnesota Press, St. Paul.

Siegel, D. I., Reeve, A. S., Glaser, P. H., and Romanowicz, E. A. (1995). Climate-driven flushing of pore water in peatlands. *Nature*, **374**, 531–33.

Sikora, L. J. and Keeney, D. R. (1983). Further aspects of soil chemistry under anaerobic conditions. In: *Ecosystems of the world. 4A. Mires: swamp, bog, fen and moor. General studies*, (ed. A. J. P. Gore), pp. 247–56, Elsevier, Amsterdam.

Silins, U. and Rothwell, R. L. (1998). Forest peatland drainage and subsidence affect soil water retention and transport properties in an Alberta peatland. *Soil Science Society of America Journal*, **62**, 1048–56.

Silins, U. and Rothwell, R. L. (1999). Spatial patterns of aerobic limit depth and oxygen diffusion rate at two peatlands drained for forestry in Alberta. *Canadian Journal of Forest Research*, **29**, 1–8.

Silvan, N., Tuittila, E.-S., Vasander, H., and Laine, J. (2004). *Eriophorum vaginatum* plays a major role in nutrient immobilisation in boreal peatlands. *Annales Botanici Fennici*, **41**, 189–99.

Silvola, J., Alm, J., Ahlholm, U., Nykänen, H., and Martikainen, P. J. (1996). CO_2 fluxes from peat in boreal mires under varying temperature and moisture conditions. *Journal of Ecology*, **84**, 219–28.

Sims, R. A. and Baldwin, K. A. (1996). *Sphagnum species in Northwestern Ontario: a field guide to their identification*. Northwestern Science and Technology Technical Report TR-101, Ontario Ministry of Natural Resources, Thunder Bay.

Sims, R. A., Towill, W. D., Baldwin, K. A., Uhlig, P. and Wickware, G. M. (1997). *Field guide to the forest ecosystem classification for Northwestern Ontario*, 2nd edn., Northwestern Science and Technology Field Guide FG-03, Ontario Ministry of Natural Resources, Thunder Bay.

Sirin, A. A., Köhler, S., and Bishop, K. (1998). Resolving flow pathways and geochemistry in a headwater forested wetland with multiple tracers. In: *Hydrology, water resources and ecology in headwaters* (ed. K. Kovar, U. Tappeiner, N. E. Peters, and R. G. Craig), pp. 337–42, IAHS Publication no. 248.

Sirin, A. A., Shumov, D. B., and Vlasova, L. S. (1997). Investigation of bog water circulation using ^{3}H analysis data. *Water Resources*, **24**, 625–33.

Sitte, P., Weiler, E. W., Kadereit, J. W., Bresinsky, A., and Körner, C. (2002). *Strasburger – Lehrbuch der Botanik für Hochschulen*. 35 edn., Spektrum Akademischer Verlag, Heidelberg.

Sjögren-Gulve, P. (1994). Distribution and extinction patterns within a northern metapopulation of the pool frog, *Rana lessonae*. *Ecology*, **75**, 1357–67.

Sjörs, H. (1948). Myrvegetation i Bergslagen. [Mire vegetation in Bergslagen, Sweden]. *Acta Phytogeographica Suecica*, **21**, 1–299.

Sjörs, H. (1949). Om *Sphagnum lindbergii* i södra Sverige. *Svensk Botanisk Tidskrift*, **43**, 568–85.

Sjörs, H. (1950). On the relation between vegetation and electrolytes in north Swedish mire waters. *Oikos*, **2**, 241–58.

Sjörs, H. (1959). Forest and peatland at Hawley Lake, Northern Ontario. *National Museum of Canada Bulletin*, **171**, 1–31.

Sjörs, H. (1961). Surface patterns in boreal peatlands. *Endeavour*, **20**, 217–24.

Sjörs, H. (1963). Bogs and fens on Attawapiskat River, northern Ontario. *National Museum of Canada Bulletin*, **186**, 45–133.

Sjörs, H. (1990). Divergent successions in mires, a comparative study. *Aquilo, Ser Botanica*, **28**, 67–77.

Sjörs, H. (1991). Phyto- and necromass above and below ground in a fen. *Holarctic Ecology*, **14**, 208–18.

Sjörs, H. (1998). Bog-pools and flark-pools – similarities and differences. In: *Patterned mires and mire pools*, (ed. V. Standen, J. H. Tallis, and R. Meade), pp. 34–36, British Ecological Society, London.

Sjörs, H. and Gunnarsson, U. (2002). Calcium and pH in north and central Swedish mire waters. *Journal of Ecology*, **90**, 650–57.

Slack, N. G. (1990). Bryophytes and ecological niche theory. *Botanical Journal of the Linnean Society*, **104**, 187–213.

Small, E. (1972). Water relations of plants in raised *Sphagnum* peat bogs. *Ecology*, **53**, 726–28.

Smith, R. I. L. and Clymo, R. S. (1984). An extraordinary peat-forming community on the Falkland Islands. *Nature*, **309**, 617–20.

Smith, R. T. and Taylor, J-A. (1989). Biopedological processes in the inception of peat formation. *International Peat Journal*, **3**, 1–24.

Smol, J. P. and Last, W. M. (series eds) (2001–2005). *Developments in paleoenvironmental research*. Springer, Berlin.

Sonesson, M. (1969). Studies on mire vegetation in the Torne Träsk area, northern Sweden. II. Winter conditions of the poor mires. *Botaniska Notiser*, **122**, 481–511.

Soro, A., Sundberg, S. and Rydin, H. (1999). Species diversity, niche width and species associations in harvested and undisturbed bogs. *Journal of Vegetation Science*, **10**, 549–60.

Sparling, J. (1967). The occurrence of *Schoenus nigricans* L. in blanket bogs: II. Experiments on the growth of *S. nigricans* under controlled conditions. *Journal of Ecology*, **55**, 15–31.

Speight, M. C. D. and Blackith, R. E. (1983). The animals. In: *Mires: swamp, bog, fen and moor*, (ed. A. J. P. Gore), pp. 349–65, Elsevier, Amsterdam.

Spink, A. J. and Parsson, A. N. (1995). An experimental investigation of the effects of nitrogen deposition to *Narthecium ossifragum*. *Environmental Pollution*, **90**, 191–98.

Spitzer, K. and Danks, H. V. (2006). Insect biodiversity of boreal peat bogs. *Annual Review of Entomology*, **51**, 137–161.

Springett, J. A. and Latter, P. M. (1977). Studies on the micro-fauna of blanket bog with particular reference to Enchytraeidae. I. Field and laboratory tests of micro-organisms as food. *Journal of Animal Ecology*, **46**, 959–74.

Standen, V., Rees, D., Thomas, C. J., and Foster, G. N. (1998). The macro-invertebrate fauna of pools in open and forested patterned mires in the Sutherland Flows, north Scotland. In: *Patterned mires and mire pools*, (ed. V. Standen, J. H. Tallis, and R. Meade), pp. 147–62, British Ecological Society, London.

Stanek, W. and Worley, I. A. (1983). A terminology of virgin peat and peatlands. In: *Proceedings of the International Symposium on Peat Utilization* (ed. C. H. Fuchsman and S. A. Spigarelli), pp. 75–102, Bemidji State University, Bemidji, Minnesota.

Starr, M. and Westman, C. J. (1978). Easily extractable nutrients in the surface peat layer of virgin sedge pine swamps. *Silva Fennica*, **12**, 65–78.

Staub, J. R. and Esterle, J. S. (1994). Peat-accumulating depositional systems of Sarawak, East Malaysia. *Sedimentary Geology*, **89**, 91–106.

Stevenson, F. J. (1986). *Cycles of soil: carbon, nitrogen, phosphorus, sulfur, micronutrients.* John Wiley & Sons, New York.

Stewart, R. B. and Rouse, W. R. (1977). Substantiation of the Priestly and Taylor parameter $\alpha = 1.26$ for potential evaporation in high latitudes. *Journal of Applied Meteorology*, **16**, 649–50.

Ström, L., Olsson, T., and Tyler, G. (1994). Differences between calcifuge and acidifuge plants in root exudation of low-molecular organic soils. *Plant and Soil*, **167**, 239–45.

Strong, L. E. (1980). Aqueous hydrochloric acid conductance from 0 to 100 °C. *Journal of Chemical and Engineering Data*, **25**, 104–06.

Stumm, W. and Morgan, J. J. (1996). *Aquatic chemistry*, 3rd edn., John Wiley & Sons, New York.

Succow, M. and Joosten, H. (ed.) (2001). *Landschaftsökologische Moorkunde*. E. Schweizerbart'sche Verlagsbuchhandlung, Stuttgart.

Sundberg, S. (2002). Sporophyte production and spore dispersal phenology in *Sphagnum*: the importance of summer moisture and patch characteristics. *Canadian Journal of Botany*, **80**, 543–56.

Sundberg, S. and Rydin, H. (1998). Spore number in *Sphagnum* and its dependence on spore and capsule size. *Journal of Bryology*, **20**, 1–16.

Sundberg, S. and Rydin, H. (2000). Experimental evidence for a persistent spore bank in *Sphagnum. New Phytologist*, **148**, 105–16.

Sundberg, S. and Rydin, H. (2002). Habitat requirements for establishment of *Sphagnum* from spores. *Journal of Ecology*, **90**, 268–78.

Sundh, I., Nilsson, M., Granberg, G., and Svensson, B. H. (1994). Depth distribution of microbial production and oxidation of methane in northern boreal peatlands. *Microbial Ecology*, **27**, 253–65.

Sundström, E., Magnusson, T., and Hånell, B. (2000). Nutrient conditions in drained peatlands along a north-south gradient in Sweden. *Forest Ecology and Management*, **126**, 149–61.

Svensson, B. M. (1995). Competition between *Sphagnum fuscum* and *Drosera rotundifolia*: a case of ecosystem engineering. *Oikos*, **74**, 205–12.

Svensson, G. (1988). Bog development and environmental conditions as shown by the stratigraphy of Store Mosse mire in southern Sweden. *Boreas*, **17**, 89–111.

Swan, J. M. A. and Gill, A. M. (1970). The origin, spread, and consolidation of a floating bog in Harvard Pond, Petersham, Massachusetts. *Ecology*, **51**, 829–40.

Swanson, D. K. and Grigal, D. F. (1988). A simulation model of mire patterning. *Oikos*, **53**, 309–14.

Szumigalski, A. R. and Bayley, S. E. (1996). Net above-ground primary production along a bog-rich fen gradient in central Alberta, Canada. *Wetlands*, **16**, 467–76.

Tahvanainen, T. (2004). Water chemistry of mires in relation to the poor-rich vegetation gradient and contrasting geochemical zones of the north-eastern Fennoscandian Shield. *Folia Geobotanica*, **39**, 353–69.

Tahvanainen, T. and Tuomaala, T. (2003). The reliability of mire water pH measurements – a standard sampling protocol and implications to ecological theory. *Wetlands*, **23**, 701–08.

Takagi, K., Tsuboya, T. Takahashi, H., and Inoue, T. (1999). Effect of the invasion of vascular plants on heat and water balance in the Sarobetsu mire, northern Japan. *Wetlands*, **19**, 246–54.

Taylor, J. A. (1983). The peatlands of Great Britain and Ireland. In: *Ecosystems of the world. 4B. Mires: swamp, bog, fen and moor. Regional studies*, (ed. A. J. P. Gore), pp. 1–46, Elsevier, Amsterdam.

Taylor, J. A. and Smith, R. T. (1972). Climatic peat – a misnomer? *Proceedings of the 4th International Peat Congress*, **1**, 471–84.

Taylor, K. (1989). The absence of mycorrhiza in *Rubus chamaemorus*. *Annales Botanici Fennici*, **26**, 421–25.

Terasmae, J. and Hughes, O. L. (1960). A palynological and geological study of Pleistocene deposits in the James Bay Lowlands, Ontario. *Geological Survey Canada Bulletin*, **62**.

Thomas, K. L., Benstead, J., Davies, K. L., and Lloyd, D. (1996). Role of wetland plants in the diurnal control of CH_4 and CO_2 fluxes in peat. *Soil Biology and Biochemistry*, **28**, 17–23.

Thompson, K. and Hamilton, A. C. (1983). Peatlands and swamps of the African continent. In: *Ecosystems of the world. 4B. Mires: swamp, bog, fen and moor. Regional studies*, (ed. A. J. P. Gore), pp. 331–74, Elsevier, Amsterdam.

Thompson, M. A., Campbell, D. I., and Spronken-Smith, R A. (1999). Evaporation from natural and modified raised peat bogs in New Zealand. *Agricultural and Forest Meteorology*, **95**, 85–98.

Thormann, M. N., Bayley, S. E., and Szumigalski, A. R. (1997). Effects of hydrologic changes on aboveground production and surface water chemistry two boreal peatlands in Alberta: Implications for global warming. *Hydrobiologia*, **362**, 171–83.

Thormann, M. N., Currah, R. S. and Bayley, S. E. (1999). The mycorrhizal status of the dominant vegetation along a peatland gradient in southern boreal Alberta, Canada. *Wetlands*, **19**, 438–50.

Thormann, M. N., Szumigalski, A. R., and Bayley, S. E. (1999). Aboveground peat and carbon accumulation potentials along a bog-fen-marsh wetland gradient in southern boreal Alberta, Canada. *Wetlands*, **19**, 305–17.

Timmermann, T. (2000). Oscillation or inundation: hydrodynamical classification of kettle-hole mires as a tool for restoration. *Proceedings of the 11th International Peat Congress*, **1**, 243–52.

Timmermann, T. (2003). The hydrological dynamics of kettle-hole mires and their significance for the development of tree stands. [In German with English abstract.] *Telma*, **33**, 85–107.

Tiner, R. W. (1998). *In search of swampland: a wetland sourcebook and field guide.* Rutgers University Press, New Brunswick.

Tiner, R. W. (1999). *Wetland indicators: a guide to wetland identification, delineation, classification, and mapping.* Lewis Publishers, Boca Raton.

Todd, D. K. and Mays, L. W. (2005). *Groundwater hydrology*, 3rd edn., John Wiley & Sons, New York.

Tolonen, K. (1983). The relationship between the calorific value and the humification of peat. [In Finnish with English summary.] *Suo*, **34**, 85–92.

Tolonen, K. (1987). Natural history of raised bogs and forest vegetation in the Lammi area, southern Finland, studied by stratigraphical methods. *Annales Academiae Scientiarum Fennicae, AIII*, **139**.

Tolonen, K. (1991). Advances in peatland palaeoecology on environmental changes. In: *Studies of mire ecosystems of Fennoscandia: Material of the Soviet-Finnish Symposium, 28–31 May 1990* (ed. M. S. Botch, O. L. Kuznetsov, and I. P. Khizova), pp. 25–34, Academy of Sciences, Institute of Biology, Petrozavodsk.

Tolonen, K. and Oldfield, F. (1986). The record of magnetic-mineral and heavy metal deposition of Regent Street Bog, Fredericton, New Brunswick, Canada. *Physics Earth Planetary Interiors*, **42**, 57–66.

Tolonen, K. and Turunen, J. (1996). Accumulation rates of carbon in mires in Finland and implications for climate change. *Holocene*, **6**, 171–78.

Tolonen, K., Warner, B. G., and Vasander, H. (1992). Ecology of Testaceans (Protozoa: Rhizopoda) in mires in southern Finland: I. Autoecology. *Archiv für Protistenkunde*, **142**, 119–38.

Tomassen, H. B. M., Smolders, A. J. P., Limpens, J., Lamers, L. P. M., and Roelofs, J. G. M. (2004). Expansion of invasive species on ombrotrophic bogs: desiccation or high N deposition. *Journal of Applied Ecology*, **41**, 139–50.

Trettin, C. C., Jurgensen, M. F., Grigal, D. F., Gale, M. R., and Jeglum, J. K. (eds) (1997). *Northern forested wetlands: ecology and management*, CRC Press, Boca Raton.

Troedsson, T. and Nykvist, N. (1973). *Marklära och markvård*. AW Läromedel, Stockholm.

Tuhkanen, S. (1980). Climatic parameters and indices in plant geography. *Acta Phytogeographica Suecica*, **67**, 1–105.

Tuhkanen, S. (1992). The climate of Tierra del Fuego from a vegetation geographical point of view and its ecoclimatic counterparts elsewhere. *Acta Botanica Fennica*, **145**, 1–64.

Tuittila, E-S., Komulainen, V-M., Vasander, H., Nykänen, H., Martikainen, P. J., and Laine, J. (2000). Methane dynamics of a restored cut-away peatland. *Global Change Biology*, **6**, 569–81.

Tuittila, E-S., Vasander, H., and Laine, J. (2004). Sensitivity of C sequestration in reintroduced *Sphagnum* to water level variation in a cut-away peatland. *Restoration Ecology* **12**, 483–93.

Turetsky, M. R., Manning, S. W., and Wieder, R. K. (2004). Dating recent peat deposits. *Wetlands*, **24**, 324–56.

Turner, R. C. and Scaife, R. G. (1995). *Bog bodies. New discoveries and new perspectives*. British Museum Press, London.

Turner, R. K., van den Bergh, J. C. J. M., and Brouwer, R. (eds) (2003). *Managing wetlands: an ecological economics approach*. Edward Elgar Publishing, Cheltenham.

Turner, S. D., Amon, J. P., Schneble, R. M., and Friese, C. F. (2000). Mycorrhizal fungi associated with plants in ground-water fed wetlands. *Wetlands*, **20**, 200–04.

Turunen, J. (2003). Past and present carbon accumulation in undisturbed boreal and subarctic mires: A review. *Suo*, **54**, 15–28.

Turunen, J. and Tolonen, K. (1996). Rate of carbon accumulation in boreal peatlands and climatic change. In: *Global peat resources* (ed. E. Lapalainen), pp. 277–80. International Peat Society. Jyskä.

Turunen, J., Tomppo, E., Tolonen, K., and Reinikainen, A. (2002). Estimating carbon accumulation rates of undrained mires in Finland – application to boreal and subarctic regions. *Holocene*, **12**, 69–80.

Tyler, G. (1999). Plant distribution and soil-plant interactions on shallow soils. *Acta Phytogeographica Suecica*, **84**, 21–32.

Ugolini, E.C. and Mann, D.H. (1979). Biopedological origin of peatlands in southeast Alaska. *Nature*, **281**, 366–8.

Umeda, Y., Shimizu, M., and Demura, M. (1986). Forming of Sarobetsu peatland – surface forms on peatland (II). [In Japanese with English Summary.] *Memoirs of the Faculty of Agriculture Hokkaido University*, **15**, 28–35.

Updegraff, K., Bridgham, S. D., Pastor, J., Weishampel, P., and Harth, C. (2004). Response of CO_2 and CH_4 emissions from peatland to warming and water table manipulation. *Ecological Applcations*, **11**, 311–26.

Väisänen, R. (1992). Distribution and abundance of diurnal Lepidoptera on a raised bog in southern Finland. *Annales Zoologici Fennici*, **29**, 75–92.

van den Bergh, J. C. J. M., Barendregt, A. G., van Herwigjnen, M. *et al.* (2003). Spatial hydro-ecological and economic modelling of land use changes in wetlands. In: *Managing wetlands – an ecological economics approach* (ed. R. K. Turner, J. C. J. M. van den Bergh, and R. Brouwer), pp. 271–300. Edward Elgar Publishing, Cheltenham.

van der Molen, P. C. and Wijmstra, T. A. (1994). The thermal regime of hummock-hollow complexes on Clara Bog, Co Offaly. *Proceedings of the Royal Irish Academy*, **94B**, 209–21.

van Duren, I. C., Boeye, D., and Grootjans, A. P. (1997). Nutrient limitation in an extant and drained poor fen: implications for restoration. *Plant Ecology*, **133**, 91–100.

van Wirdum, G. (1991). Vegetation and hydrology of floating rich-fens. PhD thesis, University of Amsterdam.

Vasander, H. (ed.) (1996). *Peatlands in Finland*. Finnish Peatland Socity, Helsinki.

Vasander, H. and Kettunen, A. (2006). Carbon in boreal peatlands. In: *Boreal peatland ecosystems*, (ed. K. Wieder and D. H. Vitt), pp. 165–194, Springer Verlag.

Vasander, H., Tuittila, E.-S., Lode, E. *et al.* (2003). Status and restoration of peatlands in Northern Europe. *Wetlands Ecology and Management*, **11**, 51–63.

Vepraskas, M. J. and Faulkner, S. P. (2001). Redox chemistry of hydric soils. In: *Wetland soils: genesis, hydrology, landscapes, and classification*, (ed. J. L. Richardson and M. J. Vepraskas), pp. 84–105, Lewis Publishers, Boca Raton.

Verhoeven, J. T. A. and Liefveld, W. M. (1997). The ecological significance of organochemical compounds in *Sphagnum. Acta Botanica Neerlandica*, **46**, 117–30.

Verhoeven, J. T. A., Keuter, A., van Logtestijn, R., van Kerkhoven, M. B., and Wassen, M. (1996a). Control of local nutrient dynamics in mires by regional and climatic factors: A comparison of Dutch and Polish sites. *Journal of Ecology*, **84**, 647–56.

Verhoeven, J. T. A., Koerselman, W., and Meuleman, A. F. M. (1996b). Nitrogen- or phosphorus-limited growth in herbaceous, wet vegetation: relations with atmospheric inputs and management regimes. *Trends in Ecology and Evolution*, **11**, 494–97.

Verhoeven, J. T. A. and Toth, E. (1995). Decomposition of *Carex* and *Sphagnum* litter in fens: effect of litter quality and inhibition by living tissue homogenates. *Soil Biology and Biochemistry*, **27**, 271–75.

Verry, E. S. (1997). Hydrological processes of natural, northern forested wetlands. In: *Northern forested wetlands: Ecology and management*, (ed. C. C. Trettin, M. F. Jurgensen, D. F. Grigal, M. R. Gale and J. K. Jeglum), pp. 163–88, CRC Press, Boca Raton.

Verry, E. S. and Timmons, D. R. (1982). Waterborne nutrient flow through an upland-peatland watershed in Minnesota. *Ecology*, **63**, 1456–67.

Verry, E. S. and Urban, N. R. (1993). Nutrient cycling at Marcell bog, Minnesota. *Suo*, **43**, 147–53.

Viereck, L. A. (1970). Forest succession and soil development adjacent to the Chena River in interior Alaska. *Arctic and Alpine Research*, **2**, 1–26.

Vilkamaa, P. (1981). Soil fauna in a virgin and two drained dwarf shrub pine bogs. *Suo*, **32**, 120–22.

Vitt, D. H. and Belland, R. J. (1995). The bryophytes of peatlands in continental western Canada. *Fragmenta Floristica et Geobotanica*, **40**, 339–48.

Vitt, D. H. and Kuhry, P. (1992). Changes in moss-dominated wetland ecosystems. In: *Bryophytes and lichens in a changing environment*, (ed. J. W. Bates and A. M. Farmer), pp. 178–210, Clarendon Press, Oxford.

Vitt, D. H. and Slack, N. G. (1975). An analysis of the vegetation of *Sphagnum*-dominated kettle-hole bogs in relation to environmental gradients. *Canadian Journal of Botany*, **53**, 332–59.

Vitt, D. H. and Slack, N. G. (1984). Niche diversification of *Sphagnum* relative to environmental factors in northern Minnesota peatlands. *Canadian Journal of Botany*, **62**, 1409–30.

Vitt, D. H., Halsey, L. A., and Zoltai, S. C. (1994). The bog landforms of continental western Canada in relation to climate and permafrost patterns. *Arctic and Alpine Research*, **26**, 1–13.

Vitt, D. H., Bayley, S. E., and Jin, T. L. (1995). Seasonal variation in water chemistry over a bog-rich fen gradient in Continental Western Canada. *Canadian Journal of Fisheries and Aquatic Sciences*, **52**, 587–606.

Vitt, D. H., Wieder, K., Halsey, L. A., and Turetsky, M. (2003). Response of *Sphagnum fuscum* to nitrogen deposition: A case study of ombrogenous peatlands in Alberta, Canada. *The Bryologist*, **106**, 235–45.

Vompersky, S. E. and Sirin, A. A. (1997). Hydrology of drained forested wetlands. In: *Northern forested wetlands: ecology and management*, (ed. C. C. Trettin, M. F. Jurgensen, D. F. Grigal, M. R. Gale, and J. K. Jeglum), pp. 189–21, CRC Press, Boca Raton.

von Post, L. (1924) Das genetische System der organogenen Bildungen Schwedens. *Comité International de Pédologie IV, Commission No.* **22**, 287–304.

von Post, L. (1937). The geographical survey of Irish bogs. *Irish Naturalists' Journal*, **6**, 210–27.

von Post, L. (1946). The prospect for pollen analysis in the study of the earth's climatic history. *New Phytologist*, **45**, 193–217.

von Post, L. and Granlund, E. (1926). Södra Sveriges torvtillgångar. *Sveriges Geologiska Undersökning Ba*, **335**

von Post, L. and Sernander, R. (1910). Pflanzenphysiognomische Studien auf Torfmooren in Närke. *Livretguide des Exc. en Suède du XIe Congrès Géologique International*, **14**, 1–48.

Waardenaar, E. C. P. (1987). A new hand tool for cutting peat profiles. *Canadian Journal of Botany*, **65**, 1772–73.

Waddington, J. M. and Roulet, N. T. (2000). Carbon balance of a boreal patterned peatland. *Global Change Biology*, **6**, 87–97.

Waksman, S. A. and Stevens, K. R. (1929). Contribution to the chemical composition of peat: V. The rôle of microörganisms in peat formation and decomposition. *Soil Science*, **28**, 315–39.

Walker, D. (1970). Direction and rate in some British post-glacial hydroseres. In: *Studies in the vegetational history of the British Isles* (ed. D. Walker and R. G. West), pp 117–139, Cambridge University Press, Cambridge.

Walker, D. and Walker, P. M. (1961). Stratigraphic evidence of regeneration in some Irish bogs. *Journal of Ecology*, **49**, 169–85.

Wallace, R. L. (1977). Distribution of sessile rotifers in an acid bog pond. *Archiv für Hydrobiologie*, **79**, 478–505.

Wallén, B. (1986). Above and below ground dry mass of the three main vascular plants on hummocks on a subarctic peat bog. *Oikos*, **46**, 51–56.

Wallén, B. (1992). Methods for studying below-ground production in mire ecosystems. *Suo*, **43**, 155–62.

Wallwork, J. A. (1991). *Ecology of soil animals*, 2nd edn., John Wiley & Sons, Chichester.

Wardle, P. (1977). Plant communities of Westland National Park (New Zealand) and neighbouring lowland and coastal areas. *New Zealand Journal of Botany*, **15**, 323–98.

Warner, B. G., Kubiw, H. J., and Hanf, K. I. (1989). An anthropogenic cause for quaking mire formation in southwestern Ontario. *Nature*, **340**, 380–4.

Warren, R. S. (1995). Evolution and development of tidal marshes. In: *Tidal marshes of Long Island Sound: ecology, history and restoration* (ed. G. D. Dreyer and W. A. Niering), pp.17–21, Connecticut College Arboretum Bulletin No. 34.

Webb, T. and McAndrews, J.H. (1976). Corresponding patterns of contemporary pollen and vegetation in central North American. *Geological Society of America Memoir*, **145**, 267–99.

Weber, C. A. (1902). *Über die Vegetation und Entstehung des Hochmoores von Augstumal im Memeldelta* Verlagsbuchhandlung Paul Parey, Berlin. [English translation: J. Couwenberg and H. Joosten, *C.A. Weber and the raised bog of Augstumal*, International Mire Conservation Group, 2002].

Weber, C. A. (1908). Aufbau und Vegetation der Moore Norddeutschlands. *Englers Botanischen Jahrbüchern* **90**.

Wells, E. D. (1996). Classification of peatland vegetation in Atlantic Canada. *Journal of Vegetation Science*, **7**, 847–78.

Wells, E. D. and Pollett, F. C. (1983). Peatlands. In: *Biogeography and ecology of the island of Newfoundland*, (ed. G. R. South), pp. 207–65, Dr W. Junk Publishers, The Hague.

Weltzin, J. F., Harth, C., Bridgham, S. D., Pastor, J., and Vonderharr, M. (2001). Production and microtopography of bog bryophytes: response to warming and water-table manipulations. *Oecologia*, **128**, 557–65.

Westman, C. J. (1987). Site classification in estimation of fertilization effects on drained mires. *Acta Forestalia Fennica*, **198**, 1–55.

Westman, C. J. and Laiho, R. (2003). Nutrient dynamics of drained peatland forests. *Biogeochemistry*, **63**, 269–98.

Wheatley, R. E., Greaves, M. P., and Inkson, R. H. E. (1996). The aerobic bacterial flora of a raised bog. *Soil Biology and Biochemistry*, **8**, 453–60.

Wheeler, B. D. and Proctor, M. C. F. (2000). Ecological gradients, subdivisions and terminology of north-west European mires. *Journal of Ecology*, **88**, 187–203.

Wheeler, B. D. and Shaw, S. C. (1995). *Restoration of damaged peatlands*. HMSO, London.

Wheeler, B. D., Shaw, S. C., Fojt, W. J., and Robertson, R. A. (1995). *Restoration of temperate wetlands*. John Wiley & Sons, Chichester.

Whiting, G. J. and Chanton, J. P. (2001). Greenhouse carbon balance of wetlands: methane emission versus carbon sequestration. *Telma*, **53B**, 521–28.

Wieder, R. K. (2001). Past, present, and future peatland carbon balance: an empirical model based on ^{210}Pb-dated cores. *Ecological Applications*, **11**, 327–42.

Wiklander, G. and Nömmik, H. (1987). Net mineralization of nitrogen in a fen peat soil, Central Sweden. *Acta Agriculturae Scandinavica*, **37**, 189–98.

Wikner, B. (1983). Distribution and mobility of boron in forest ecosystems. *Communicationes Instituti Forestalis Fenniae*, **116**, 131–40.

Wilcox, D. A., Sweat, M. J., Carlson, M. A., and Kowalski, K. P. (2006). A water-budget approach to restoring a sedge fen affected by diking and ditching. *Journal of Hydrology*, **320**, 501–17.

Williams, B. L., Buttler, A., Grosvernier, P. *et al.* (1999). The fate of NH_4NO_3 added to *Sphagnum magellanicum* carpets at five European mire sites. *Biogeochemistry*, **45**, 73–93.

Williams, B. L. and Sparling, G. P. (1988). Microbial biomass carbon and readily mineralized nitrogen in peat and forest humus. *Soil Biology and Biochemistry*, **20**, 579–81.

Williams, R. T. and Crawford, R. L. (1983). Microbial diversity of Minnesota USA peatlands. *Microbial Ecology*, **9**, 201–14.

Wilson, H. D. (1987). *Vegetation of Stewart Island, New Zealand: a supplement to the New Zealand Journal of Botany*. DSIR Science Information Publishing Centre, Wellington.

Wolff, W. J. (1993). Netherlands-Wetlands. *Hydrobiologia*, **265**, 1–14.

World Commission on Environment and Development (1987). *Our common future*, Oxford University Press, Oxford.

Wright Jr., H. E. (ed.) (1983). *Late-quaternary environment of the United States, vol 2, the Holocene*, University of Minnesota Press, St. Paul.

Wright Jr., H. E., Coffin, B. A. and Asseng, N. E. P. (eds) (1992). *The patterned peatlands of Minnesota*, University of Minnesota Press, St. Paul.

Wright Jr, H. E., Mann, D. H., and Glaser, P. H. (1984). Piston corers for peat and lake sediments. *Ecology*, **65**, 657–59.

Yu, Z., Vitt, D. H., Campbell, I. D., and Apps, M. J. (2003). Understanding Holocene peat accumulation pattern of continental fens in western Canada. *Canadian Journal of Botany*, **81**, 267–82.

Yung, Y.-K., Stokes, P., and Gorham, E. (1986). Algae of selected continental and maritime bogs in North America. *Canadian Journal of Botany*, **64**, 1825–33.

Zhang, X. and Andrews, J. H. (1993). Evidence for growth of *Sporothrix schenckii* on dead but not on living sphagnum moss. *Mycopathologia*, **123**, 87–94.

Zobel, M. (1990). Soil oxygen conditions in paludifying boreal forest sites. *Suo*, **41**, 81–9.

Zoltai, S. C. (1972). Palsas and peat plateaus in central Manitoba and Saskatchewan. *Canadian Journal of Forest Research*, **2**, 291–302.

Zoltai, S. C. (1995). Permafrost distribution in peatlands of west-central Canada during the Holocene warm period 6000 years BP. *Géographie physique et Quaternaire*, **49**, 45–54.

Zoltai, S. C. and Vitt, D. H. (1990). Holocene climatic change and the distribution of peatlands in western interior Canada. *Quaternary Research*, **33**, 231–40.

Zoltai, S.C., Taylor, S. Jeglum, J.K., Mills, G.F., and Johnson, J.D. (1988). Wetlands of boreal Canada. In: *Wetlands of Canada* (ed. National Wetlands Working Group, Canada), pp. 97–154, Environment Canada, Ottawa.

Glossary

Aapa mire A Finnish term denoting a large patterned fen or mixed string mire. Often used in a wider meaning for vast boreal mire complexes dominated by minerotrophic mire expanse vegetation in their central parts.

Acrotelm The upper layer in a peatland, down to the level at which the peat is always water saturated.

Anoxic Environment without oxygen, allowing only anaerobic biological processes.

Bog, ombrotrophic peatland Peatland in which the surface peat and vegetation only receives water and nutrient from precipitation, dust, sea-spray, and airborne deposition.

Boundary layer *see* **Recurrence surface**

Brown mosses An ecological group of bryophyte species characteristic of rich fens. Mostly from the Amblystegiaceae family, but some other species are included (e.g. *Tomentypnum nitens* and *Paludella squarrosa*). Most of them have brown, reddish, yellowish brown, or even golden colour.

Carpet Peatland vegetation with a bryophyte dominance that is so soft that a footprint remains visible for a long time. With a sparse cover of cyperaceous plants. Often from 5 cm below to 5 cm above the water table.

Catotelm The lower layer in a peatland, the zone which is always water saturated.

Dy Aquatic sediment in oligotrophic lakes rich in humic substances. Consists mainly of amorphous precipitated humus colloids. Dark brown, in dry state almost black. In more nutrient-rich lakes the sediment is **gyttja**, not dy.

Feathermosses An ecological category of bryophytes characteristic of upland coniferous forests. In peatlands they occur in coniferous swamp forests and on wooded bog hummocks, and they expand after drainage. Examples: *Hylocomium splendens, Pleurozium schreberi, Ptilium crista-castrensis*.

Fen, minerotrophic peatland Peatland receiving inflow of water and nutrients from the mineral soil. Distinguished from **swamp forest** by a lack of tree cover or with only a sparse ($< 25\%$) crown cover. Indistinctly separated from **marsh** (which is always beside open water, and usually has a mineral substrate).

Flark (Swedish), **rimpi** (Finnish) Elongated wet depressions with sparse vegetation (**mud-bottom**) in fen. For most of the time a flark is waterlogged or even flooded. The water level is maintained by the damming strings on the downslope side, and the flarks and the strings are arranged perpendicularly to the slope.

Floating mat *see* **Quagmire**

Gley Soil type formed under reduced waterlogged conditions, with grey to blue-grey colour caused by reduced iron.

Grenzhorizont *see* **Recurrence surface**

Groundwater Water beneath the **water table**. The groundwater matrix is essentially saturated, but trapped air or biogenic gasses are sometimes present.

Gyttja Aquatic sediment composed of a mixture of organic and minerogenic materials, including plant detritus, plankton, and pollen. Grey brown to black, becomes lighter in colour on drying. In oligotrophic lakes the sediment is more often **dy**.

Hollow The lower feature (depression) in peatlands with an alternating microtopograhy, a hummock-hollow pattern.

Humification The degree to which peat has been modified by decomposition, that is, how much of the original fibrous structure of the vegetation remains and how much of the peat consists of humified material in which the original plant organs are no longer recognizable. Often given by the von Post humification scale.

Hummock Peatland vegetation raised 20–50 cm above the lowest surface level, characterized by drier-occurring mosses, lichens, and dwarf shrubs.

Hydraulic conductivity The permeability of peat (or other material), a measure of how easily water can flow through the peat.

Hydromorphology The shape of a peatland, and the patterns within it, as determined by the interactions between underlying terrain form, climate, and hydrology.

Infilling, terrestrialization The process whereby peat develops on the margins and into the centres of ponds, lakes, or slow-flowing rivers.

Lagg A narrow fen or swamp surrounding a bog, receiving water both from the bog and from the surrounding mineral soil.

Lawn Peatland vegetation with graminoid (cyperaceous plants, grasses etc) dominance and with a diversified bryophyte layer. Due to the abundance of roots and rhizomes, the lawns are are so firm that foot prints rapidly disappear. Most of the time 5 to 20 cm above the water table

Marl Whitish, grey, blue-grey, or yellow sediment in lakes or calcareous fens. Consists of calcium carbonate often mixed with gyttja.

Marsh Mostly on mineral soil, but could be a peatland. Beside open water, with standing or flowing water, or flooded seasonally. Submerged, floating, emergent, or tussocky vegetation.

Mineral soil water limit The limit between minerotrophic and ombrotrophic, often established in the field by plants indicating some minerotrophic influence.

Minerotrophic peatland Peatlands receiving nutrients through an inflow of water that has filtered through mineral soil.

Mire Wetland with at least some peat, dominated by living peat-forming plants. Different

from the definition of peatland (which requires a minimum peat depth).

Mire complex An area consisting of several hydrologically connected, but often very different mire types. Sometimes separated by mineral soil uplands, but in regions dominated by peatlands there are vast continuous mire complexes.

Mixed mire A mire type with **bog** and **fen** features or sites in close connection.

Mud-bottom Patch in peatland which is often inundated. Mostly with exposed bare peat with very loose consistency, often with a thin cover of algae and scattered bryophytes.

Ombrotrophic peatland *see* **Bog.**

Oxic Environment with oxygen, allowing aerobic biological processes.

Paludification Peatland expansion caused by gradual raising of the water table as peat accumulation impedes drainage, or encroachment of peatland onto adjacent, drier mineral soil.

Peat Remains of plant and animal litter accumulating under more or less water-saturated conditions through incomplete decomposition. It is the result of anoxic conditions, low temperatures, low decomposability of the material and other complex causes.

Peatland Peat-covered terrain. A minimum depth of peat is required for a site to be classified as peatland (e.g. 30 or 40 cm).

Pool Permanently water-filled basin in bog (bog-pool) or fen (flark-pool), often with some vegetation at its edge. Pools were initiated and deepened after the peatland was formed.

Primary peat formation The process whereby peat is formed directly on freshly exposed, wet mineral soil.

Quagmire (quaking mat, floating mat, Schwingmoor) Peat-forming vegetation floating on water. Often with *Sphagnum* or brown moss cover, but held together and kept afloat by the roots and rhizomes of graminoid species.

Recurrence surface, boundary horizon, Grenzhorizont Transitions between peat layers, from darker, highly humified peat interpreted as laid down under dry conditions, to lighter, poorly humified peat which indicates higher humidity.

Schwingmoor *see* **Quagmire**

Sedentary deposits   Deposits of organic material that has formed in place, such as **peat**.

Sedimentary deposits   Deposits in springs, ponds, lakes, or sea embayments, either by settling from above, washing in, or chemical or biological precipitation. Includes wide range of compositions from predominantly organic to predominantly mineral. See dy, gyttja, and marl.

Soak   A minerotrophic, often narrow seepage of moving water, within a bog or crossing a bog (cf. water track). (Sometimes flush has been used as a synonym for soak.)

Soligenous   Minerogenous peatlands that are sloping, and with directional water flow through the peat or on the surface.

String   Elongated ridges in patterned fens and bogs arranged perpendicularly to the slope with a **hummock** or **lawn** level vegetation.

Subsidence   The sinking of the peat surface after drainage (or drought) caused by shrinkage (by loss of water and collapse of pore spaces), and increased aerobic decomposition.

Swamp, swamp forest   Forested wetland on mineral soil or minerotrophic peat. In the latter case it is separated from wooded fen by having a denser tree canopy (e.g., > 25% crown cover)

Terrestrialization *see* **infilling**

Topogenous   Minerogenous peatlands with a virtually horizontal water table, located in basins.

Water table   The level to which water will rise in a hole in the peatland, i.e. the upper surface of the groundwater.

Water track   Broad soligenous fen drainage ways bordered by bogs, swamp forests, or uplands in large peatland complexes.

Waterlogging   Where the water table is situated near the surface of the soil.

Wetland   Land with the water table near the surface. Inundation lasts for such a large part of the year that the dominant organisms must be adapted to wet and reducing conditions. Includes shore, marsh, swamp, fen, and bog.

Index

Numbers in bold refer to figures, tables or captions